DEVELOPMENTS IN BOUNDARY ELEMENT METHODS—1

THE DEVELOPMENTS SERIES

Developments in many fields of science and technology occur at such a pace that frequently there is a long delay before information about them becomes available and usually it is inconveniently scattered among several journals.

Developments Series books overcome these disadvantages by bringing together within one cover papers dealing with the latest trends and developments in a specific field of study and publishing them within *six months* of their being written.

Many subjects are covered by the series, including food science and technology, polymer science, civil and public health engineering, pressure vessels, composite materials, concrete, building science, petroleum technology, geology, etc.

Information on other titles in the series will gladly be sent on application to the publisher.

DEVELOPMENTS IN BOUNDARY ELEMENT METHODS—1

Edited by

P. K. BANERJEE

B.C.E., Ph.D., C.Eng., M.I.C.E.

Senior Lecturer, Department of Civil and Structural Engineering, University College Cardiff, UK

and

R. BUTTERFIELD

D.Sc., D.I.C., C.Eng., M.I.C.E., M.I.Struct.E.

Professor, Department of Civil Engineering, University of Southampton, UK

APPLIED SCIENCE PUBLISHERS LTD

LONDON

APPLIED SCIENCE PUBLISHERS LTD
RIPPLE ROAD, BARKING, ESSEX, ENGLAND

British Library Cataloguing in Publication Data

Developments in boundary element methods.—
(Developments series).
1
1. Boundary value problems—Numerical
solutions
I. Banerjee, P K II. Butterfield, R
III. Series
515′.353 QA379

ISBN 0-85334-845-6

WITH 14 TABLES AND 81 ILLUSTRATIONS

Printed in Great Britain by Galliard (Printers) Ltd, Great Yarmouth

PREFACE

This book is an attempt to demonstrate the power and versatility of Boundary Element Methods by illustrating and explaining recent developments in their application to a wide range of complex practical problems.

These range through potential flow and electromagnetism, into two- and three-dimensional elasticity, plates and shells, thermoelasticity, two- and three-dimensional elastoplasticity, fracture mechanics, wave propagation and, finally, contributions to the analysis of some inhomogeneous and non-linear systems together with algorithms for coupling BEM with other numerical methods.

We hope that the diversity of problems covered will convince the uninitiated reader that here is a method which is better than most and worthy of study. For those already persuaded, we would mention that the material in the book is being presented for the first time in this form and, we believe, fairly represents the current state of development of BEM, since each chapter has been written by an author who is already well established in the field. As Boundary Element Methods are relatively new we thought it would be worthwhile to include the following brief outline of their main features.

The numerical methods most widely used at present tackle the governing differential equations directly in the form in which they were derived without any further mathematical manipulation. This is usually done either by approximating the differential operators in the equations by simpler, localised algebraic ones valid at a series of nodes within the region (e.g. finite difference methods) or by representing the region itself by non-infinitesimal (i.e. finite) elements of material which are assembled to provide an approximation to the real system. All such methods involve whole-body discretisation schemes which require, finally, the solution of very large systems of algebraic equations. These methods unavoidably generate the solution at all the internal 'nodes' used, whether or not this information is required.

The Finite Element Method, which epitomises the latter approach, has in recent years reached such a stage of development that many workers would doubt whether any equivalent, let alone superior, technique could ever appear. However, we could return to our set of differential equations and, as an alternative approach, try to integrate them analytically in some way before either proceeding to any discretisation scheme or introducing any approximations. We are, of course, attempting to integrate the differential equations to find a solution whatever method we use, but the essence of integral equation techniques is the transformation of the differential equations into equivalent sets of integral ones as the first step in their solution. Intuitively one would expect from such an operation a set of equations which would involve only values of the variables at the extremes of the range of integration (i.e. on the boundaries of the region) with the implication that any discretisation scheme needed subsequently would only involve subdivision of the bounding surface of the body. This is exactly what happens and has led to the name Boundary Element Methods being coined† for the quite general, practical, numerical technique which has developed from the basic integral equation approach. Some of the attractive features of BEM are that, by using a boundary discretisation scheme, the effective dimensionality of any problem is reduced by one, which, especially in three dimensions, leads to an appreciable reduction in the number of algebraic equations generated for solution, as well as much simplified data preparation. In addition, the solution variables vary continuously throughout the region and therefore detailed output can be obtained at subsequently chosen internal points. The solution is, in principle, exact, and approximations need only be introduced if they are required either to describe the boundary geometry or to evaluate the various integrals involved. At the very heart of the method is the fact that infinitely distant boundaries can be automatically accounted for without discretisation, a property unique to BEM.

We hope that this book will encourage not only the use of Boundary Element Methods, but also their development, for, although many

† In *Finite Elements in Geomechanics* (1977), Chapter 16, Ed. Gudehus, G., Wiley, New York.

computer algorithms, immediately useful to the practising analyst, now exist, they are by no means fully explored and still offer the exciting possibility of solving, efficiently, problems which are currently quite intractable.

P. K. BANERJEE
R. BUTTERFIELD

CONTENTS

LIST OF CONTRIBUTORS

P. K. BANERJEE, B.C.E., Ph.D., C.Eng., M.I.C.E.
*Senior Lecturer Department of Civil and Structural Engineering, University College Cardiff, CF*1 1*XL*, *UK*.

R. BUTTERFIELD, D.Sc., D.I.C., C.Eng., M.I.C.E., M.I.Struct.E.
*Professor, Department of Civil Engineering, University of Southampton, SO*9 5*NH*, *UK*.

D. N. CATHIE, B.Sc., Ph.D., M.I.C.E.
*Lecturer, Department of Civil Engineering and Building Technology, University of Wales Institute of Science and Technology, Cardiff, CF*1 3*NU*, *UK*.

T. A. CRUSE, B.S., M.S., Ph.D.
Project Engineer, Structures Technology, Pratt & Whitney Aircraft Group, Commercial Products Division East Hartford, Connecticut 06108, *USA*.

T. G. DAVIES, B.Sc., Ph.D.
*Researcher, Department of Civil and Structural Engineering, University College Cardiff, CF*1 1*XL*, *UK*.

M. FRIEDMAN, M.Sc., Ph.D.
Senior Scientist, Nuclear Research Center-Negev, Beer-Sheva, Israel; also, *Lecturer, Department of Electrical Engineering, Ben Gurion University, Beer-Sheva, Israel.*

D. W. KELLY, Ph.D.
Lecturer, Department of Civil Engineering, University of Wales, Swansea, SA2 8PP, UK.

M. H. LEAN, B.Sc.
Graduate Student, Department of Electrical Engineering, University of Manitoba, Winnipeg, R3T 2N2, Canada.

G. G. W. MUSTOE, B.Sc.
Research Assistant, Department of Civil Engineering, University of Wales, Swansea, SA2 8PP, UK.

F. J. RIZZO, B.S., M.S., Ph.D.
Professor, Department of Engineering Mechanics, University of Kentucky, Lexington, Kentucky 40506, *USA.*

R. P. SHAW, B.S., M.S., Ph.D.
Professor, Department of Engineering Science, Aerospace Engineering and Nuclear Engineering, State University of New York at Buffalo, Buffalo, New York 14214, *USA.*

D. J. SHIPPY, B.S., M.S., Ph.D.
Professor, Department of Engineering Mechanics, University of Kentucky, Lexington, Kentucky 40506, *USA.*

H. TOTTENHAM, M.A., C.Eng., F.I.C.E., F.I.Struct.E.
Partner, Hume, Tottenham and Bennett, Consulting Engineers, Broadwater House, Romsey, Hants, SO5 8GT, UK; also, *Visiting Professor, Department of Civil Engineering, The University, Southampton, SO9 5NH, UK.*

J. O. WATSON, B.Sc.(Eng.), Ph.D.
Advanced Fellow, Interdepartmental Rock Mechanics Project, Imperial College of Science and Technology, London SW7 2AZ, UK.

A. WEXLER, B.Sc., Ph.D., F.R.S.A.
Professor, Department of Electrical Engineering, University of Manitoba, Winnipeg, R3T 2N2, Canada.

O. C. ZIENKIEWICZ, F.R.S.
Professor, Department of Civil Engineering, University of Wales, Swansea, SA2 8PP, UK.

Chapter 1

NEW CONCEPTS ILLUSTRATED BY OLD PROBLEMS

R. BUTTERFIELD

Department of Civil Engineering, University of Southampton, UK

SUMMARY

A simple introduction to Boundary Element Methods is provided by illustrating the application of both Indirect and Direct Methods to the 'beam on elastic foundation' problem. The close correspondence between the one-dimensional algorithms and their three-dimensional counterparts is noted.

The beam example is then used to demonstrate the consequences of using the homogeneous, free space Green's function, in lieu of the specific one, for a completely inhomogeneous beam on elastic foundation system. This leads to a new BEM algorithm and also emphasises the value of exploring the solution of simple one-dimensional systems.

1.1 INTRODUCTION

This chapter is primarily concerned with a demonstration of the application of Boundary Element Methods (BEM) to a class of elasticity problems involving only one space dimension. The material covered has not been published previously although it has been used on introductory BEM lecture courses.

There are, the author believes, at least two sufficient reasons for presenting the work here,

(a) As the BEM become more widely recognised as one of the most attractive and powerful tools of engineering analysis, so does the problem of introducing them to a wider engineering audience become more important. A large number of engineers, very often with a background in structural mechanics, are interested in BEM, particularly in relation to elasticity, but shy away due to the apparent sophistication of the underlying mathematics without realising that they are in fact already familiar with the basic ideas involved.

The first part of this chapter considers both the Indirect (IBEM) and Direct (DBEM) solutions of the well known problem of a uniform beam on a elastic (Winkler) spring foundation, and shows, using a very simple analysis, how the solution algorithms are closely related to those for the general three-dimensional elasticity problem.

(b) Apart from a previous article by the author (Butterfield, 1978), dealing with potential flow in inhomogeneous bodies, the application of BEM to one-dimensional problems does not appear to have been studied. In addition to their value as teaching tools, notwithstanding the fact that the BEM are not shown to advantage in such simple problems, the one-dimensional cases also provide an invaluable means of exploring the application of BEM to non-linear and completely inhomogeneous systems. This is because analytical solutions are available for a number of one-dimensional inhomogeneous field problems on which new BEM techniques can be tested. The second part of this chapter extends the basic BEM to explore the possibility of analysing continuous inhomogeneity without introducing internal *zonal* subdivisions.

1.2 THE HOMOGENEOUS BEAM ON AN ELASTIC FOUNDATION

The simple one-dimensional problem of an elastic beam on an elastic spring foundation will be used to illustrate the basic concepts underlying both Indirect BEM and Direct BEM and to develop algorithms which are essentially similar to those arising in the general study of elastostatics.

The example we are to study is shown in Fig. 1 which illustrates a simple unit width beam (length $= L$, second moment of cross-sectional area for in-plane bending $= I$, material Young's modulus $= E$) supported on a bed of uniform elastic springs of stiffness k per unit length of beam. Our 'boundary' for such a one-dimensional system is merely the two end-points of the beam (P_1, P_2) and only two 'boundary elements', one at each of them, will arise. In general, co-ordinates x_i would locate 'field points'—here $x_1 = x$, and u is the beam displacement in the x_2 direction—and co-ordinates $y_i \equiv y$ locate 'load points' on the beam. The loading considered will be a distributed transverse one, $b(x)$ only, although distributed moment loading, or both types applied simultaneously, can be dealt with similarly. Such loading is exactly analogous in the general elasticity problem to a body force distribution $b(x_i)$, say, throughout a region R with

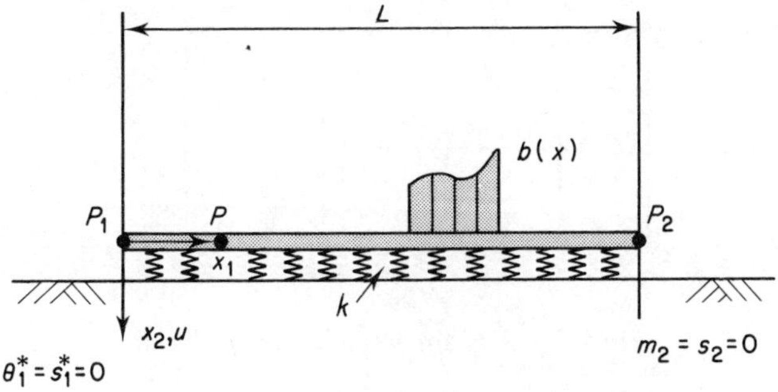

FIG. 1. A loaded beam on an elastic foundation.

surface S. Here R is merely the field length (L), and S the system end-points (P_1, P_2).

Boundary conditions will be specified over S: for example, that the beam slope (θ) and shear force (s) could both be zero at P_1, and the shear force and bending moment (m) zero at P_2. Two items of boundary data at each 'boundary element' are, of course, necessary for a properly posed problem governed by a fourth-order differential equation as this one is. Furthermore a 'mixed' set of boundary values has been chosen in order to emphasise that such problems present no especial difficulty with BEM.

The governing differential equation is well known to be,

$$EI\frac{\mathrm{d}^4u}{\mathrm{d}x^4} = b(x) - ku(x) \tag{1}$$

and deflections, slopes, bending moments and shear forces are interrelated by

$$s = \frac{\mathrm{d}m}{\mathrm{d}x} = -EI\frac{\mathrm{d}^2\theta}{\mathrm{d}x^2} = -EI\frac{\mathrm{d}^3u}{\mathrm{d}x^3} \tag{2}$$

We shall now demonstrate the solution of the problem shown in Fig. 1 by both Indirect and Direct BEM.

1.2.1 Indirect BEM Solution

The classical Green's function, or influence function, approach would be to obtain solutions for a unit load acting alone on the system and then, by using the principle of superposition for the linear system, combine multiples of this solution for loading applied at any number of specified points Q, (Fig. 2). However, the Green's function obtained is entirely dependent on the particular boundary conditions specified, i.e., it changes

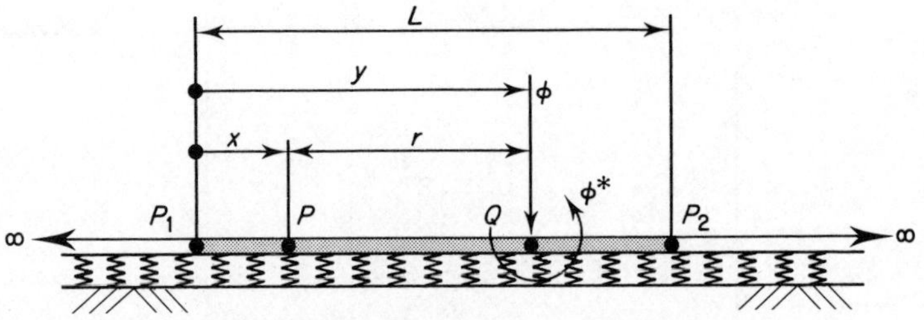

FIG. 2. A beam within an infinite field.

if the boundary conditions are altered for an otherwise identical problem. For complicated systems the determination of the influence functions is essentially as difficult as solving the problem itself, and this approach is not generally useful.

It is much more rewarding to consider the free space Green's function, variously called the infinite region unit singular solution or the fundamental solution, etc. (i.e., the solution obtained for a load $\phi \equiv 1$ (Fig. 2) on an *infinitely long beam* in our case). Such solutions are well known for all the classical field equations—unit point sources in infinite regions in potential problems, unit point loads applied to elastic bodies of infinite extent in elasticity, etc.

We now conceptually embed our real region within an infinite one (Fig. 2) where for convenience, corresponding points are marked identically in Figs. 1 and 2.

The infinite region unit solution means that we have available the value of the displacement $u(x)$ at any point P (and thereby $\theta(x)$, $m(x)$ and $s(x)$ from eqn. (2)) due to any specified loading $\phi(y)$ at any point Q.

For example due to $\phi(y)$ (Hetenyi, 1946),

$$u(x) = \hat{u}(x, y)\phi(y) \tag{3a}$$

or

$$u(r) = \hat{u}(r)\phi, \qquad r = |y - x| \tag{3b}$$

With $\beta = \sqrt[4]{(k/EI)}$

$$\hat{u}(r) = \frac{\beta}{2k} \exp(-\beta r)(\cos \beta r + \sin \beta r) \tag{4}$$

The corresponding expressions for $\theta(r)$, $m(r)$ and $s(r)$ are then, using eqn. (2),

$$\begin{aligned} \theta(r) &= \hat{\theta}(r)\phi, & \hat{\theta}(r) &= \frac{\beta^2}{k} \exp(-\beta r) \sin \beta r \, \mathrm{sgn}(y - x) \\ m(r) &= \hat{m}(r)\phi, & \hat{m}(r) &= \frac{\exp(-\beta r)}{4\beta}(\cos \beta r - \sin \beta r) \\ s(r) &= \hat{s}(r)\phi, & \hat{s}(r) &= \frac{\exp(-\beta r)}{2} \cos \beta r \, \mathrm{sgn}(y - x) \end{aligned} \tag{5}$$

where $\mathrm{sgn}(y - x) = 1$ when $y > x$, $\mathrm{sgn}(y - x) = -1$ when $y < x$ and $\mathrm{sgn}(y - x)$ is undefined when $y = x$.

Although the form of eqn. (3b) is simpler than (3a) this latter equation does emphasise that all of $\hat{u}, \hat{\theta}, \hat{m}, \hat{s}$ are 'two-point functions' which generate the effect of ϕ, at a load point y, on any field point x. The significance of this distinction will become more evident as the direct BEM algorithm is developed subsequently. The function $\hat{u}(r)$ is, of course, the free-space Green's function for $u(r)$, $(\phi = 1)$.

There is also a similar Green's function ($u^*(r)$, say), for the case of a unit concentrated moment ($\phi^* \equiv 1$, Fig. 2) for which (Hetenyi, 1946),

$$u(r) = u^*(r)\phi^*, \qquad u^* = \frac{2\beta^2}{k}\hat{\theta}(r) \tag{6}$$

and from which, using eqn. (2), equations for θ^*, m^*, s^* analogous to (5) can be obtained as,

$$\begin{aligned} \theta(r) &= \theta^*(r)\phi^*, & \theta^* &= -\frac{4\beta^4}{k}\hat{m}(r) \\ m(r) &= m^*(r)\phi^*, & m^* &= \hat{s}(r) \\ s(r) &= s^*(r)\phi^*, & s^* &= \frac{k\hat{u}}{\beta}(r) \end{aligned} \tag{7}$$

Hereafter the symbol $\boldsymbol{\phi}$ will be used to represent the two-component column vector (ϕ, ϕ^*).

We shall be very much concerned with the situation where Q is located on the boundary for example at P_1, and the field point P also approaches P_1, (i.e., $r \to 0$). All the expressions in eqns. (5), (6) containing the sgn function undergo a step change of sign as P passes through Q (as $y - x$ changes sign)—they are multi-valued at $r = 0$. Such multi-valued functions abound in BEM but once their presence has been recognised and the fundamental notion appreciated that our field points will always approach the boundary load point, to within an infinitesimal distance $\varepsilon, \varepsilon \to 0$, from *inside* the region of interest (i.e., here from inside P_1, P_2) then the values of u, θ, s, m will always be defined unambiguously. Exactly analogous multi-valued functions arise in all potential flow and elasticity problems.

The final step is to consider Fig. 3 which is Fig. 2 once more but with the known loading $b(x)$ applied together with two (one per 'boundary element')

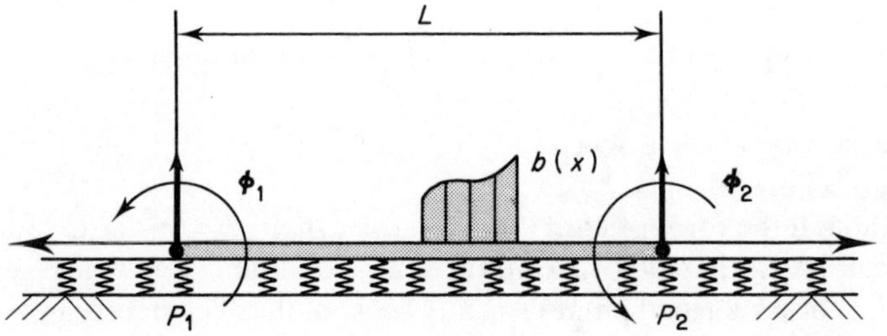

FIG. 3. The fictitious system with boundary element constraints.

unknown 'fictitious' loading vectors, $\boldsymbol{\phi}_1$, and $\boldsymbol{\phi}_2$ applied at P_1 and P_2 respectively. We now determine the unique values of $\boldsymbol{\phi}_1$ and $\boldsymbol{\phi}_2$ which will generate the real (Fig. 1) boundary conditions at P_1 and P_2 within the 'fictitious' problem domain (Fig. 2). Once $\boldsymbol{\phi}_1$ and $\boldsymbol{\phi}_2$ have been found our initial problem is solved for we can then calculate, say $u(x)$, at any internal point (x) merely by superimposing the effects of $\boldsymbol{\phi}_1, \boldsymbol{\phi}_2$ and $b(x)$ using eqns. (3a) and (6).

The distributed loading $b(x)$ is catered for by either approximating it as a closely spaced series of point loads or by using the uniformly distributed load equivalent of $\hat{u}(r)$ (again well known) to approximate $b(x)$ as a series of load steps.

In a properly posed problem there will always be specified exactly the number of boundary values, here four, required to determine the unknown $\boldsymbol{\phi}$ components.

As an example we shall consider the simplified problem of Fig. 4 in which the loading $b(x)$ is reduced to the point excitation vector $\mathbf{b}$, the column vector (b, b^*). Thus if we take the origin of co-ordinates at P_1, we have, using the Fig. 1 boundary conditions (i.e., a free end at P_1, $m_1 = s_1 = 0$ and a 'symmetrically loaded beam centreline condition' $\theta_2 = s_2 = 0$ at P_2), from eqns. (4) and (5).

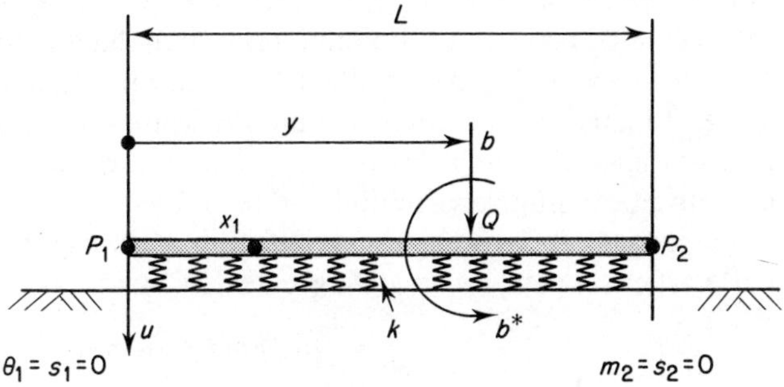

FIG. 4. A beam on elastic foundation example with specified boundary conditions.

For $s_1 = 0$, here $x = +\varepsilon$

$$s_1 = \phi_1 \hat{s}(\varepsilon, 0)(-1) + \phi_1^* s^*(\varepsilon, 0) + \phi_2 \hat{s}(\varepsilon, L)(1) + \phi_2^* s^*(\varepsilon, L) + b\hat{s}(\varepsilon, y) + b^* s^*(\varepsilon, y) = 0 \qquad (8)$$

and for $s_2 = 0$, here $x = (L - \varepsilon)$

$$s_2 = \phi_1 \hat{s}(L-\varepsilon, 0)(-1) + \phi_1^* s^*(L-\varepsilon, 0) + \phi_2 \hat{s}(L-\varepsilon, L)(1) + \phi_2^* s^*(L-\varepsilon, L) + b\hat{s}(L-\varepsilon, y) + b^* s^*(L-\varepsilon, y) = 0$$

Two similar equations are easily assembled to express the remaining boundary conditions $\theta_1 = m_2 = 0$. The 'unit solution' equations (5) for ϕ_1 and their counterparts for ϕ^* will provide all the $\hat{u} \cdots \hat{s}$ and $u^* \cdots s^*$ values, b and b^* are specified loads and for any such problem a matrix equation of the following form will always be generated

$$\mathbf{G\Phi} + \mathbf{Hb} = \mathbf{F} \qquad (9)$$

where $\mathbf{\Phi} = (\boldsymbol{\phi}_1, \boldsymbol{\phi}_2)^{\mathrm{T}}$ and **G** is a (4×4) matrix whose components are the coefficients of the ϕ terms in eqns. (7), and **F** is the vector of specified boundary values (here nul). There will clearly be one (**Hb**) term for each loaded point on the beam. Equation (9) determines $\mathbf{\Phi}$ and thereafter, for example, the shear force $s(x)$ at any point (x) can be calculated from say the first of eqns. (8) by substituting x for ε. Similar equations can obviously be written down for $u(x), \theta(x), m(x)$. It is important to appreciate that the dimensions of the matrix to be inverted, **G**, are not in any way affected by the number of internal nodes (Q) selected to represent the loading applied to the beam.

It is also worth noting that Hetenyi provides similar, closed form, unit solutions for a semi-infinite beam for which $\theta_1 = s_1 = 0$ at one end automatically. Thus by using this solution to provide the 'kernel functions' equivalent to $\hat{u}$, $\hat{\theta}$, $\hat{m}$, $\hat{s}$ we only have one 'boundary element' (P_2) at which we need to apply ϕ_2 to satisfy the boundary conditions. The size of the problem is thereby halved and, although this is a trivial case, the point that any particularly convenient singular solution can be utilised with BEM is an important one which can be used to advantage in, for example, three-dimensional elasticity problems involving extensive, plane traction-free

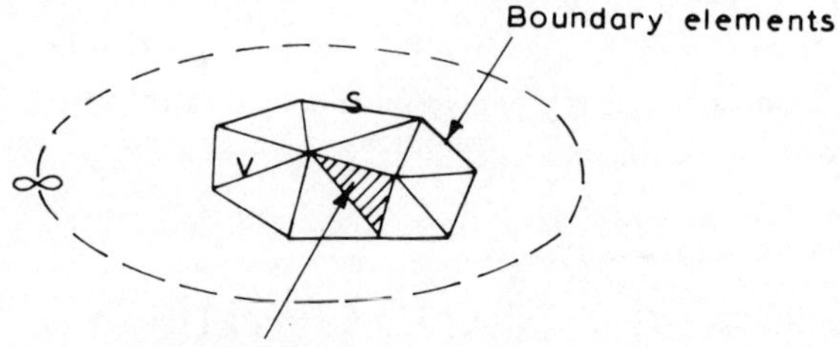

FIG. 5. A more general system.

surfaces. In such cases the use of Mindlin's point load solution as the kernel function in lieu of the usual Kelvin one will ensure that the traction-free surface boundary condition is satisfied automatically without requiring any boundary discretisation whatsoever. In the analysis of many foundation engineering problems this produces a major reduction in the dimensions of the final matrix equations.

To conclude this introductory section on the indirect BEM, the following is a summary of the very few basic steps involved in the solution algorithm, steps which are used identically in two- and three-dimensional problems in potential flow, elastostatics, etc.

(i) Obtain the most convenient singular solution to the governing field equations.
(ii) Distribute this at the specified intensity over internal body cells to represent internal (body) forces (Fig. 5).
(iii) Distribute the solution at initially unknown intensities over a set of arbitrarily chosen boundary elements (Fig. 5).
(iv) Compile equations incorporating (i), (ii) and (iii) which evaluate boundary data around the domain of interest within an infinite or semi-infinite space (i.e. eqns. (8) and (9)) and, by requiring the boundary values to be identical to the specified ones use these equations to calculate the intensities introduced in (iii).

(v) Incorporate these distributions, together with those in (ii), into equations exactly similar to those in (iv), but applicable to internal rather than boundary points, in order to evaluate solution variables at any subsequently selected internal points.

The basic idea underlying IBEM is not a new one at all, indeed Hetenyi solves general 'beam on elastic foundation' problems this way. What is new is its generalisation in recent years to encompass virtually all problems in classical engineering science.

1.2.2 Direct BEM Solution

In contrast to the rather intuitive approach adopted for the indirect method, which nevertheless is equally powerful and in some cases more efficient than the direct one, the direct method is best approached more formally.

We are still to solve the same governing equation (1).

$$EI\frac{\mathrm{d}^4u(x)}{\mathrm{d}x^4} = b(x) - ku(x) \tag{10}$$

We also require the solution to the equation

$$EI\frac{\mathrm{d}^4\hat{U}(x,y)}{\mathrm{d}x^4} = \delta(x,y) - k\hat{U}(x,y) \tag{11}$$

in which $\delta(x,y)$ is the Dirac delta function and, corresponding to the derivatives of $u(x)$ in eqn. (2), we define analogous derivatives of $\hat{U}(x,y)$. Since we shall show below that $\hat{U}(x,y) \equiv \hat{u}(x,y)$ we have,

$$\hat{s} = \frac{\mathrm{d}\hat{m}}{\mathrm{d}x} = -EI\frac{\mathrm{d}^2\hat{\theta}}{\mathrm{d}x^2} = -EI\frac{\mathrm{d}^3\hat{u}}{\mathrm{d}x^3} = -EI\frac{\mathrm{d}^3\hat{U}}{\mathrm{d}x^3} \tag{12}$$

A Dirac delta function is associated with two points (x,y) and takes the value zero when $x \neq y$ but becomes infinite when $x = y$ in such a way that

$$\int_{-\infty}^{\infty} \delta(x,y)\,\mathrm{d}x = 1(y) \tag{13}$$

and therefore represents a unit load at the point y. Consequently the solution to (11) is once more our infinite region Green's function (i.e., eqn. (3), with $\phi = 1$) and therefore we have,

$$\hat{U}(x,y) \equiv \hat{u}(x,y) \tag{14}$$

Thus $\hat{u}(x)$ is the displacement influence function for $\phi = 1$ which justifies the notation used in eqn. (12) where the hatted variables $\hat{s}$, $\hat{m}$, $\hat{\theta}$ are again the shears, moments and slopes given by eqn. (5).

The basic operation, common to the formal development of all direct BEM formulations, is to multiply both sides of the governing equation (10) by $\hat{U}$ and integrate by parts (here four times) over L until all the derivatives of u have been removed and replaced by derivatives of $\hat{U}$, thus

$$\int_0^L EI\frac{\mathrm{d}^4u}{\mathrm{d}x^4}\hat{U}\,\mathrm{d}x = \int_0^L (b-ku)\hat{U}\,\mathrm{d}x$$

$$\therefore \left[\hat{U}EI\frac{\mathrm{d}^3u}{\mathrm{d}x^3}\right]_0^L - \int_0^L EI\frac{\mathrm{d}^3u}{\mathrm{d}x^3}\frac{\mathrm{d}\hat{U}}{\mathrm{d}x}\mathrm{d}x = \int_0^L (b-ku)\hat{U}\,\mathrm{d}x$$

or from (2) and (12)

$$[-\hat{u}(x)s(x)]_0^L + \int_0^L s(x)\hat{\theta}(x)\,\mathrm{d}x = \int_0^L (b-ku)\hat{U}\,\mathrm{d}x$$

Integrating by parts again, three times, produces

$$[-\hat{u}(x)s(x)]_0^L + [\hat{\theta}(x)m(x)]_0^L + \int_0^L \frac{m(x)\hat{m}(x)}{EI}\mathrm{d}x = \int_0^L (b-ku)\hat{U}\,\mathrm{d}x$$

$$\therefore [-\hat{u}(x)s(x) + \hat{\theta}(x)m(x)]_0^L - [\hat{m}(x)\theta(x)]_0^L + \int_0^L \theta(x)\hat{s}(x)\,\mathrm{d}x$$

$$= \int_0^L (b-ku)\hat{U}\,\mathrm{d}x$$

and finally

$$[-\hat{u}(x)s(x) + \hat{\theta}(x)m(x) - \hat{m}(x)\theta(x)]_0^L + [s^*(x)u(x)]_0^L$$

$$+ \int_0^L u(x)\left(EI\frac{\mathrm{d}^4\hat{U}}{\mathrm{d}x^4}\right)\mathrm{d}x = \int_0^L (b-ku)\hat{U}\,\mathrm{d}x$$

i.e., from eqn. (11)

$$[-\hat{u}\,.\,s + \hat{\theta}\,.\,m - \hat{m}\,.\,\theta + \hat{s}\,.\,u]_0^L + \int_0^L (\delta(x,y) - k\hat{U})u(x)\,\mathrm{d}x$$

$$= \int_0^L (b-ku)\hat{U}\,\mathrm{d}x \qquad (15)$$

and using the basic property of the δ function (13) that

$$\int_0^L \delta(x,y)u(x)\,\mathrm{d}x = u(y)$$

we arrive at

$$u(y) = [\hat{u}\,.\,s - \hat{\theta}\,.\,m + \hat{m}\,.\,\theta - \hat{s}\,.\,u]_0^L + \int_0^L b(x)\hat{u}(x, y)\,\mathrm{d}x \qquad (16)$$

An equation which provides the deflection u at any point, now (y), in terms of known parameters $\hat{u}(x, y), \hat{\theta}(x, y), \hat{m}(x, y), \hat{s}(x, y)$ at $x = 0, L$; an integral of the specified loading $(b(x)\,.\,\hat{u}(x, y))$ over L, which may be evaluated either numerically or analytically, and the *complete* set of boundary values $u(x), \theta(x), m(x)$ and $s(x)$ at $x = 0, L$, (i.e., at the boundary elements P_1, P_2).

It should be added that eqn. (16) is well known and could have been derived more directly from either a virtual work approach or Betti's reciprocal theorem. Analogous equations are also well known for potential flow and elasticity problems and therefore the foregoing derivation is *not* a necessary preliminary to the use of direct BEM although successive integration by parts is a basic procedure which can be applied to other differential operators.

Initially we shall know only half of the set of boundary values and the next step is to derive the remainder, noting that once this is done eqn. (16) can generate $u(y)$ directly at any selected interior point along the beam.

The values of $\hat{u}, \hat{\theta}, \hat{m}, \hat{s}(x, y)$ are obtained from precisely† the same equations (3) used in the indirect method. It should be noted, however, that in eqn. (16) the roles of x and y are reversed. This point becomes significant if we wish to calculate the slope of the beam $\theta(y)$ at y. It is permissible to differentiate under the integral signs throughout the development of this equation, and therefore, since $\theta(y) = \mathrm{d}u(y)/\mathrm{d}y$, we have

$$\theta(y) = \left[\frac{\mathrm{d}}{\mathrm{d}y}(\hat{u}(x, y)s(x) - \hat{\theta}(x, y)m(x) + \hat{m}(x, y)\theta(x) - \hat{s}(x, y)u(x))\right]_0^L + \int_0^L b(x)\frac{\mathrm{d}\hat{u}(x, y)}{\mathrm{d}y}\mathrm{d}x \qquad (17)$$

Only the $\hat{u}, \hat{\theta}, \hat{m}, \hat{s}$ terms involve y and these can be differentiated very simply. However although $\hat{\theta}(x, y) = \mathrm{d}\hat{u}(x, y)/\mathrm{d}x$ by definition, $\hat{\theta}(x, y) \neq \mathrm{d}\hat{u}(x, y)/\mathrm{d}y$ unless $\hat{u}(x, y)$ is symmetrical in x and y (which it is not). This point does not raise any particular problems and is only mentioned here

† Should the loading function include distributed moments $b^*(x)$, say, then clearly eqn. (16) would have to be augmented by terms involving the relevant fundamental solution $u^*(x, y)$ and its derivatives, eqns. (6) and (7).

since it illustrates simply the underlying cause of the subtly different kernel functions which arise in direct and indirect BEM generally. We shall use the symbols $\hat{u}', \hat{\theta}', \hat{m}', \hat{s}'$ to denote $d\hat{u}/dy = \hat{u}'$, etc.

Reverting to the determination of the unknown boundary values in eqn. (16) we do this by taking our field point (y) as before to $y = +\varepsilon$, and $y = (L - \varepsilon)$ within the bounding elements at $x = 0, L$. Writing this out fully in the form of (16) we have the matrix equation.

$$\begin{Bmatrix} u(L-\varepsilon) \\ u(0+\varepsilon) \end{Bmatrix} = \begin{bmatrix} \hat{u}(L, L-\varepsilon), -\hat{\theta}(L, L-\varepsilon), \hat{m}(L, L-\varepsilon), -\hat{s}(L, L-\varepsilon) \\ \hat{u}(L, 0+\varepsilon), -\hat{\theta}(L, 0+\varepsilon), \hat{m}(L, 0+\varepsilon), -\hat{s}(L, 0+\varepsilon) \end{bmatrix} \begin{Bmatrix} s(L) \\ m(L) \\ \theta(L) \\ u(L) \end{Bmatrix}$$

$$- \begin{bmatrix} \hat{u}(0, L-\varepsilon), -\hat{\theta}(0, L-\varepsilon), \hat{m}(0, L-\varepsilon), -\hat{s}(0, L-\varepsilon) \\ \hat{u}(0, 0+\varepsilon), -\hat{\theta}(0, 0+\varepsilon), \hat{m}(0, 0+\varepsilon), -\hat{s}(0, 0+\varepsilon) \end{bmatrix} \begin{Bmatrix} s(0) \\ m(0) \\ \theta(0) \\ u(0) \end{Bmatrix}$$

$$+ \begin{Bmatrix} \int_0^L b(x)\hat{u}(x, L-\varepsilon)\,dx \\ \int_0^L b(x)\hat{u}(x, 0+\varepsilon)\,dx \end{Bmatrix} \tag{18}$$

Once again these equations are algebraically cumbersome, but from a computational point of view only involve systematic assembly of the matrices by inserting specific numerical co-ordinate values in the singular solution equations. The last term in eqn. (18) involves the specified loading on the beam and this is usually handled most conveniently by treating the load discretely, as say n lumped contributions and writing for example,

$$\int_0^L b(x)\hat{u}(x, L-\varepsilon)\,dx = \sum_{i=1}^{n} b(x_i)\hat{u}(x_i, L-\varepsilon)\Delta x_i \tag{19}$$

Alternatively sections of uniformly distributed loading may be incorporated by using the relevant Green's function as mentioned in 1.2.1.

Equation (17) can be used to write a further set of two equations exactly analogous to (18) which provides four equations in all interrelating the eight boundary values. Four of the boundary values will always be specified and these equations serve to calculate the remaining four. Thus we always arrive at an identity interrelating the boundary data which takes the final form.

$$\mathbf{D}_1\mathbf{F}_1 + \mathbf{D}_2\mathbf{F}_2 = \mathbf{Hb} \tag{20}$$

where $\mathbf{F}_1$ and $\mathbf{F}_2$ are vectors of known and unknown boundary values respectively and, as before, $\mathbf{b}$ is a known loading vector. Once all of $\mathbf{F}$ have been calculated using (20), eqns. (16) and (17) can be used to obtain $u(y), \theta(y)$ at any point (y) as mentioned previously. It is of interest at this point to recall the indirect BEM eqn. (9) which in terms of the known $\mathbf{F}_1$ becomes, say

$$\mathbf{G}_1\mathbf{\Phi} + \mathbf{H}_1\mathbf{b} = \mathbf{F}_1 \tag{21}$$

A similar equation could obviously be assembled relating $\mathbf{\Phi}$ to the unspecified boundary data $\mathbf{F}_2$ as,

$$\mathbf{G}_2\mathbf{\Phi} + \mathbf{H}_2\mathbf{b} = \mathbf{F}_2 \tag{22}$$

eliminating $\mathbf{\Phi}$ between (22) and (23) leads to

$$\mathbf{G}_1^{-1}\mathbf{F}_1 - \mathbf{G}_2^{-1}\mathbf{F}_2 = (\mathbf{G}_1^{-1}\mathbf{H}_1 - \mathbf{G}_2^{-1}\mathbf{H}_2)\mathbf{b} \tag{23}$$

which has to be equivalent to (20).

Therefore in practice the direct and indirect methods are essentially similar despite the rather different presentations used here. Whereas the indirect BEM algorithm varies very little in its form no matter to what problem it is applied, the direct method algorithm, in effect, translates a known identity between the boundary data into numerical form. The relevant identities are well known and do not have to be derived uniquely each time. Thus from a computational point of view a direct BEM solution starts from eqns. (16) and (17) expressed as a boundary identity in the form of (18). Once this equation has been used to obtain all the boundary values they are then inserted in (16) and (17) to evaluate displacements and slopes at selected internal points.

It may be helpful here to outline the close analogy which exists between the solution presented here for a one-dimensional 'biharmonic' problem with the full three-dimensional elasticity problem. If, for example, eqns. (16) and (17) are written together as,

$$u_i(y) = \begin{Bmatrix} u \\ \theta \end{Bmatrix}_y = \left[\begin{bmatrix} \hat{u}, & -\hat{\theta} \\ \hat{u}', & -\hat{\theta}' \end{bmatrix}\begin{Bmatrix} s \\ m \end{Bmatrix} - \begin{bmatrix} \hat{s}, & -\hat{m} \\ \hat{s}', & -\hat{m}' \end{bmatrix}\begin{Bmatrix} u \\ \theta \end{Bmatrix}\right]_0^L + \int_0^L \begin{bmatrix} \hat{u}, & -\hat{\theta} \\ \hat{u}', & -\hat{\theta}' \end{bmatrix}\begin{Bmatrix} b \\ b^* \end{Bmatrix} \mathrm{d}x$$

where loading terms due to applied moments b^* have been added in and if $t_j = \begin{Bmatrix} s \\ m \end{Bmatrix}$ represents the boundary traction terms, this equation can be written compactly as, say

$$u_i(y) = [\mathrm{U}_{ij}t_j - \mathrm{T}_{ij}u_j]_0^L + \int_0^L \mathrm{U}_{ij}b_j\,\mathrm{d}x, \qquad (i,j = 1,2) \tag{24}$$

The corresponding equation for three-dimensional elasticity, $(i,j = 1,2,3)$, is, (see eqn. (12) Chapter 3 or eqn. (23) Chapter 4)

$$u_i(y) = \int_s (\mathrm{U}_{ij}t_j - \mathrm{T}_{ij}u_j)\,\mathrm{d}s + \int_v \mathrm{U}_{ij}b_j\,\mathrm{d}v \tag{25}$$

We see that eqns. (24) and (25) are essentially identical in form apart from the increase in spatial dimensions, by two, in each of the integral terms. (Comparable expressions for potential flow problems are given in Butterfield (1978)). The different solutions therefore reflect simply the increase in dimensionality of the elastic system without any conceptual change in the underlying analysis.

1.2.3 The Continuously Inhomogeneous Beam System

Many practical situations involve bodies composed of zones of different, but homogenous, materials ('piecewise' homogeneous bodies). Such problems can be solved quite simply by discretising the zonal boundaries and treating each zone as a separate body with boundary conditions satisfying continuity across the zone interfaces. (Tomlin & Butterfield, 1974).

We shall now consider the more general problem in which the properties of the body vary continuously in some specified way throughout its extent. For example, in Foundation Engineering a model of the ground with elastic moduli increasing linearly with depth can provide practically useful results. One approach to such problems is to model the ground by layers (zones) arranged to approximate the continuous variation in modulus by step changes (Banerjee & Davies, 1978).

This section explores the implications of solving, by DBEM, the beam on elastic foundation problem in which the section modulus $C \equiv EI$ varies continuously along the length of the beam but the spring stiffness k remains constant.

If now the inhomogeneous beam modulus is $(C\alpha)$ with C constant and

$\alpha = \alpha(x)$ the governing differential equations become

$$\frac{d^2}{dx^2}\left(C\alpha\frac{d^2u}{dx^2}\right) = b(x) - ku(x) \tag{26}$$

and

$$s = \frac{dm}{dx} = -\frac{d}{dx}\left(C\alpha\frac{d}{dx}\right) = -\frac{d}{dx}\left(C\alpha\frac{d^2u}{dx^2}\right) \tag{27}$$

which reduce to (1) and (2) for $\alpha = 1$.

If we were fortunate enough to know the fundamental free space Green's function related to eqn. (26) we could use precisely the solution procedure outlined in Section 1.2.2.

Therefore if $\bar{U}(x, y)$ say, were the solution of

$$\frac{d^2}{dx^2}\left(C\alpha\frac{d^2\bar{U}}{dx^2}\right) = \delta(x, y) - k\bar{U} \tag{28}$$

then $\bar{U}$ would be our fundamental solution with related slopes, moments and shears $(\bar{\theta}, \bar{m}, \bar{s}$, say). If the integration by parts procedure is now repeated as in Section 1.2.2 with eqn. (11) and $\hat{U}$ replaced by eqn. (28) and $\bar{U}$ it will be found that eqn. (16) is generated identically except that $(\hat{U}, \hat{u}, \hat{\theta}, \hat{m}, \hat{s})$ are replaced by $(\bar{U}, \bar{u}, \bar{\theta}, \bar{m}, \bar{s})$ respectively. This is, of course, necessarily so since the fundamental solution of the governing equation has been used. However the fundamental solutions for general inhomogeneous bodies are neither known nor ever likely to be and an alternative method of attack is needed.

Since we do know the free space Green's function $(\hat{U})$ (eqn. (11)) for the homogeneous body it is interesting to explore the consequences of carrying out the integration by parts operations using $\hat{U}$ in lieu of $\bar{U}$ to multiply eqn. (28), i.e. to integrate by parts three times the equation

$$\int_0^L \frac{d^2}{dx^2}\left(C\alpha\frac{d^2u}{dx^2}\right)\cdot \hat{U}\,dx = \int_0^L (b - ku)\,.\,\hat{U}\,dx \tag{29}$$

The result is the following

$$(\alpha u)(y) = [\hat{u}s - \hat{\theta}m + (\alpha\theta)\hat{m} - \hat{t}(\alpha u)]_0^L + \int_0^L b(x)\hat{u}(x, y)\,dx + \int_0^L u\left(2\frac{d\alpha}{dx}\hat{s} + \frac{d^2\alpha}{dx^2}\cdot\hat{m}\right)dx \tag{30}$$

The function

$$\hat{t} = \frac{1}{\alpha}\frac{d(\alpha m)}{dx}$$

arises in eqn. (30) from the integration by parts, where although the term

$$\left(-\hat{u}\frac{\mathrm{d}}{\mathrm{d}x}\left(\alpha\frac{\mathrm{d}^2u}{\mathrm{d}x^2}\right)\right) = \hat{u}s,$$

the 'symmetrical' one

$$\left(u\frac{\mathrm{d}}{\mathrm{d}x}\left(\alpha\frac{\mathrm{d}^2\hat{u}}{\mathrm{d}x^2}\right)\right)$$

only reduces to

$$\left(-u\frac{\mathrm{d}(\alpha\hat{m})}{\mathrm{d}x}\right).$$

If eqns. (16) and (30) are compared, the difference (apart from the trivial changes of u to αu, θ to $\alpha\theta$, etc.) is seen to be the appearance of an extra integral on the right-hand side of (30). This integral involves the known parameters $\hat{s}$, $\hat{m}$; known derivatives of α and the initially unknown value of $u(x)$. A possible iterative solution procedure would clearly be to solve the boundary equations corresponding to (30) with the final integral (I_0, say) set at zero. This corresponds exactly to the DBEM procedure set out in eqns. (16) to (20) which can be used to provide a first approximation to $u(x)$ and therefore allow an approximate value of I_0 to be calculated. The procedure could then be repeated to produce progressively closer approximations to the correct values of both I_0 and $u(x)$.

A similar technique has been used successfully in a potential flow problem in which the conductivity was proportional to x^m (Butterfield, 1978) and compared with the known analytical solution. The same paper also discusses the consequences of using the homogeneous space Green's function to solve inhomogeneous potential flow problems in two and three dimensions. If the development of eqn. (16) is followed through it will be seen that the all important $u(y)$ term is generated at the stage (eqn. (15)) where the Dirac delta function is reintroduced using the fact that $\hat{u}$ satisfies the original differential equation (11). Therefore $\hat{u}$ can be any solution of (11), not necessarily the free space Green's function that we have used.

We can apparently take this point one stage further. If we postulate U to be the free space Green's function for a simple homogeneous beam, without spring support, (i.e., $C(\mathrm{d}^4U/\mathrm{d}x^4) = \delta(x, y)$), and use this function in lieu of u we will find that the $\int_0^L(kuU)\,\mathrm{d}x$ term on the right-hand side of eqn. (15)

has to be retained and the resulting expression, corresponding to (30) becomes,

$$(\alpha u)(y) = [Us - \Theta m + (\alpha\theta)M - T(\alpha u)]_0^L + \int_0^L bU\,\mathrm{d}x + \int_0^L u\left(2\frac{\mathrm{d}\alpha}{\mathrm{d}x}S + \frac{\mathrm{d}^2\alpha}{\mathrm{d}x^2}M - kU\right)\mathrm{d}x \qquad (31)$$

where, $U \cdots T$ are the simple beam functions corresponding to $\hat{u} \cdots \hat{t}$.

Although eqn. (31) has an additional term in the right-hand side integral the $U \cdots T$ functions are simpler than $\hat{u} \cdots \hat{t}$ and furthermore there is now no restriction on k in this development. Thus eqn. (31) may offer a means of solving, by BEM, an inhomogeneous beam on an inhomogeneous spring foundation problem, utilising merely the simple $U \cdots T$ and $U' \cdots T'$ functions for a homogeneous free beam with an iterative procedure in which the main system matrix would only have to be reduced once. From the previous discussion, the successful implementation of such a scheme could have far reaching consequences in the solution of very general inhomogeneous elastic systems.

1.3 THE SIMPLE BEAM

Since the free space Green's function for an infinite simple beam has now become important, it and the related functions are listed below,

$$\begin{aligned}
U &= \tfrac{1}{12}\{2l^3 + |x-y|^3 - 3l|x-y|^2\} & U' &= -\tfrac{1}{4}|x-y|(|x-y| - 2l)\,\mathrm{sgn}(x-y)\\
\Theta &= \tfrac{1}{4}|x-y|(|x-y| - 2l)\,\mathrm{sgn}(x-y) & \Theta' &= \tfrac{1}{2}(l - |x-y|)\\
M &= -\tfrac{1}{2}(|x-y| - l) & M' &= \tfrac{1}{2}\,\mathrm{sgn}(x-y)\\
S &= -\tfrac{1}{2}\,\mathrm{sgn}(x-y) & S' &= 0
\end{aligned} \qquad (32)$$

where at $|x-y| = l$, $u = \theta = 0$.

1.3.1 An Example

Consider the simple ($c = 1$) homogeneous cantilever in Fig. 6(a) with the boundary conditions, $u_2 = \theta_2 = 0$ and $m_1 = 0, u_1 = 2000$. Direct substitution of the kernel functions into the boundary equations (18), and the corresponding $\theta(y)$ equations from (17), together with the above boundary conditions when rearranged, leads to,

$$\begin{bmatrix} 500/3 & 0 & 0 & 0 \\ 0 & -500/3 & 25 & -5 \\ 0 & 25 & -5 & 1/2 \\ 25 & 0 & 0 & -1/2 \end{bmatrix} \begin{Bmatrix} s_2 \\ s_1 \\ m_2 \\ \theta_1 \end{Bmatrix} = \begin{Bmatrix} -1/2 \\ 1/2 \\ 0 \\ 0 \end{Bmatrix} . (2000) \tag{33}$$

and hence the exact, analytical DBEM solution: $s_1 = s_2 = -6$: $m_2 = -60$: and $\theta_1 = -300$, which is easily checked by direct analysis. Considering now the strongly inhomogeneous cantilever of Fig. 6(b) with $c\alpha = x^4$. The free space Green's function, and related functions, for such a beam extending from $x = 0$ to $x = \infty$, with $u = \theta = 0$ at infinity, are easily shown to be,

$$y < x \leq \infty$$

$$\bar{U} = \frac{1}{2x}\left(1 - \frac{y}{3x}\right) \qquad \bar{U}' = -\frac{1}{6x^2}$$

$$\bar{\Theta} = \frac{1}{2x^2}\left(\frac{2y}{3x} - 1\right) \qquad \bar{\Theta}' = \frac{1}{3x^3}$$

$$\bar{M} = -(x - y) : \bar{S} = -1 \qquad \bar{M}' = 1 : \bar{S}' = 0$$

$$0 \leq x < y$$

$$\bar{U} = \frac{1}{2y}\left(1 - \frac{x}{3y}\right) \qquad \bar{U}' = \frac{1}{y^2}\left(\frac{x}{3y} - \frac{1}{2}\right)$$

$$\bar{\Theta} = -\frac{1}{6y^2} \qquad \bar{\Theta}' = \frac{1}{3y^3}$$

$$\bar{M} = \bar{S} = 0 \qquad \bar{M}' = \bar{S}' = 0 \tag{34}$$

Precisely the same operations, using eqns. (17) and (18), which built up eqn. (33), but using $\bar{U}$, $\bar{U}'$, etc., will generate the equation,

$$\begin{bmatrix} 48 & -68 & 2 & 0 \\ 68 & -288 & 16/3 & 0 \\ -2 & 16/3 & -1/3 & 0 \\ -2 & 72 & -72 & -1 \end{bmatrix} \begin{Bmatrix} s_2 \\ s_1 \\ m_2 \\ \theta_1 \end{Bmatrix} = \begin{Bmatrix} 0 \\ 1 \\ 0 \\ 0 \end{Bmatrix} . (2000) . (1728) \tag{35}$$

with a solution: $s_1 = s_2 = -20\,736$: $m_2 = -207\,360$: $\theta_1 = -800$: which is

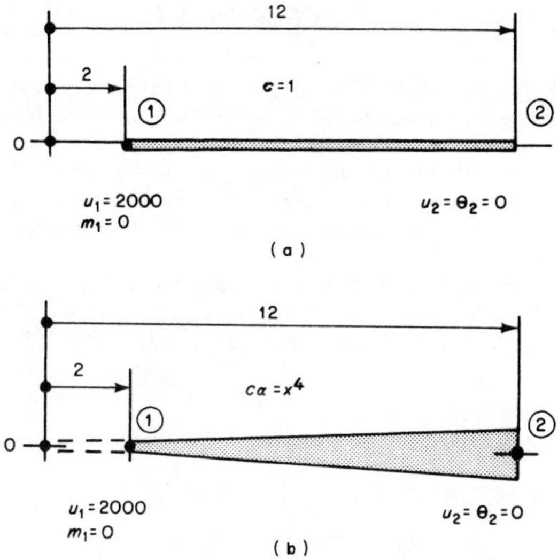

FIG. 6. Cantilevers with specified end displacement (a) homogeneous (b) inhomogeneous member.

an exact analytic DBEM result for a continuously inhomogeneous beam problem using the relevant Green's function.

These two analytical solution examples serve to demonstrate very simply the basic DBEM solution procedure. The reader will find it instructive to compile the matrices involved and to re-solve problem (a) augmented by elastic spring support. The efficient solution of the elastically supported inhomogeneous beam (b) is more of a challenge.

1.4 CONCLUSION

This chapter was intended to provide simple, new, introductory material to illustrate the underlying philosophy of BEM, whilst at the same time indicating how inhomogeneous systems can be solved by the Direct Boundary Element Method. The one-dimensional examples used serve also to show how, at the cost of an iterated solution, the free space Green's function for a free homogeneous body may be useful in solving the rather intractable problems of inhomogeneous, restrained elastic bodies. Additionally, such one-dimensional problems are a useful source of closed form analytical solutions against which numerical algorithms can be checked.

REFERENCES

BANERJEE, P. K. & DAVIES, T. G. (1978). The behaviour of axially and laterally loaded single piles embedded in non-homogenous soils, *Geotechnique*, **28,** 309–27.

BUTTERFIELD, R. (1978). An application of the boundary element method to potential flow problems in generally inhomogeneous bodies. In: *Recent Advances in Boundary Element Methods*, Ed. C. A. Brebbia, Pentech Press, London.

HETENYI, M. (1946). *Beams on Elastic Foundations*, Univ. of Michigan Press.

TOMLIN, G. R. & BUTTERFIELD, R. (1974). Elastic analysis of zoned orthotropic continua, *Proc. ASCE, Eng. Mech. Div.*, **100,** EM3, 511–29.

Chapter 2

NON-LINEAR PROBLEMS OF POTENTIAL FLOW

P. K. BANERJEE

Department of Civil and Structural Engineering,
University College Cardiff, UK

SUMMARY

Boundary element formulations are developed for dealing with non-linear potential flow problems. The problem of non-linearities due to boundary conditions as well as those arising from the changes in the material properties are discussed. A typical solution to a difficult interface problem in coastal ground water flow is described.

2.1 INTRODUCTION

In problems of heat conduction, seepage through porous media, fluid flow, the distribution of electrical or magnetic potential, lubrication of pad bearings, and indeed in many other problems of engineering. it is often necessary to obtain a numerical solution to the differential equation

$$\frac{\partial}{\partial x_i}\left(C\frac{\partial H}{\partial x_i}\right) = Q \tag{1}$$

where H is the total potential at a point; C is the conductivity or the permeability; and Q is the specified sources in the interior of the domain; subject to boundary conditions:

$$H(x_i) = f(x_i) \text{ over the part of the boundary } S_1$$

and

$$V_n(x_i) = V_i n_i = -C\frac{\partial H}{\partial x_i}n_i = h(x_i) \text{ over the part of the boundary } S_2$$

where V_i is the velocity vector; n is the outward normal to the boundary; $S_1 + S_2 = S$, the complete boundary of the region; and V_n is the velocity normal to S_2.

Direct and indirect boundary integral representations for eqn. (1) have been quite well explored in classical texts. Butterfield (1972), Cruse & Rizzo (1975), Hess & Smith (1964), Hess (1974), Jaswon & Symm (1977) and Tomlin (1972) have used these formulations for the numerical solution of linear boundary value problems involving homogeneous and piece-wise homogeneous bodies.

However, in any realistic idealisation of practical problems several non-linear features may arise:

(i) a part of the boundary S may not be known initially,
(ii) the right-hand side of eqn. (1) may be a function of H or V_i,
(iii) the material property C may be dependent on H or V_i.

In this chapter we shall explore each of these non-linearities and develop the necessary integral formulations to deal with them.

2.2 BASIC INTEGRAL FORMULATIONS FOR LINEAR PROBLEMS

By utilising the reciprocal integral identity corresponding to the differential equation (1) with constant C, we can write the direct boundary integral equation as

$$\alpha H(\xi) = \int_S [V_n(x)G(x,\xi) - H(x)F(x,\xi)]\,ds + \int_V Q(x)G(x,\xi)\,dv \quad (2)$$

where

$$\begin{aligned} \alpha &= 1 && \text{if } \xi \text{ is inside } V \\ &= \tfrac{1}{2} && \text{if } \xi \text{ is on } S \\ &= 0 && \text{if } \xi \text{ is outside } V \end{aligned}$$

The corresponding indirect formulation is represented by the two boundary integral equations

$$H(x) = \int_S \phi(\xi)G(x,\xi)\,ds + \int_V Q(\xi)G(x,\xi)\,dv \quad (3)$$

$$V_n(x) = \alpha\phi(x) + \int_S \phi(\xi)F(x,\xi)\,ds + \int_V Q(\xi)F(x,\xi)\,dv \quad (4)$$

where

$$\begin{aligned} \alpha &= 0 && \text{if } x \text{ is inside or outside } V \\ &= \tfrac{1}{2} && \text{for the interior problem} \\ &= -\tfrac{1}{2} && \text{for the exterior problem} \\ \phi &= \text{a boundary source density} \end{aligned}$$

The kernel functions involved in the integral representation given above are:

$G(x, \xi) = -(1/2\pi C) \ln r$ for two dimensions and $1/4\pi Cr$ for three dimensions and $F(x, \xi) = -C(\partial G/\partial x_i)n_i$, where the normal n is evaluated at x.

The corresponding velocity vector V_i at any interior point can be obtained by differentiating eqns. (2) and (3) with respect to the field variables ξ and x respectively.

2.3 NON-LINEAR BOUNDARY CONDITIONS

In many problems of fluid flow such as the seepage through a porous media with a free surface, seepage during pumping from a well, flow over a spillway crest, etc., the free surface of flow is not known *a priori*. Although at the steady state the flow is governed by the linear differential equation (1), the boundary conditions are non-linear. On the known part of the boundary either V_n or H is known, but on the unknown free surface both conditions must be satisfied simultaneously.

Boundary element formulations for these problems already exist in the current literature (Niwa *et al.*, 1974; Liggett, 1977). These authors utilised numerical algorithms developed for the finite element analyses of such problems (Liam Finn, 1967; Taylor & Brown, 1967; France *et al.*, 1971) and described solutions to the problem of seepage through an earthen dam.

In this section we shall describe the solution to a more complex interface problem of coastal ground water flow as shown in Fig. 1. The region ABCDEFGHA is the solution domain for which the exact locations of the free surface CD and the sea water–fresh water interface FG are not known (see Polubarinova-Kochina, 1962).

By considering the fresh water–sea water interface we have

$$H_w = Z + \frac{p}{\rho_w} \tag{5}$$

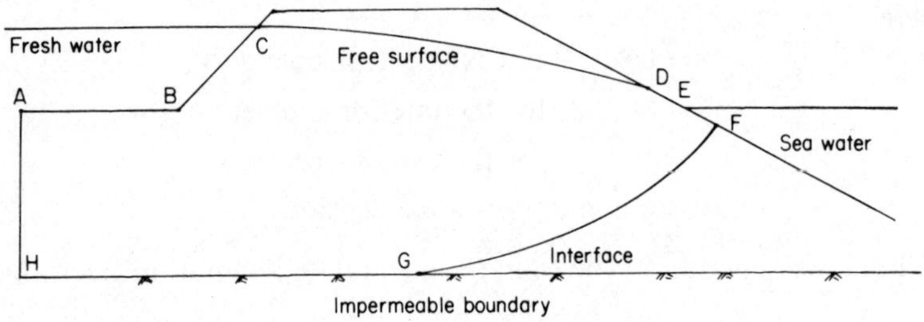

FIG. 1. A typical coastal ground water flow problem.

and

$$H_s = Z + \frac{p}{\rho_s} \tag{6}$$

where H_w, ρ_w are the total head and specific weight of fresh water respectively; H_s, ρ_s are the total head and specific weight of sea water respectively; p is the pressure at the interface; Z is the elevation above a reference datum.

Eliminating p between (5) and (6) we have

$$H_w = Z\left[\frac{\rho_w - \rho_s}{\rho_w}\right] + \frac{\rho_s}{\rho_w} H_s \tag{7}$$

Since in the steady state the sea water may be considered as static, only the fresh water flows; H_s may be regarded as constant. Therefore the total head $H = H_w$ of the fresh water at the interface FG varies as a linear function of the elevation Z.

At the air–water interface CD (assuming the specific weight of air as zero), we have:

$$H_w = Z \tag{8}$$

Thus in the region under consideration, ABC is an equipotential boundary (H specified) and GHA is an impermeable boundary ($V_n = 0$). Along FG and CD the total head varies according to eqns. (7) and (8) respectively. Additionally, since these are also stream lines we must have $V_n = 0$. Part of the boundary DEF is a seepage surface where H is specified unambiguously with $V_n \neq 0$.

In the absence of any specified sources Q in the interior of the domain (no pumping, for example) we can discretise eqn. (2) in the usual manner by dividing the boundary into N straight-line segments and assuming uniform V_n and H over them. Thus:

$$\alpha H(\xi^p) = \sum_{q=1}^{N} \left\{ V_n(x^q) \int_{\Delta S_q} G(x^q, \xi^p)\,\mathrm{d}s - H(x^q) \int_{\Delta S_q} F(x^q, \xi^p)\,\mathrm{d}s \right\} \quad (9)$$

By taking ξ^p onto all the boundary nodes in turn and separating the known and unknown components of V_n and H we can form the following system of equations

$$\mathbf{Ax} = \mathbf{b} \quad (10)$$

where $\mathbf{A}$ is a fully populated $N \times N$ matrix (usually non-symmetric); $\mathbf{x}$ is an $N \times 1$ vector containing the unknown boundary values of V_n and H; and $\mathbf{b}$ is an $N \times 1$ vector compiled from the prescribed boundary values of V_n and H.

It is also possible to use eqns. (3) and (4) as a basis for constructing the final system of equations (10) as discussed by Butterfield (1972), Banerjee & Butterfield (1979), for example. However, for problems of free surface flow the direct formulations involving physical boundary variables are generally more efficient.

Computations for the non-linear solution of the coastal ground water flow problem may be described as follows:

(i) The locations of CD and FG are guessed initially and the total head H at the collocation points on each of the boundary elements (mid-points, usually) are calculated by using eqns. (7) and (8) for FG and CD respectively. The boundary conditions along the remaining part of the boundary are of course defined unambiguously.

(ii) Solution of eqn. (10) will yield the unknown values of V_n along CD and FG, which will not be zero unless the assumed locations are correct. With these values of V_n as the known boundary data along CD and FG (with boundary conditions elsewhere remaining unchanged) solve eqn. (10) again. This time the solution will yield the unknown values of H along CD and FG.

(iii) Using these values of H, an improved location (elevation) of the free surface and the interface may be obtained via eqns. (8) and (7) respectively.

(iv) Repeat the process until Z^n and Z^{n+1} agree as closely as required. Usually 5 iterations provide satisfactory results.

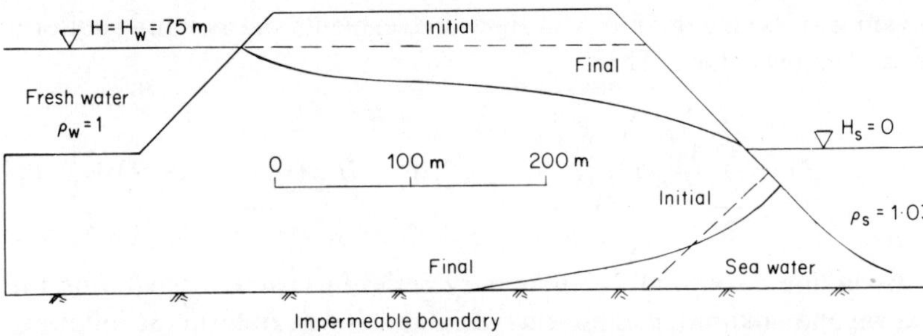

FIG. 2. Positions of the initial and the final free surfaces and interfaces.

Figure 2 shows an example of a typical coastal ground water problem. The initially guessed locations and the final positions of the free surface and the interface are also shown in this figure.

The analysis described above can be easily extended to deal with movements of the free surface or the interface due to rapid draw-down as discussed by France *et al.* (1971) with reference to the finite element method.

2.4 NON-LINEARITIES DUE TO THE SOURCE BEING DEPENDENT ON POTENTIAL

Problems of compressible flow (Coulmy, 1976; Luu & Coulmy, 1977), elasto-hydrodynamics of lubrication (Herrebrugh, 1968), torsion of an elasto-plastic shaft (Mendelson, 1973), etc., can be formulated in such a way that the source Q is dependent on the potential H. Indeed many problems involving material non-linearities (i.e., C being dependent on H or V_i) can also be treated in the same way.

The computational procedure for these problems would involve discretising the boundary into N boundary elements, and the interior into a convenient number of interior cells. This will lead to the final system of equations (see eqns. (2) and (10)):

$$\mathbf{Ax} = \mathbf{b} - \mathbf{GQ} \tag{11}$$

The computational steps can then be developed as follows:

(i) Solve eqn. (11) with $Q = 0$ as a first approximation. From the boundary values of V_n and H calculate the interior values of V_i and H.

(ii) From the known relationship between Q and H (or V_n) calculate the values of the sources $\mathbf{Q}$ for the interior cells. Solve eqn. (11) with these $\mathbf{Q}$ (essentially this means that the same system of equations is solved for a different right-hand side).
(iii) Calculate a revised set of interior H and V_i and thus a revised source density Q within the domain.
(iv) Repeat the process until the changes in Q between two successive steps are less than an acceptable norm.

The procedure listed above usually converges if the non-linearities involved are not severe. Coulmy (1976), Luu & Coulmy (1977) adopted an essentially similar procedure for calculating the compressible flow around an aerofoil up to the shock-free transonic range by using an indirect boundary element formulation.

2.5 MATERIAL NON-LINEARITIES

In problems of flow through porous media, electrical and magnetic fields, elasto-hydrodynamics of lubrication, etc., the material parameter C is often dependent on the values of V_i or H. We may express C as:

$$C = C^{\circ} + C^{\mathrm{d}} \tag{12}$$

where C° is the constant part
C^{d} is the part dependent on V_i or H.

The governing differential equation (1), for convenience with $Q = 0$, can be expressed as:

$$C^{\circ}\frac{\partial^2 H}{\partial x_i \partial x_i} + \frac{\partial}{\partial x_i}\left(C^{\mathrm{d}}\frac{\partial H}{\partial x_i}\right) = 0$$

or

$$C^{\circ}\frac{\partial^2 H}{\partial x_i \partial x_i} + Q^* = 0 \tag{13}$$

Over the boundary the value of potential remains unaffected, however, the normal velocity must be modified to

$$V_n' = V_n + C^{\mathrm{d}}\frac{\partial H}{\partial x_i}n_i \tag{14}$$

The problem is then reduced to the solution of eqn. (13) subject to a modified normal velocity at the boundary given by eqn. (14).

The direct integral formulation for (13) can then be expressed as

$$\alpha H(\xi) = \int_S [V_n'(x)G(x,\xi) - H(x)F(x,\xi)]\,\mathrm{d}s - \int_V \frac{\partial}{\partial x_i}\left[C^{\mathrm{d}}\frac{\partial H(x)}{\partial x_i}\right]G(x,\xi)\,\mathrm{d}v \qquad (15)$$

By applying divergence theorems the volume integral can be expressed as

$$-\int_S C^{\mathrm{d}}\frac{\partial H(x)}{\partial x_i}n_i G(x,\xi)\,\mathrm{d}s + \int_V C^{\mathrm{d}}\frac{\partial H(x)}{\partial x_i}\frac{\partial G(x,\xi)}{\partial x_i}\,\mathrm{d}v$$

By substituting this result for the volume integral in (15) and using eqn. (14) we obtain

$$\alpha H(\xi) = \int_S [V_n(x)G(x,\xi) - H(x)F(x,\xi)]\,\mathrm{d}s + \int_V C^{\mathrm{d}}\frac{\partial H(x)}{\partial x_i}\frac{\partial G(x,\xi)}{\partial x_i}\,\mathrm{d}v \qquad (16)$$

the boundary integral equation we require for material non-linearities. It is of interest to note that eqn. (16) does not require any modification to the normal velocities at the boundary.

A numerical computation scheme can be developed by assuming $C^{\mathrm{d}} = 0$ initially. As the field establishes the values of C^{d} in the interior, the contribution from the volume integral modifies the original solution. Chapter 4 of this volume describes complete details of such a scheme of numerical calculation for much more complex problems of two- and three-dimensional elasto-plasticity. The boundary integral used there is essentially of identical form.

In some problems the value of C may change by orders of magnitude (often exponentially with respect to the field H), then it is convenient to work with a new set of transformed variables in place of V_n and H. This is usually achieved by expressing the governing differential equation and the boundary conditions in terms of these new variables.

2.6 CONCLUDING REMARKS

In this chapter we have explored methods for dealing with non-linearities in potential flow problems and reviewed the current state of knowledge that exists in this area.

Methods of numerical calculation for treating non-linearities arising due to boundary conditions, material properties, etc., are discussed. A typical example of an application to a problem in coastal ground water flow is presented.

Obviously most of these developments are in their early stages and a considerable research effort will be necessary before it becomes possible to use these solutions for routine analyses of practical problems.

REFERENCES

BANERJEE, P. K. & BUTTERFIELD, R. (1979). *Boundary Element Methods in Engineering Science*, Mc-Graw Hill, London (in press).

BUTTERFIELD, R. (1972). *The application of the integral equation methods to continuum problems in soil mechanics*, Stress–strain behaviour of soils, Roscoe Memorial Symposium, Cambridge, Foulis, pp. 573–87.

COULMY, G. (1976). *Calculation of flow field by means of panel methods*, Euromech Colloquium '75, Braunschweig/Rhode, May, 1976.

CRUSE, T. A. & RIZZO, F. J., Eds (1975). *Proc. ASME Conference on boundary integral methods*, AMD 11, ASME, New York.

FRANCE, P. W., PAREKH, C. J., PETERS, J. C. & TAYLOR, C. (1971). Numerical analysis of free surface seepage problems. *Jour. Irrigation and Drainage Division, ASCE*, 165–79.

HERREBRUGH, K. (1968). Solving the incompressible and isothermal problem in elasto-hydrodynamic lubrication through an integral equation. *Trans. ASME, F.*, **90,** 262–8.

HESS, J. L. (1974). The problem of three-dimensional lifting potential flow and its solution by means of surface singularity method, *Computer Methods in Applied Mechanics and Engineering*, **4,** 283–319.

HESS, J. L. & SMITH, A. M. O. (1964). Calculations of nonlifting potential flow about arbitrary three-dimensional bodies, *Jour. Ship Research*, **8,** (2), 22–44.

JASWON, M. A. & SYMM, G. T. (1977). *Integral Equation Methods in Potential Theory and Elasto-statics*, Academic Press, London.

LIAM FINN, W. D. (1967). Finite element analysis of seepage through dams, *Jour. of Soil Mech. and Foundation Engineering, ASCE*, **SM 6,** 41–8.

LIGGETT, J. A. (1977). Location of free-surface in porous media, *Jour. Hydraulics Division, ASCE*, **HY4,** 353–65.

LUU, T. S. & COULMY, G. (1977). Method of calculating the compressible flow around an aerofoil or a cascade up to the shockfree transonic range, *Computers and Fluids*, **5,** 261–75.

MENDELSON, A. (1973). *Boundary integral methods in elasticity and elasto-plasticity*, NASA Tech. Note, No. T.N. D 7416.

NIWA, Y., KOBAYASHI, S. & FUKUI, T. (1974). An application of the integral equation method to seepage problems, *Proc. 24th Japanese National Congress for Applied Mechanics*, pp. 479–86.

POLUBARINOVA-KOCHINA, P. YA. (1962). *Theory of Ground Water Movement*, Princeton Univ. Press, Princeton, NJ.

TAYLOR, R. L. & BROWN, C. B. (1967). Darcy flow solutions with a free surface, *Jour. Hydraulics Division*, *ASCE*, **HY2,** 25–32.

TOMLIN, G. R. (1972). *Numerical analysis of zoned continuum problem in anisotropic media*, Ph.D. Thesis, Southampton University.

Chapter 3

ADVANCED IMPLEMENTATION OF THE BOUNDARY ELEMENT METHOD FOR TWO- AND THREE-DIMENSIONAL ELASTOSTATICS

J. O. WATSON

Interdepartmental Rock Mechanics Project,
Imperial College, London, UK

SUMMARY

Boundary integral equations of elastostatics are derived by the direct formulation, for finite and infinite domains in two and three dimensions. Quadratic isoparametric elements for the representation of boundaries of finite and infinite extent in two and three dimensions are described. A system of simultaneous equations approximating to the boundary integral equation is presented, and it is shown how the coefficients of these equations may accurately and economically be evaluated by using Gaussian quadrature formulae. The accuracy of the numerical implementation is assessed by reference to simple examples, and suggestions are made for future work.

3.1 INTRODUCTION

The basic feature of boundary element formulations is the derivation, for each homogeneous zone or subregion of a continuum, of an integral equation over the boundary. To make such a derivation possible, it is necessary first to express the solution as the sum of a particular integral, which is a displacement field satisfying the partial differential equation of equilibrium everywhere within the subregion, and of a complementary function, which satisfies the homogeneous partial differential equation corresponding to zero body force, subject to certain boundary conditions.

In the indirect boundary element formulation, the complementary function is represented by an integral in terms of an arbitrary function over the boundary; this function, which may be visualised as a fictitious force distribution acting on the boundary embedded in the infinite elastic space, is found to be the solution of an integral equation. This integral equation may be of the first or second kind, depending upon the boundary conditions and choice of integral representation. It has been shown by Kupradze (1963) that, subject to certain conditions of functional and surface smoothness rarely attained in engineering practice, the solution of the integral equation obtained by the indirect formulation exists, and varies smoothly over the boundary. However, the case in which the boundary has edges and corners has not been considered by mathematicians, and there is no guarantee that in general the proposed fictitious force distribution exists; if it does exist, it is likely to be unbounded near corners and difficult to represent numerically.

In the direct formulation (Jaswon & Ponter, 1963; Symm, 1963; Rizzo, 1967; Cruse, 1969), the partial differential equation for the complementary function is transformed by use of Betti's theorem to an integral equation in terms of boundary values of displacement and traction. If the boundary value problem itself has a solution, then so does the integral equation. In addition, the variation of displacement and traction near corners is well understood, and so can with confidence be represented numerically. In both indirect and direct formulations, displacement and stress at points inside a subregion are calculated after the integral equation has been solved, by integration over the boundary.

Only the direct formulation is considered in this chapter. The boundary integral equation for the three-dimensional case is derived first for a finite subregion, then for one of infinite extent. Corresponding results are given for plane strain and plane stress. The numerical implementation presented here is that of Lachat & Watson (1975, 1976), with extensions for infinite bodies. The surface is represented by elements whose geometry is defined in terms of the quadratic shape functions of intrinsic co-ordinates proposed by Ergatoudis (1968), and these same functions are used for the interpolation of displacement and traction in terms of nodal values. Coefficients of the system of simultaneous equations approximating the integral equation are evaluated using Gaussian quadrature formulae. In devising integration schemes, special attention is paid to the accurate evaluation of integrals of unbounded functions, and to the attainment of computational efficiency by choosing quadrature formulae according to the estimated rapidity of variation of the integrand.

3.2 INTEGRAL EQUATION FOR A FINITE SUBREGION

The partial differential equation of equilibrium for subregion $V^{(k)}$ is

$$\frac{\partial \sigma_{ij}^{(k)}[w]}{\partial x_j} + b_i^{(k)}(x) = 0 \tag{1}$$

where $b_i^{(k)}(x)$ is body force, and $\sigma_{ij}^{(k)}[w]$ is the stress field corresponding to the displacements $w_i^{(k)}(x)$. For an isotropic material, $\sigma_{ij}^{(k)}[w]$ is given in terms of $w_i^{(k)}(x)$ by the constitutive relationship

$$\sigma_{ij}^{(k)}(x) = \lambda^{(k)}\delta_{ij}\frac{\partial w_s^{(k)}}{\partial x_s} + \mu^{(k)}\left[\frac{\partial w_i^{(k)}}{\partial x_j} + \frac{\partial w_j^{(k)}}{\partial x_i}\right] \tag{2}$$

where the Lamé constants $\lambda^{(k)}$ and $\mu^{(k)}$ are given in terms of Young's modulus $E^{(k)}$ and Poisson's ratio $\nu^{(k)}$ by

$$\begin{aligned} \lambda^{(k)} &= \frac{\nu^{(k)}E^{(k)}}{(1+\nu^{(k)})(1-2\nu^{(k)})} \\ \mu^{(k)} &= \frac{E^{(k)}}{2(1+\nu^{(k)})} \end{aligned} \tag{3}$$

The displacement field $w^{(k)}$ is taken to be the sum of a particular integral $v^{(k)}$ and complementary function $u^{(k)}$. The particular integral is any function satisfying

$$\frac{\partial \sigma_{ij}^{(k)}[v]}{\partial x_j} + b_i^{(k)}(x) = 0 \tag{4}$$

for all $x \in V^{(k)}$. Let $n_i^{(k)}(x)$ be the unit outward normal at a point $x \in S^{(k)}$, the surface of $V^{(k)}$. Then, denoting by $t_i^{(k)}(x)$ the traction $n_j^{(k)}(x)\sigma_{ij}^{(k)}[u]$, the boundary conditions for the complementary function may be calculated from

$$\begin{aligned} u_i^{(k)}(x) &= w_i^{(k)}(x) - v_i^{(k)}(x) \\ t_i^{(k)}(x) &= \{\sigma_{ij}^{(k)}[w] - \sigma_{ij}^{(k)}[v]\}n_j^{(k)}(x) \end{aligned} \tag{5}$$

At a point x on the interface between subregions $V^{(k)}$ and $V^{(l)}$, the compatibility and equilibrium conditions are

$$\begin{aligned} u_i^{(k)}(x) + v_i^{(k)}(x) &= u_i^{(l)}(x) + v_i^{(l)}(x) \\ t_i^{(k)}(x) + \sigma_{ij}^{(k)}[v]n_j^{(k)}(x) &= -\{t_i^{(l)}(x) + \sigma_{ij}^{(l)}[v]n_j^{(l)}(x)\} \end{aligned} \tag{6}$$

Calculation of the particular integral presents no problem. In the general case, it may be taken to be

$$v_i^{(k)}(x) = \int_{V^{(k)}} U_{ij}^{(k)}(x, y) b_j^{(k)}(y) \, dS_y \tag{7}$$

where $U_{ij}^{(k)}(x, y)$ is the Kelvin solution, corresponding to a concentrated force acting at the point y in the infinite elastic space (Lachat, 1975):

$$U_{ij}^{(k)}(x, y) = \frac{(1 + \nu^{(k)})}{8\pi E^{(k)}(1 - \nu^{(k)})r} \left\{ (3 - 4\nu^{(k)})\delta_{ij} + \frac{(x_i - y_i)(x_j - y_j)}{r^2} \right\} \tag{8}$$

In eqn. (8), r is the distance between x and y. In commonly occurring special cases such as self-weight or centrifugal force, simple polynomial expressions exist for $v_i^{(k)}(x)$, so the integration of eqn. (7) can be avoided.

The boundary integral equation for the complementary function is obtained from Betti's theorem. Let $\alpha_s(y)$ and $\beta_s(y)$ be continuously differentiable displacement fields, and let $\sigma_{js}[\alpha]$ and $\sigma_{js}[\beta]$ be the corresponding stress fields. Then (Lachat, 1975)

$$\int_R \left\{ \beta_j(y) \frac{\partial \sigma_{js}[\alpha]}{\partial y_s} - \alpha_j(y) \frac{\partial \sigma_{js}[\beta]}{\partial y_s} \right\} dR_y$$

$$= \int_S \{\beta_j(y)\sigma_{js}[\alpha] - \alpha_j(y)\sigma_{js}[\beta]\} n_s(y) \, dS_y \tag{9}$$

where S is the surface of the finite domain R. Now let us replace $\alpha_j(y)$ by $u_j^{(k)}(y)$, and $\beta_j(y)$ by $U_{ji}^{(k)}(y, x)$, $x \in V^{(k)}$. The latter function is not continuously differentiable at x, so the domain R is taken to be $V^{(k)} - \tau(x; \varepsilon)$, where (see Fig. 1) $\tau(x; \varepsilon)$ is a sphere of radius ε centred at x.

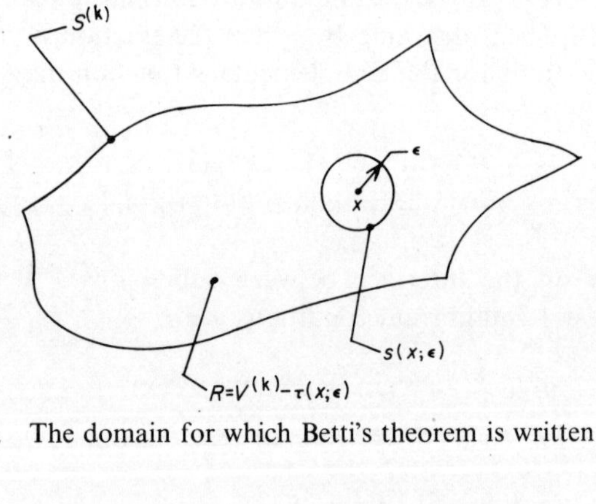

FIG. 1. The domain for which Betti's theorem is written, $x \in V^{(k)}$.

Then, noting that $U_{ji}^{(k)}(y,x) = U_{ij}^{(k)}(x,y)$ and that the volume integral vanishes,

$$\int_{S^{(k)}+s(x;\varepsilon)} \{U_{ij}^{(k)}(x,y)t_j^{(k)}(y) - T_{ij}^{(k)}(x,y)u_j^{(k)}(y)\}\,\mathrm{d}S_y = 0 \tag{10}$$

where $s(x;\varepsilon)$ is the surface of $\tau(x;\varepsilon)$, and (Lachat, 1975)

$$\begin{aligned} T_{ij}^{(k)}(x,y) &= \sigma_{js}^{(k)}[U]n_s^{(k)}(y) \\ &= \frac{1}{8\pi(1-\nu^{(k)})r^2}\left\{(1-2\nu^{(k)})\left[n_i^{(k)}(y)\frac{(x_j-y_j)}{r} - n_j^{(k)}(y)\frac{(x_i-y_i)}{r}\right]\right. \\ &\quad \left. + \left[(1-2\nu^{(k)})\delta_{ij} + 3\frac{(x_i-y_i)(x_j-y_j)}{r^2}\right]n_s^{(k)}(y)\frac{(x_s-y_s)}{r}\right\} \end{aligned} \tag{11}$$

Letting $\varepsilon \to 0$ in eqn. (10), it is found that

$$u_i^{(k)}(x) = \int_{S^{(k)}} U_{ij}^{(k)}(x,y)t_j^{(k)}(y)\,\mathrm{d}S_y - \int_{S^{(k)}} T_{ij}^{(k)}(x,y)u_j^{(k)}(y)\,\mathrm{d}S_y \tag{12}$$

Equation (12) is known as Somigliana's identity, and is used to compute results at points inside the subregion once the boundary integral equation has been solved.

The boundary integral equation is obtained by carrying out the same limiting process, but for $x \in S^{(k)}$. The domain R is taken to be $V^{(k)} - \tau(x;\varepsilon)$ where (see Fig. 2) $\tau(x;\varepsilon)$ is that part of a sphere of radius ε centred at x which is contained within $V^{(k)}$. Then

$$\int_{S^{(k)}-S(x;\varepsilon)+s(x;\varepsilon)} \{U_{ij}^{(k)}(x,y)t_j^{(k)}(y) - T_{ij}^{(k)}(x,y)u_j^{(k)}(y)\}\,\mathrm{d}S_y = 0 \tag{13}$$

where (see Fig. 2) $S(x;\varepsilon)$ is that part of $S^{(k)}$ contained within the sphere of radius ε, and $s(x;\varepsilon)$ is that part of the surface of the sphere contained within

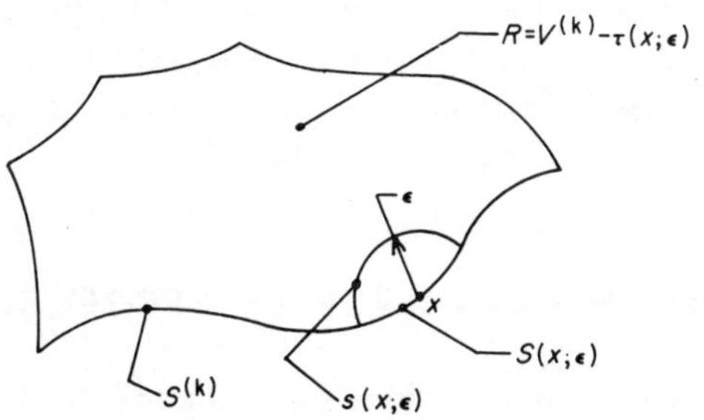

FIG. 2. The domain for which Betti's theorem is written, $x \in S^{(k)}$.

$V^{(k)}$. Letting $\varepsilon \to 0$, it is found that

$$c_{ij}^{(k)}(x)u_j^{(k)}(x) + \int_{S^{(k)}} T_{ij}^{(k)}(x,y)u_j^{(k)}(y)\,\mathrm{d}S_y = \int_{S^{(k)}} U_{ij}^{(k)}(x,y)t_j^{(k)}(y)\,\mathrm{d}S_y \tag{14}$$

where

$$c_{ij}^{(k)}(x) = \lim_{\varepsilon \to 0} \int_{s(x;\varepsilon)} T_{ij}^{(k)}(x,y)\,\mathrm{d}S_y \tag{15}$$

As is implied by the shape of the region of exclusion $\tau(x;\varepsilon)$, the Cauchy principal value of the integral on the left side of eqn. (14) is taken. Note that, since $T_{ij}^{(k)}(x,y) \in o(1/r^2)$, only a principal value corresponding to a specified shape of $\tau(x;\varepsilon)$ which does not change as $\varepsilon \to 0$, could exist. If the tangent plane at x is continuous, $c_{ij}^{(k)}(x) = \frac{1}{2}\delta_{ij}$.

The foregoing derivations may also be carried out for the problems of plane strain and plane stress. The kernels $U_{ij}^{(k)}(x,y)$ and $T_{ij}^{(k)}(x,y)$ for these cases are (Lachat, 1975)

$$U_{ij}^{(k)}(x,y) = \frac{(1+v_1^{(k)})}{4\pi E_1^{(k)}(1-v_1^{(k)})}\left\{(3-4v_1^{(k)})\delta_{ij}\log\frac{1}{r} + \frac{(x_i-y_i)(x_j-y_j)}{r^2}\right\} \tag{16}$$

and

$$T_{ij}^{(k)}(x,y) = \frac{1}{4\pi(1-v_1^{(k)})r}\left\{(1-2v_1^{(k)})\left[n_i^{(k)}(y)\frac{(x_j-y_j)}{r} - n_j^{(k)}(y)\frac{(x_i-y_i)}{r}\right]\right.$$
$$\left. + \left[(1-2v_1^{(k)})\delta_{ij} + 2\frac{(x_i-y_i)(x_j-y_j)}{r^2}\right]n_s^{(k)}(y)\frac{(x_s-y_s)}{r}\right\} \tag{17}$$

where, for plane strain, $E_1^{(k)} = E^{(k)}$ and $v_1^{(k)} = v^{(k)}$; for plane stress, $E_1^{(k)} = E^{(k)}(1-v_1^{(k)2})$ and $v_1^{(k)} = v^{(k)}/(1+v^{(k)})$. The kernel $T_{ij}^{(k)}(x,y) \in o(1/r)$, and as for the three-dimensional case the Cauchy principal value of the integral on the left side of eqn. (14) is taken.

3.3 INTEGRAL EQUATION FOR AN INFINITE SUBREGION

It is assumed in the case of an infinite subregion that the limiting state of stress at infinity is known. The solution is again taken as the sum of a particular integral and complementary function, and now the particular integral must satisfy not only eqn. (4) but also the conditions

$$v_i^{(k)}(x) - w_i^{(k)}(x) \in o\left(\frac{1}{r_0}\right)$$

$$\sigma_{ij}^{(k)}[v] - \sigma_{ij}^{(k)}[w] \in o\left(\frac{1}{r_0^2}\right) \tag{18}$$

where r_0 is the distance from x to an arbitrary reference point x_0. It follows from eqn. (5) that

$$u_i^{(k)}(x) \in o\left(\frac{1}{r_0}\right)$$

$$t_i^{(k)}(x) \in o\left(\frac{1}{r_0^2}\right) \tag{19}$$

Let us now consider the derivation of a boundary integral equation for the complementary function, and let us suppose the argument x of the free term to be located within a sphere of radius r_1 centred at the arbitrary reference point x_0. The equation is derived by considering the infinite subregion $V^{(k)}$ to be the limit as $r_0 \to \infty$ of the finite region $V_0^{(k)}(r_0)$, which is that part of $V^{(k)}$ contained within a sphere of radius r_0, where $r_0 \geq 2r_1$, centred at x_0. The surface of $V_0^{(k)}(r_0)$ consists of two parts: that part of $S^{(k)}$ contained within the sphere, denoted $S_0^{(k)}(r_0)$, and that part of the surface of the sphere contained within $V^{(k)}$, denoted $\Gamma_0(r_0)$. Then eqn. (14) may be re-written

$$c_{ij}^{(k)}(x)u_j^{(k)}(x) + \int_{S_0^{(k)}+\Gamma_0} T_{ij}^{(k)}(x,y)u_j^{(k)}(y)\,\mathrm{d}S_y = \int_{S_0^{(k)}+\Gamma_0} U_{ij}^{(k)}(x,y)t_j^{(k)}(y)\,\mathrm{d}S_y \tag{20}$$

On the surface $\Gamma_0(r_0)$

$$T_{ij}^{(k)}(x,y) \in o\left(\frac{1}{r_0^2}\right)$$

$$U_{ij}^{(k)}(x,y) \in o\left(\frac{1}{r_0}\right) \tag{21}$$

From this and eqn. (19) it follows that the integrals over the surface of the sphere tend to zero as $r_0 \to \infty$. Assuming now that, for $r_0 \geq 2r_1$, the area $A_0^{(k)}(r_0)$ of $S_0^{(k)}(r_0)$ satisfies the inequality

$$\left|\frac{\mathrm{d}A_0^{(k)}}{\mathrm{d}r_0}\right| \leq Kr_0 \tag{22}$$

it can similarly be shown that

$$\int_{S^{(k)}-S_0^{(k)}} T_{ij}^{(k)}(x,y)u_j^{(k)}(y)\,\mathrm{d}S_y \in o\left(\frac{1}{r_0}\right)$$
$$\int_{S^{(k)}-S_0^{(k)}} U_{ij}^{(k)}(x,y)t_j^{(k)}(y)\,\mathrm{d}S_y \in o\left(\frac{1}{r_0}\right) \tag{23}$$

and so the integrals over $S_0^{(k)}(r_0)$ both converge to a limit as $r_0 \to \infty$. Therefore, for x within a sphere of radius r_1 centred at the arbitrary reference point x_0, eqn. (14) is valid for the infinite subregion $V^{(k)}$. Since r_1 may be chosen arbitrarily, this is indeed the case for all $x \in S^{(k)}$. It may similarly be shown that eqn. (12), Somigliana's identity, is also valid for an infinite subregion.

The derivation for an infinite subregion may also be carried out for the problems of plane strain and plane stress, provided conditions (18) are satisfied and

$$\left|\frac{\mathrm{d}s_0^{(k)}}{\mathrm{d}r_0}\right| \leq K \tag{24}$$

where $s_0^{(k)}(r_0)$ is the length of the boundary $S_0^{(k)}(r_0)$.

3.4 NUMERICAL REPRESENTATION OF GEOMETRY AND FUNCTIONS

In the formulation presented here, the boundary is represented by elements whose geometry is defined using the quadratic shape functions of intrinsic co-ordinates proposed by Ergatoudis (1968). The boundary elements for plane problems and for the three-dimensional problem are shown in Figs. 3(a) and 3(b) respectively. The cartesian co-ordinates $x_i(\xi)$ of an arbitrary point of an element are given in terms of the nodal co-ordinates x_i^a by

$$x_i(\xi) = N^a(\xi)x_i^a \tag{25}$$

where, for plane problems, the superscript a ranges from 1 to 3 and (Lachat, 1975)

$$N^1(\xi) = \tfrac{1}{2}\xi(\xi + 1)$$
$$N^2(\xi) = 1 - \xi^2$$
$$N^3(\xi) = \tfrac{1}{2}\xi(\xi - 1) \tag{26}$$

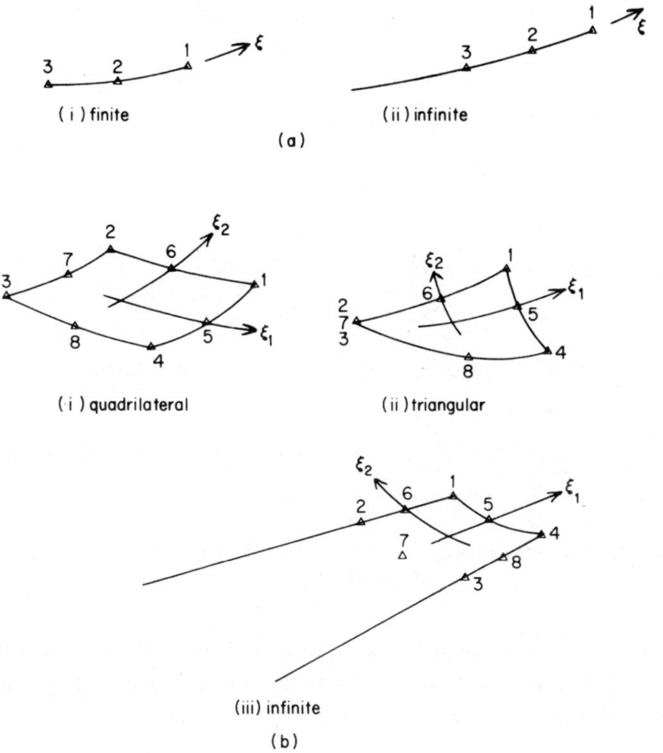

FIG. 3. Quadratic boundary elements. Boundary elements for (a) the plane problems; (b) the three-dimensional problem.

For three-dimensional problems, the superscript a ranges from 1 to 8, and

$$
\begin{aligned}
N^1(\xi) &= \tfrac{1}{4}(\xi_1 + 1)(\xi_2 + 1)(\xi_1 + \xi_2 - 1),\\
N^5(\xi) &= \tfrac{1}{2}(\xi_1 + 1)(1 - \xi_2^2), \text{ etc.}
\end{aligned} \tag{27}
$$

Equation (27) applies to both quadrilateral and triangular elements: in the latter case, nodes 2, 7 and 3 are coincident in space (see Fig. 3(b)). This quadratic variation of geometry can, as experience with finite elements has already shown, adequately represent surfaces occurring in practice.

Displacement and traction at an arbitrary point of a finite boundary element are given in terms of nodal values u_i^a and t_i^a by

$$
\begin{aligned}
u_i(\xi) &= N^a(\xi)u_i^a\\
t_i(\xi) &= N^a(\xi)t_i^a
\end{aligned} \tag{28}
$$

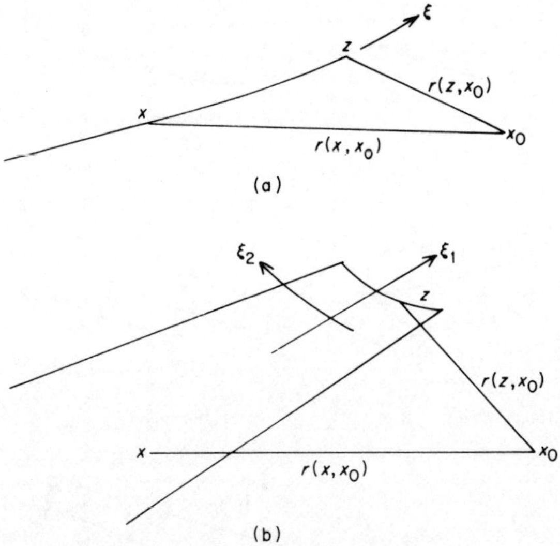

FIG. 4. Asymptotic functions over infinite element. (a) Plane problems; (b) three-dimensional problem.

Note that from here onwards, the superscript (k) is omitted in the interest of clarity. Over an infinite line element (see Fig. 4(a)), displacement and traction at a point $x_i(\xi)$ are taken to be

$$u_i(\xi) = u_i^1 \frac{r(z, x_0)}{r(x, x_0)}$$

$$t_i(\xi) = t_i^1 \left[\frac{r(z, x_0)}{r(x, x_0)} \right]^2 \tag{29}$$

where z is node 1, $r(z, x_0)$ is the distance from node 1 to an arbitrary reference point x_0, and $r(x, x_0)$ is the distance from x to x_0. In eqn. (29) it is assumed that the traction $t_i(x)$, which as will be seen later is in general taken to be discontinuous across element boundaries, is continuous at node 1. The asymptotic behaviour implied by eqn. (29) satisfies conditions (19). Over an infinite surface element (see Fig. 4(b)), displacement and traction at a point $x_i(\xi)$ are taken to be

$$u_i(\xi_1, \xi_2) = u_i(1, \xi_2) \frac{r(z, x_0)}{r(x, x_0)}$$

$$t_i(\xi_1, \xi_2) = t_i(1, \xi_2) \left[\frac{r(z, x_0)}{r(x, x_0)} \right]^2 \tag{30}$$

where z is the point $(1, \xi_2)$, $r(z, x_0)$ is the distance from that point to an arbitrary reference point x_0, and $r(x, x_0)$ is the distance from x to x_0. The traction $t_i(x)$ is taken to be continuous across the line $\xi_1 = +1$, the boundary between the infinite element and the adjoining finite element. The asymptotic behaviour implied by eqn. (30) satisfies conditions (19), and corresponds to the variation of displacement and traction due to a concentrated load at x_0.

3.5 EQUATIONS OF NODAL COLLOCATION

It is now possible to write systems of simultaneous equations in values of displacement and traction at nodes of finite elements, approximating to the boundary integral equation (14). In general some areas of the surface are either partially or completely fixed, and in the case of partial restraint the directions in which a point is fixed and free to move are not necessarily parallel with the global co-ordinate axes. Let $l_{\alpha i}(x)$ be the direction cosines of the directions in which the point x is fixed or free to move. The integral equation for the complementary function is written in terms of the components of displacement and traction in these directions, which are

$$\begin{aligned} u'_\beta(x) &= l_{\beta j}(x) u_j(x) \\ t'_\beta(x) &= l_{\beta j}(x) t_j(x) \end{aligned} \tag{31}$$

Let there be p elements, let the element b have $n(b)$ nodes which are nodes of finite elements, and let there be a total of q distinct nodes of finite elements on the surface S. Let there be constructed a global numbering system for nodes which are nodes of finite elements, the number of node c of element b being denoted $\mathrm{d}(b, c)$. Writing eqn. (14) at each of the q nodes of finite elements, and substituting the parametric functional representations of eqn. (28), and eqn. (29) or (30),

$$c'_{\alpha\beta}(x^a) u'_\beta(x^a) + \sum_{b=1}^{p} \sum_{c=1}^{n(b)} u'_\beta(x\,\mathrm{d}(b, c)) \int_{S_b} T'_{\alpha\beta}(x^a, y(\xi)) N^c(\xi) J(\xi)\, \mathrm{d}\xi$$

$$= \sum_{b=1}^{p} \sum_{c=1}^{n(b)} t'_\beta(b, c) \int_{S_b} U'_{\alpha\beta}(x^a, y(\xi)) N^c(\xi) J(\xi)\, \mathrm{d}\xi \tag{32}$$

where

$$c'_{\alpha\beta}(x) = \lambda_{\alpha i}(x)\lambda_{\beta j}(x)c_{ij}(x)$$
$$T'_{\alpha\beta}(x, y) = \lambda_{\alpha i}(x)\lambda_{\beta j}(y)T_{ij}(x, y)$$
$$U'_{\alpha\beta}(x, y) = \lambda_{\alpha i}(x)\lambda_{\beta j}(y)U_{ij}(x, y) \tag{33}$$

In eqn. (33), $\lambda_{\alpha i}l_{\alpha j} = \delta_{ij}$. In eqn. (32), S_b is the element b, $J(\xi)$ is the jacobian $\mathrm{d}S/\mathrm{d}\xi$ and the superscript a ranges from 1 to q. $t'_\beta(b, c)$ is the limiting value of traction acting on element S_b as the node c is approached. There are in general as many different limiting values of traction at a node x^a as there are elements adjacent to it; all of these may be known boundary data or alternatively some or all of them may be equal to the value $t'_\beta(x^a)$ of a traction unknown associated with the node, and any others known. By $N^c(\xi)$ in eqn. (32) is meant the function defined by eqn. (28), (29) or (30), depending on the type of problem and type of element. When all known boundary data are substituted, eqns. (32) become a system of $2q$ or $3q$ simultaneous equations, depending on the type of problem, in terms of unknowns which number either 2 or 3 per surface node, and either 4 or 6 per interface node.

The order of the system of eqns. (32) can be reduced in cases in which there is symmetry of structure and loading with respect to one or more planes. In Fig. 5(a), for example, is shown a plane problem with axes of symmetry XX and YY. It is possible to represent the two axes of symmetry as boundaries (Fig. 5(b)), as would be done if the finite element method were used. But it might be more efficient to take account of either one (Fig. 5(c)) or both (Fig. 5(d)) of the symmetries during construction of the equations. These alternatives require integration over reflected elements in the other half or quadrants of the structure, so they are not always more economical.

Equations (32) should be written in non-dimensional form, to avoid excessive round-off error that can occur during solution if, due to numerically either very large or very small values of Young's modulus or of typical dimensions of the structure, matrix coefficients of displacement unknowns and of traction unknowns are of a very different order of magnitude. In practice, this requires the use in calculations, not of the units of length and force chosen for the specification of input data, but of, for example, the greatest dimension of the structure as the unit of length, and a unit of force such that Young's modulus equals unity.

After the equations have been solved, the stress tensor at surface points may be determined by substituting into the constitutive relationship (eqn. (2)) the calculated values of tangential strain (obtained by differentiating

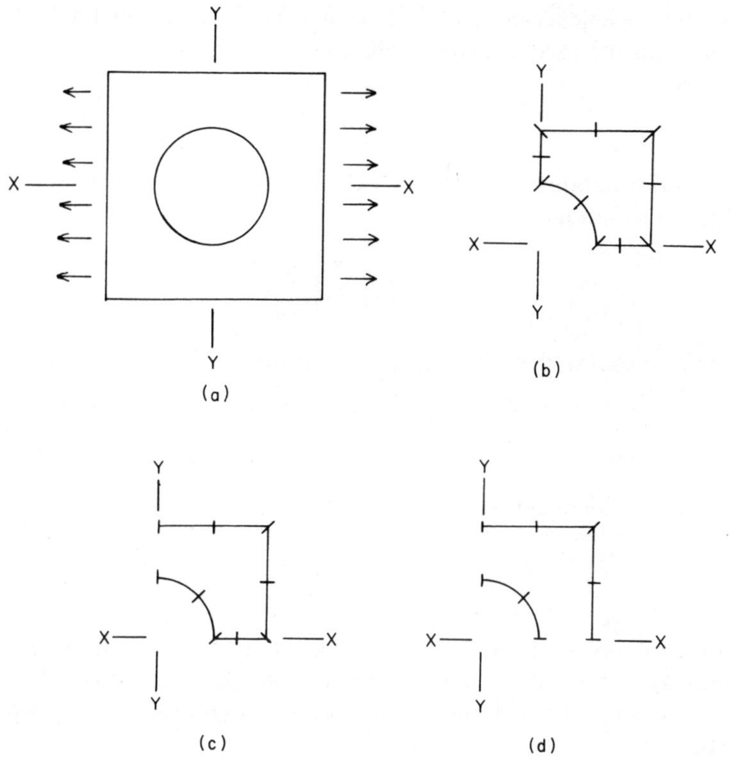

FIG. 5. Example of representation of symmetries. (a) Structure to be analysed; (b) boundary elements, XX and YY represented as boundaries; (c) boundary elements, XX represented as boundary; (d) boundary elements, neither XX nor YY represented as boundary.

eqn. (28)) and traction. Displacement and stress at interior points may be calculated by integration over the surface elements (see eqn. (12)).

3.6 INTEGRATION OF KERNEL–SHAPE FUNCTION PRODUCTS

The kernel–shape function products appearing in eqn. (32) may accurately be evaluated using Gaussian quadrature formulae, provided due precautions are taken where the integrand tends to infinity. In order that formulae for finite intervals such as $(-1, +1)$ may be used, integrations over infinite elements are carried out with respect to co-ordinates $\bar{\eta}$, where, for the elements used in plane problems,

$$\xi = \frac{3\bar{\eta} - 1}{1 + \bar{\eta}} \tag{34}$$

This transformation maps the infinite element onto the $\bar{\eta}$ interval $(-1, +1)$, and the jacobian is

$$\frac{\mathrm{d}\xi}{\mathrm{d}\bar{\eta}} = \frac{4}{(1 + \bar{\eta})^2} \tag{35}$$

For the three-dimensional case, the transformation is

$$\xi_1 = \frac{3\bar{\eta}_1 - 1}{(1 + \bar{\eta}_1)^2}, \qquad \xi_2 = \bar{\eta}_2 \tag{36}$$

for which the jacobian is

$$\frac{\mathrm{d}\xi}{\mathrm{d}\bar{\eta}} = \frac{4}{(1 + \bar{\eta}_1)^2} \tag{37}$$

For plane problems, where $x^a \notin S_b$ and S_b is a finite boundary element, kernel–shape function products may reasonably be approximated by polynomials and so may be integrated according to the formula with weight function unity;

$$\int_{-1}^{+1} f(\eta)\,\mathrm{d}\eta \simeq \sum_{i=1}^{m} A_i^m f(\eta_i^m) \tag{38}$$

Where $x^a \notin S_b$ and S_b is an infinite element, the jacobian

$$\frac{\mathrm{d}S}{\mathrm{d}\bar{\eta}} = \frac{\mathrm{d}S}{\mathrm{d}\xi} \cdot \frac{\mathrm{d}\xi}{\mathrm{d}\bar{\eta}} \in o(r^2)$$

but, by virtue of conditions (19), which are satisfied by the shape functions defined by eqn. (29),

$$T'_{\alpha\beta}(x^a, y(\xi))N^c(\xi) \in o\left(\frac{1}{r^2}\right)$$

$$U'_{\alpha\beta}(x^a, y(\xi))N^c(\xi) \in o\left(\frac{1}{r^2}\log r\right) \tag{39}$$

where $N^c(\xi)$ is taken to represent those functions. The product $T'_{\alpha\beta}(x^a, y(\xi))N^c(\xi)(\mathrm{d}S/\mathrm{d}\bar{\eta})$ therefore remains bounded everywhere in $|\bar{\eta}| \leq 1$, and formula (38) may be used. The product

$$U'_{\alpha\beta}(x^a, y(\xi))N^c(\xi)(\mathrm{d}S/\mathrm{d}\bar{\eta}) \in o(\log r),$$

and may reasonably be approximated by the product of a polynomial in $\bar{\eta}$ and the function $\log(1/(1+\bar{\eta}))$. It is therefore integrated using the Gaussian quadrature formula with weight function $\log 1/\eta$ (Stroud & Secrest, 1966):

$$\int_0^1 f(\eta)\log\frac{1}{\eta}\,\mathrm{d}\eta \simeq \sum_{i=1}^{m} B_i^m f(\eta_i^m) \tag{40}$$

Where $x^a \in S_b$, certain integrands tend to infinity as x^a is approached. Let us consider the case in which x^a is the second node of a finite boundary element S_b (see Fig. 3(a)). Then it is convenient to divide the element into two subelements, as shown in Fig. 6, and to integrate over each in turn. Where $\mathrm{d}(b,c) \neq a$ (see eqn. (32)), formula (38) may be used because although $U'_{\alpha\beta}(x^a, y(\xi)) \in o(\log 1/r)$ and $T'_{\alpha\beta}(x^a, y(\xi)) \in o(1/r)$, the shape function $N^c(\xi) \in o(r)$ and so the product to be integrated tends to a finite limit as x^a is approached, and may reasonably be approximated by a polynomial. Where $\mathrm{d}(b,c) = a$, the shape function $N^c(\xi)$ equals unity at x^a. The product $U'_{\alpha\beta}(x^a, y(\xi))N^c(\xi)$ may reasonably be approximated by the product of a polynomial and the function $\log 1/r$, and so its integral may be evaluated according to formula (40). Where x^a is at the extremities of two elements rather than within an element as shown in Fig. 6, two elements instead of one require special consideration, but it is unnecessary to divide either into subelements. Where one of these elements is an infinite element, integration of $U'_{\alpha\beta}(x^a, y(\xi))N^c(\xi)$ over it is carried out using a double application of formula (40).

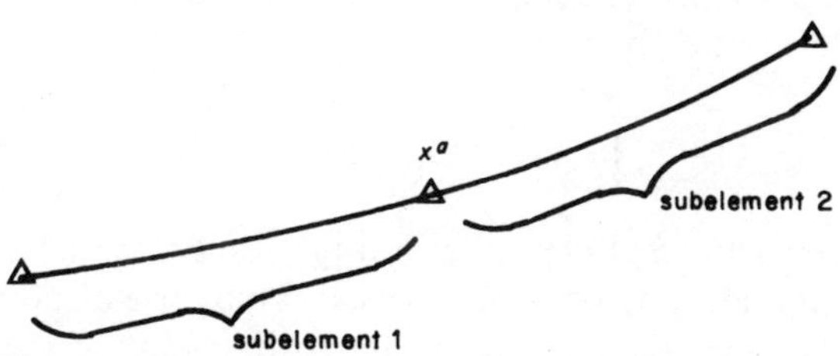

FIG. 6. Subelements for integration, plane problems.

This leaves integrals of $T'_{\alpha\beta}(x^a, y(\xi))N^c(\xi)$, $d(b, c) = a$, and the coefficient of the free term. The integral of $T'_{\alpha\beta}(x^a, y(\xi))N^c(\xi)$ over one subelement or adjacent element alone does not exist when $d(b, c) = a$: the principal value of the integral over the two subelements or adjacent elements taken together does. It is not, however, necessary to evaluate separately either this remaining integral or the coefficient $c'_{\alpha\beta}(x^a)$ of the free term, both of which contribute to the same 2×2 leading diagonal submatrix of coefficients of eqns. (32). Instead, the sum of the two components of this submatrix may be evaluated by considering rigid body translations in the directions of the coordinate axes (Lachat & Watson, 1976). In the case of an infinite subregion, the rigid body translation is considered of the finite region $V_0^{(k)}(r_0)$ with surface $S_0^{(k)}(r_0) + \Gamma_0(r_0)$ (see eqn. (20)). Writing eqn. (32) in terms of global components $u_i(x)$ and $t_i(x)$, substituting $t_i(x) = 0$, and in the case of an infinite subregion letting $r_0 \to \infty$ and taking account of eqn. (19),

$$c_{ij}(x^a) + \sum_{b=1}^{p}\sum_{c=1}^{n(b)} I(a, b, c) \int_{S_b} T_{ij}(x^a, y(\xi))N^c(\xi)J(\xi)\,d\xi$$

$$= -\left\{\sum_{b=1}^{p}\sum_{c=1}^{n(b)} (1 - I(a, b, c)) \int_{S_b} T_{ij}(x^a, y(\xi))N^c(\xi)J(\xi)\,d\xi + \lim_{r_0 \to \infty} \int_{\Gamma_0(r_0)} T_{ij}(x_0, y)\,dS_y\right\} \tag{41}$$

where $I(a, b, c) = 1$ if $d(b, c) = a$, zero otherwise. The implication of this result is that, in order to evaluate the leading diagonal submatrix, it is sufficient to sum all explicitly calculated integrals of $T_{ij}(x, y)$ (global components), add the limiting value of the integral over $\Gamma_0(r_0)$ (the azimuthal integral) in the case of an infinite subregion, change sign, and transform to components in the directions $l_{\alpha i}(x^a)$ (see eqn. (31)). The azimuthal integral is easy to compute. If the subregion comprises the entire elastic space less some finite cavities,

$$\lim_{r_0 \to \infty} \int_{\Gamma_0(r_0)} T_{ij}(x_0, y)\,dS_y = -\delta_{ij} \tag{42}$$

This procedure illustrates clearly the arbitrary nature of the definition of the Cauchy principal value: however that value and the corresponding value of the coefficient of the free term are defined (by choice of the region of exclusion: see eqns. (14) and (15)), the numerical values of the coefficients of the system of simultaneous equations approximating the boundary integral equation are the same.

For three-dimensional problems, the procedure is similar. Where $x^a \notin S_b$ and S_b is a finite surface element, kernel–shape function products are integrated according to the two-dimensional Gaussian formula with weight function unity:

$$\int_{-1}^{+1}\int_{-1}^{+1} f(\eta_1,\eta_2)\,\mathrm{d}\eta_1\,\mathrm{d}\eta_2 \simeq \sum_{i=1}^{m_1}\sum_{j=1}^{m_2} A_i^{m_1} A_j^{m_2} f(\eta_i^{m_1},\eta_j^{m_2}) \tag{43}$$

Where $x^a \notin S_b$ and S_b is an infinite element, the jacobian

$$\frac{\mathrm{d}S}{\mathrm{d}\bar{\eta}} = \frac{\mathrm{d}S}{\mathrm{d}\xi}\cdot\frac{\mathrm{d}\xi}{\mathrm{d}\bar{\eta}} \in o(r^3)$$

but, by virtue of conditions (19), which are satisfied by the shape functions defined by eqn. (30),

$$T'_{\alpha\beta}(x^a, y(\xi))N^c(\xi) \in o\left(\frac{1}{r^3}\right)$$

$$U'_{\alpha\beta}(x^a, y(\xi))N^c(\xi) \in o\left(\frac{1}{r^3}\right) \tag{44}$$

where $N^c(\xi)$ is taken to represent those functions. The integrands therefore remain bounded everywhere in $|\bar{\eta}_i| \leq 1$, and formula (43) may be used.

Now let us consider the case $x^a \in S_b$. A quadrilateral element (see Fig. 3(b)) is divided into either two or three triangular subelements as shown in Fig. 7(a), and integration over each is carried out with respect to the system of co-ordinates, obtained by a linear transformation of the intrinsic co-ordinates of S_b, shown in Fig. 8. It may be seen that, since the lines $\bar{\eta}_1 = +1$ and $\bar{\eta}_1 = -1$ converge to a point at x^a, the jacobian $J(\bar{\eta}) = (\mathrm{d}S/\mathrm{d}\bar{\eta}) \in o(r)$. Where $\mathrm{d}(b,c) \neq a$ (see eqn. (32)), formula (43) may be used because, although $U'_{\alpha\beta}(x^a, y(\xi)) \in o(1/r)$ and $T'_{\alpha\beta}(x^a, y(\xi)) \in o(1/r^2)$, the shape functions $N^c(\xi) \in o(r)$, so the integrand tends to a finite limit as x^a is approached. Where $\mathrm{d}(b,c) = a$, the shape function $N^c(\xi)$ equals unity at x^a; the product $U'_{\alpha\beta}(x^a, y(\xi))N^c(\xi)J(\bar{\eta})$ tends to a finite limit as x^a is approached, and so may be integrated using formula (43). A triangular element (see Fig. 3(b)) need only be divided into subelements if x^a is a midside node (see Fig. 7(b)). For the triangular element described here, a degenerate quadrilateral, the co-ordinate system $\bar{\eta}$ is in certain

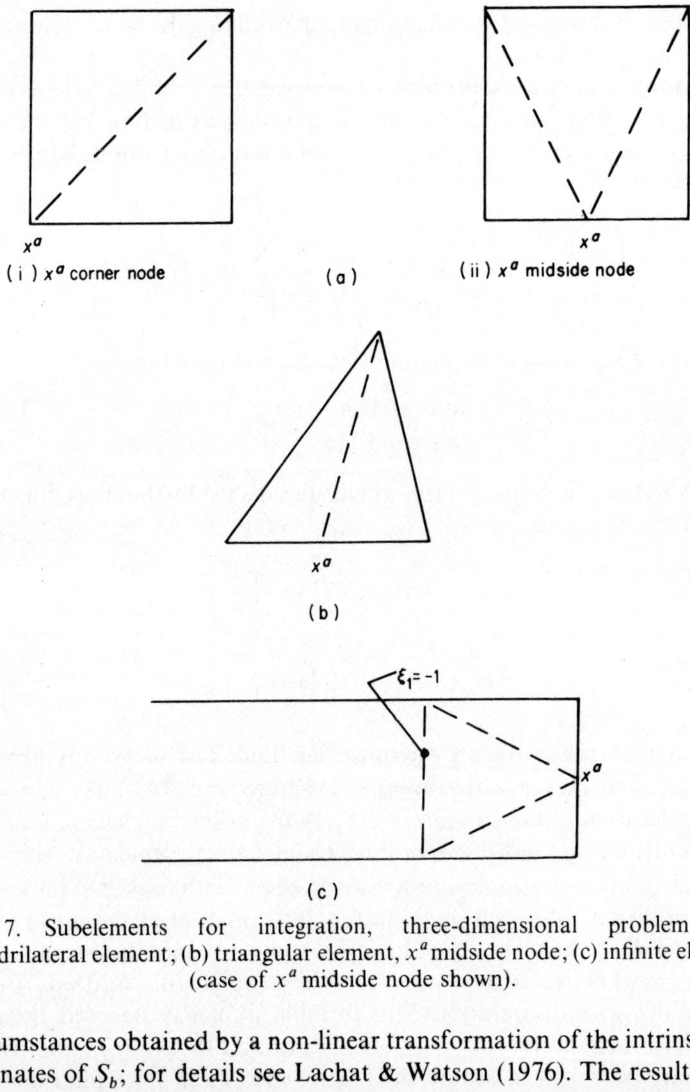

FIG. 7. Subelements for integration, three-dimensional problem. (a) Quadrilateral element; (b) triangular element, x^a midside node; (c) infinite element (case of x^a midside node shown).

circumstances obtained by a non-linear transformation of the intrinsic co-ordinates of S_b; for details see Lachat & Watson (1976). The result in all cases is, however, a system similar to that shown in Fig. 8. An infinite element (see Fig. 3(b)) is divided into one infinite quadrilateral and either two or three finite triangular subelements, as shown in Fig. 7(c). Integrations over the triangular subelements are carried out in the same manner as for a quadrilateral element (and without reference to the co-ordinate system defined by eqn. (36)), whereas for the infinite quadrilateral,

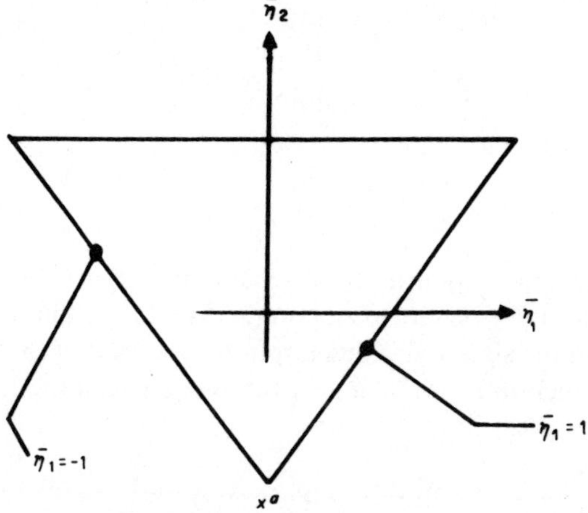

FIG. 8. Co-ordinates for integration over triangular subelement.

formula (43) is used to integrate with respect to the co-ordinates defined by eqn. (36). This leaves only the Cauchy principal value of the integrals of $T'_{\alpha\beta}(x^a, y(\xi))N^c(\xi)J(\xi)$ for $d(b, c) = a$ and the coefficient of the free term: as for the plane problems, the sum of these two contributions is evaluated by considering rigid body translations (see eqn. (41)).

In the foregoing discussion, nothing is said about orders of Gaussian quadrature formulae (the parameters m, m_1 and m_2). If the order of the formulae is to be the same for all S_b and for all x^a, then a reasonable value is 4. However, an invariant integration formula is wasteful of central processing time, because a value of m that gives reasonable accuracy over a large element S_b near x^a is more accurate than necessary for a small element S_b distant from x^a: the wastage is squared in the case of the three-dimensional formulation. Since especially for three-dimensional problems it can take longer to construct the equations than it does to solve them, it is worthwhile to incorporate in the algorithm an automatic choice of integration formula, designed to assure roughly uniform precision of integration. A convenient error bound upon which to base such a choice is that in terms of the $2m$th derivative of the integrand given by Stroud & Secrest (1966). For the one-dimensional formula with weight function unity (eqn. (38)) this is

$$\left| \int_{-1}^{+1} f(\eta)\,d\eta - \sum_{i=1}^{m} A_i^m f(\eta_i^m) \right| \leq eH \tag{45}$$

where

$$\left|\frac{d^{2m}f}{d\eta^{2m}}\right| \le H \tag{46}$$

and, very nearly,

$$e = \frac{4}{2^{2m}(2m)!} \tag{47}$$

It would be uneconomical to calculate the upper bound H for the integrands in question, so to obtain a formula for choosing integration schemes, certain simplifying assumptions are made (Lachat & Watson, 1976). The function $1/r$ is taken to be representative of the actual integrands

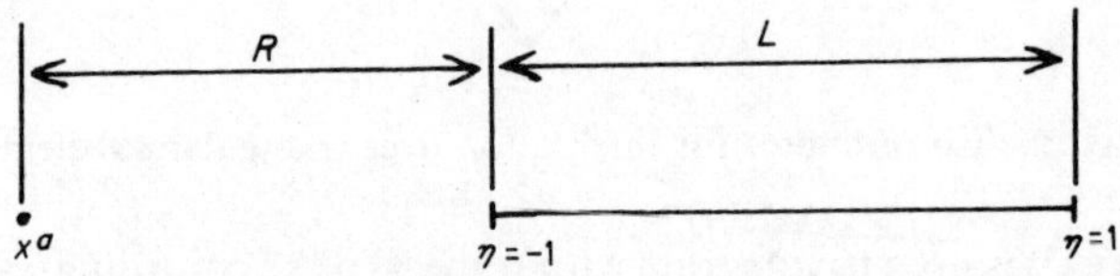

FIG. 9. Special case of integration over one-dimensional region.

(which are products of shape functions and the kernels given in eqns. (16) and (17)), and the case is considered of a region of integration which is a straight line of length L and constant jacobian, the singularity x^a being situated on the continuation of that line, at a distance R from one end (see Fig. 9). For this simple case the upper bound is easy to calculate. Proposing now that a limit be imposed on the ratio of the upper bound eH (eqn. (45)) to the upper bound L/R of the integral it is intended to evaluate, m should be chosen such that

$$\left(\frac{L}{4R}\right)^{2m} \le KL \tag{48}$$

where K is a constant. For the case $x^a \notin S_b$, where S_b is a finite boundary element, m is chosen by taking R in inequality (48) to be the minimum distance from x^a to the element. Where S_b is an infinite element, L is taken to be the length of the interval $|\xi| \le 1$. For the case $x^a \in S_b$, R is arbitrarily taken equal to L. There is no rigorous justification for this, especially as it is on occasion used to choose the order of a formula with logarithmic weight function, but it gives reasonable results. The computational procedure for a given element and point x^a is as follows.

(1) Calculate approximately L and R.

(2) By choosing a suitable value of m, and if necessary by subdividing the element into more than one region of integration, satisfy inequality (48).

For the two-dimensional formula required for integration over surface elements in three dimensions (eqn. (43)),

$$\left| \int_{-1}^{+1} \int_{-1}^{+1} f(\eta_1, \eta_2)\, \mathrm{d}\eta_1\, \mathrm{d}\eta_2 - \sum_{i=1}^{m_1} \sum_{j=1}^{m_2} A_i^{m_1} A_j^{m_2} f(\eta_i^{m_1}, \eta_j^{m_2}) \right| \leq 2 \sum_{r=1}^{2} e_r H_r \tag{49}$$

where

$$\left| \frac{\partial^{2m_r} f}{\partial \eta^{2m_r}} \right| \leq H_r \tag{50}$$

and, very nearly (Stroud & Secrest, 1966),

$$e_r = \frac{4}{2^{2m_r}(2m_r)!} \tag{51}$$

As for the plane problems, simplifying assumptions are made to obtain a formula for choosing integration schemes (Lachat & Watson, 1976). The function $1/r^2$ is taken to be representative of the actual integrands, and the case is considered of a rectangular region of integration of dimensions $L_1 \times L_2$ and constant jacobian, the singularity x^a being situated at a distance R from one side (see Fig. 10). Proceeding as before, it is found that m_r should be chosen such that

$$(2m_r + 1)\left(\frac{L_r}{4R}\right)^{2m_r} \leq KL_1L_2 \tag{52}$$

where K is a constant. For the case $x^a \notin S_b$, where S_b is a quadrilateral or triangular element, m_r is chosen by taking R in inequality (52) to be the minimum distance from x^a to the element. Where S_b is an infinite element, L_1, L_2 are taken to be the dimensions of the region $|\xi_i| \leq 1$. For the case $x^a \in S_b$, R is arbitrarily taken to be the minimum distance from x^a to the opposite side of a triangular element or subelement, L_1 is the length of that side, and $L_2 = R$. As for the plane problem, there is no rigorous justification for this procedure, but it is probably not greatly in error. The imposition of inequality (52) sets a limit on each of the two terms on the right side of inequality (49), and so assures comparable precision of

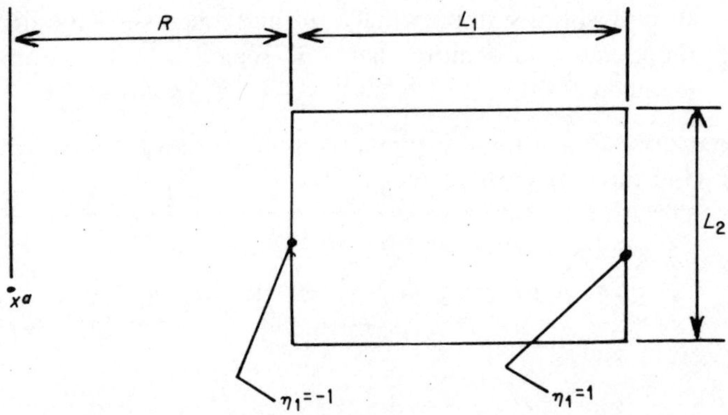

FIG. 10. Special case of integration over two-dimensional region ($r = 1$ shown).

integration in each intrinsic co-ordinate direction. This in turn allows the use in surface discretisation of elements with high aspect ratio (length/breadth), where these are suggested by the expected variation of surface displacement, without loss of accuracy. Inequalities similar to (48) and (52) may be deduced for integration formulae to be used when calculating stress and displacement at points inside a subregion (see eqn. (12)).

3.7 EXAMPLE OF THREE-DIMENSIONAL ANALYSIS

The example is that of the 1/10 scale prototype prestressed concrete nuclear reactor vessel, tested by Taylor Woodrow (see Fig. 11). The elastic constants are taken to be Young's modulus = 41 400 N/mm^2 and Poisson's ratio = 0·15, except in the perforated region of the cap, where Young's modulus = 29 000 N/mm^2 (Langan *et al.*, 1972). Due to symmetry, it is only necessary to consider a $22\frac{1}{2}°$ segment. In addition, the vessel is nearly symmetric with respect to the equatorial plane, and is assumed to be so for the purpose of numerical analysis. The effect of the steel liner is ignored. The load cases are

(1) a band of prestress 127 mm wide centred 114 mm above the equator, exerting a pressure of 7·32 N/mm^2 on the outside of the vessel;
(2) pressure of 4·42 N/mm^2 inside the reactor void, gas duct and boiler penetration.

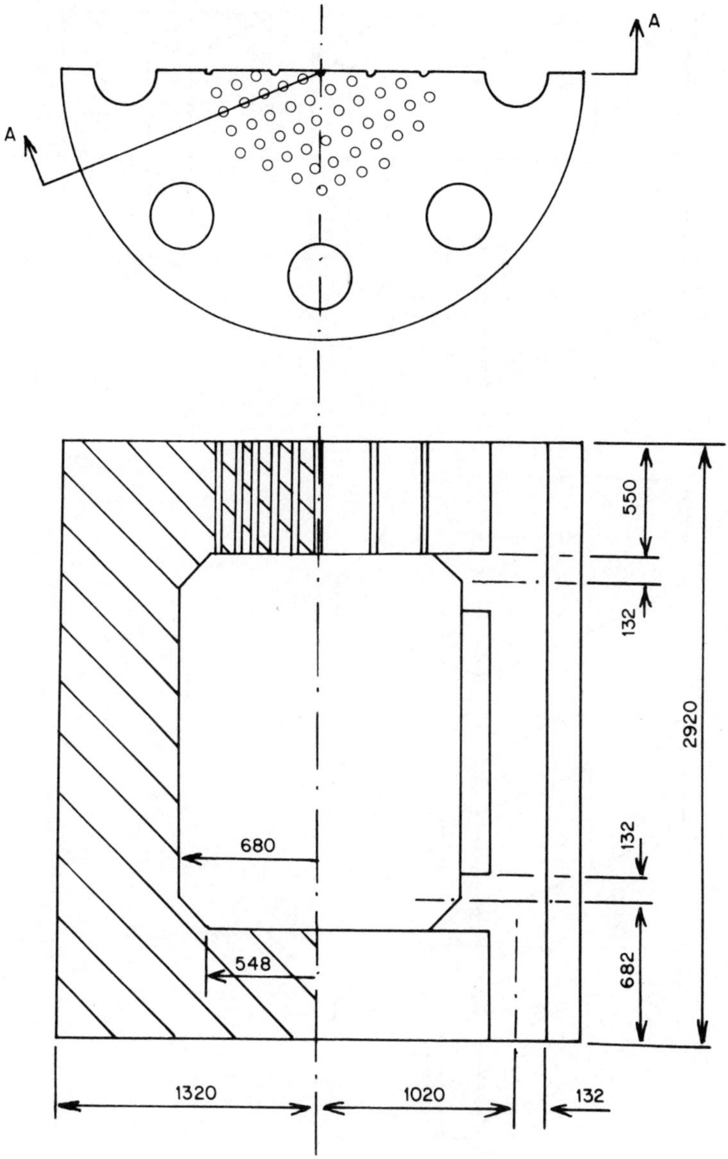

FIG. 11. Plan and section of pressure vessel.

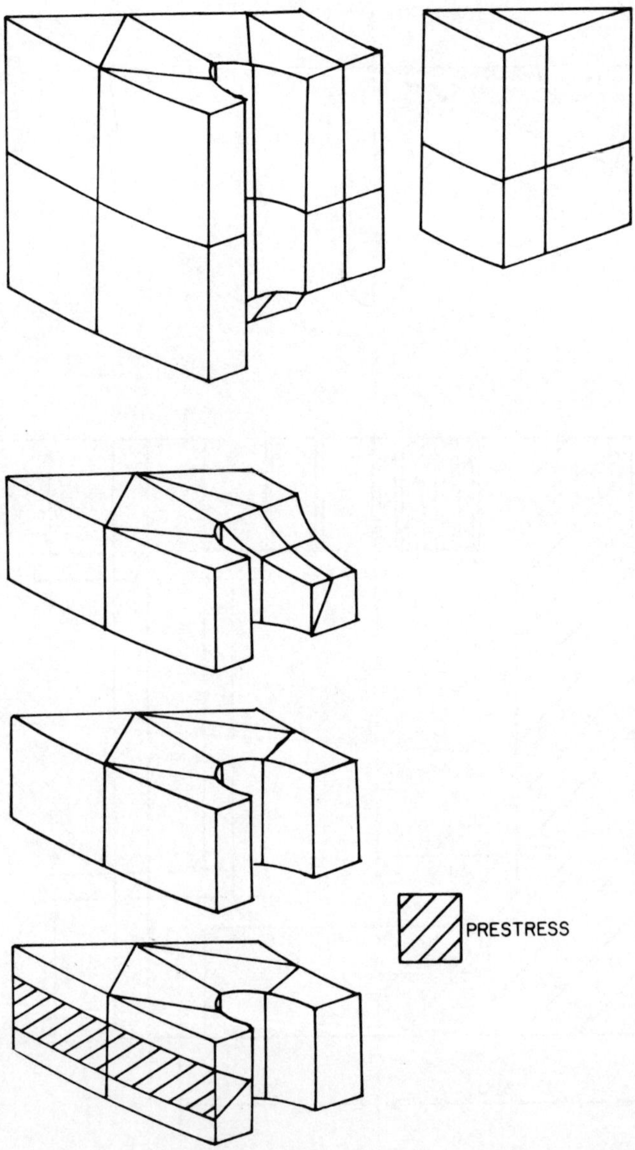

FIG. 12. Fine discretisation of surface of pressure vessel.

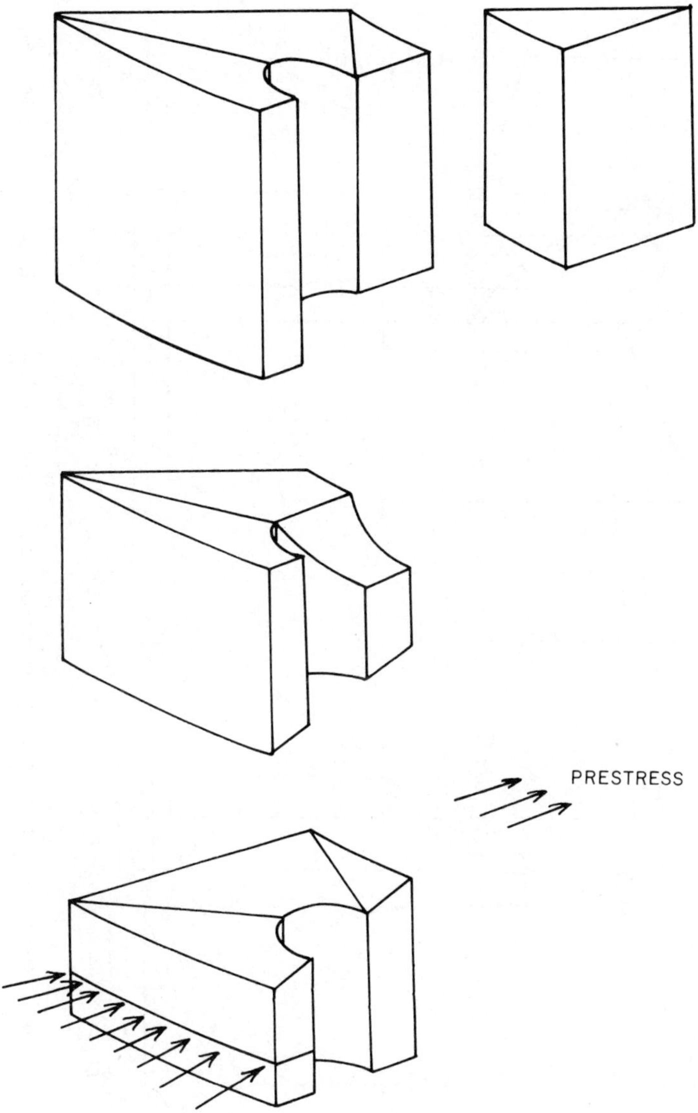

FIG. 13. Coarse discretisation of surface of pressure vessel.

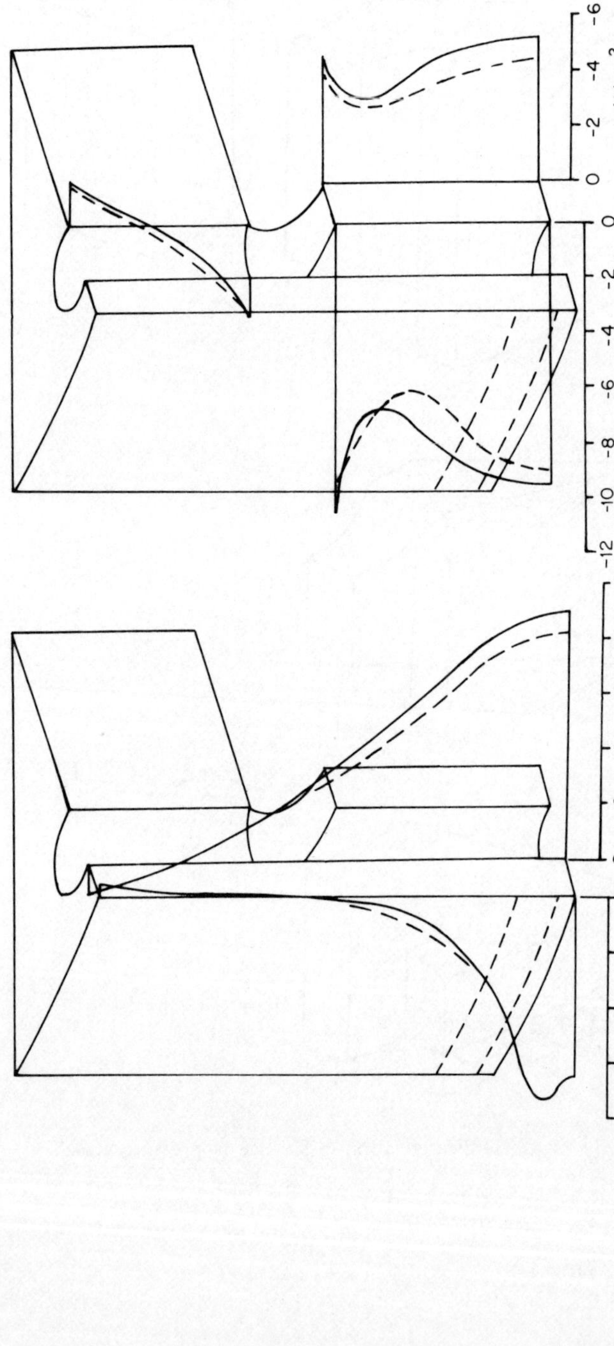

FIG. 14. Calculated variation of circumferential stress due to prestress. —— Fine, quadratic; – – – coarse, cubic.

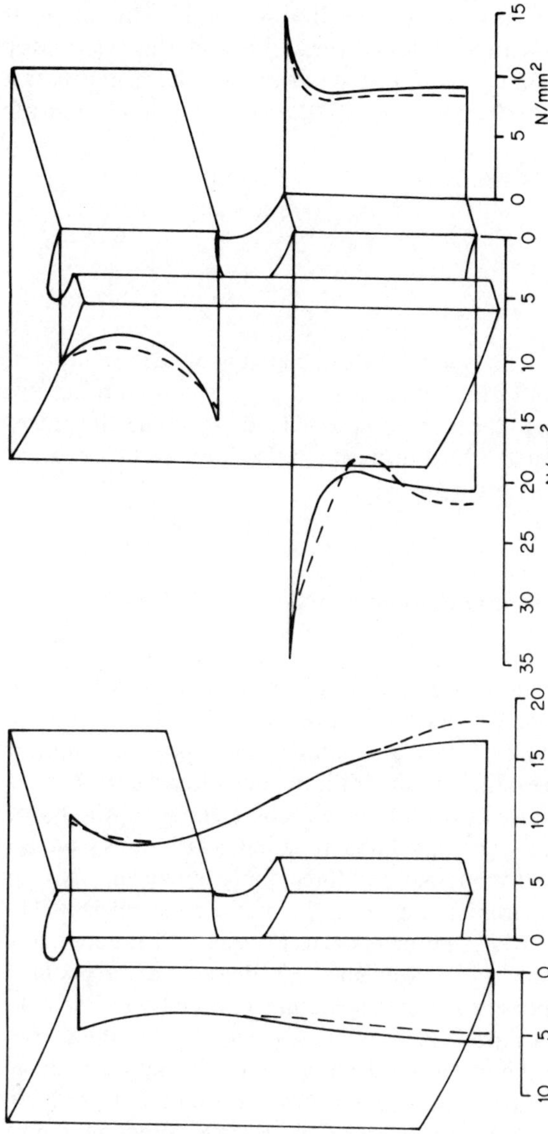

FIG. 15. Calculated variation of circumferential stress due to internal pressure. —— Fine, quadratic; – – – coarse, cubic.

Two discretisations are considered. The first is a division of the body into 5 subregions with a total of 124 surface and interface elements (see Fig. 12), with quadratic functional variation. The second is a division of the body into 4 subregions with a total of 42 surface and interface elements (see Fig. 13), with cubic functional variation over each element (see Lachat & Watson, 1976). Computer (CDC 7600) execution times were as follows.

Discretisation	Central processing	Input–output
Coarse	71·8 s	4·2 s
Fine	159·8 s	12·6 s

In Figs. 14 and 15 are shown the calculated variations of circumferential stress, along lines in the plane of symmetry passing through two boiler penetrations, due to the first and second load cases. The difference between the results calculated using the two discretisations nowhere exceeds 10 % of the calculated maximum stress.

3.8 COMMENTS AND CONCLUSION

The numerical treatment described here has successfully been used in the analysis of practical problems of two- and three-dimensional elasticity (Lachat, 1975; Lachat & Watson, 1976). The original implementation allowed the choice of linear, quadratic or cubic Ergatoudis shape functions for the representation of displacement and traction. Although for plane problems both quadratic and cubic functional variations have performed well, it has been found that for three-dimensional analyses, quadratic variation, the isoparametric case, is usually the most efficient. The reason probably is that, whilst an increase in the order of functional variation reduces the number of parameters or degrees of freedom required adequately to represent a given displacement or traction field, in the case of the Ergatoudis shape functions, it also results in a less favourable distribution over the surface of the nodes at which collocation takes place.

The difficulty of poor distribution of collocation points for higher order functional variations may in principle be overcome by replacing the procedure of nodal collocation by the Galerkin formulation (Jeng & Wexler, 1977), in which the product of eqn. (14) with each of q functions

$f^a(x)$ formed from the shape functions is integrated over S:

$$\int_S f^a(x)c_{ij}(x)u_j(x)\,\mathrm{d}S_x + \int_S\int_S f^a(x)T_{ij}(x,y)u_j(y)\,\mathrm{d}S_y\,\mathrm{d}S_x$$

$$= \int_S\int_S f^a(x)U_{ij}(x,y)t_j(y)\,\mathrm{d}S_y\,\mathrm{d}S_x \qquad (53)$$

Work on this alternative is not yet sufficiently advanced for conclusions on its effectiveness to be drawn. In practice the Galerkin formulation requires collocation at the Gauss integration points for the integration with respect to x, followed by the construction of a system of equations in terms of nodal values of displacement and traction as before. This presents two major problems. First, it is difficult to devise criteria for the choice of the order of Gauss formula to use for the integration with respect to x over each element, such that whilst there is no danger of a singular matrix, the total number of collocation points on S and therefore the cost of constructing the system of equations stays within reasonable limits. There must always be at least as many Galerkin collocation points as there are nodes; usually there will be considerably more, and for a given surface discretisation the Galerkin formulation will be correspondingly more costly than nodal collocation: the question is whether superior accuracy for a given discretisation will allow sufficiently fewer elements to be used to offset this disadvantage. The second difficulty is that, for problems of practical size, the construction from the equations of collocation at Gauss integration points of the system of equations necessitates a large random access file handling operation, which may become a significant additional overhead.

From experience with the isoparametric elements, simple guidelines for the discretisation of surfaces have been developed. If the part of a boundary between two re-entrant corners is represented by three elements (see Fig. 16(a)), and the central element is twice as long as each of the others, accuracy will be good everywhere except near the corners, where theoretically there are singularities that cannot be represented by the smoothly varying shape functions. A 90° arc of a circle may usually be represented by just two elements (see Fig. 16(b)). As an example of the accuracy obtained by following this rule, let us consider the calculation of the stress field near a spherical cavity in an infinite elastic medium with Poisson's ratio 0·3, when subjected to the uniaxial stress $\sigma_{33} = 1{\cdot}0$ at infinity. If one-eighth of the surface is represented by four elements as shown in Fig. 17(a) then according to the three-dimensional analysis, the stress σ_{33} on the equator of the sphere is 2·02 at a corner node and 2·01 at a

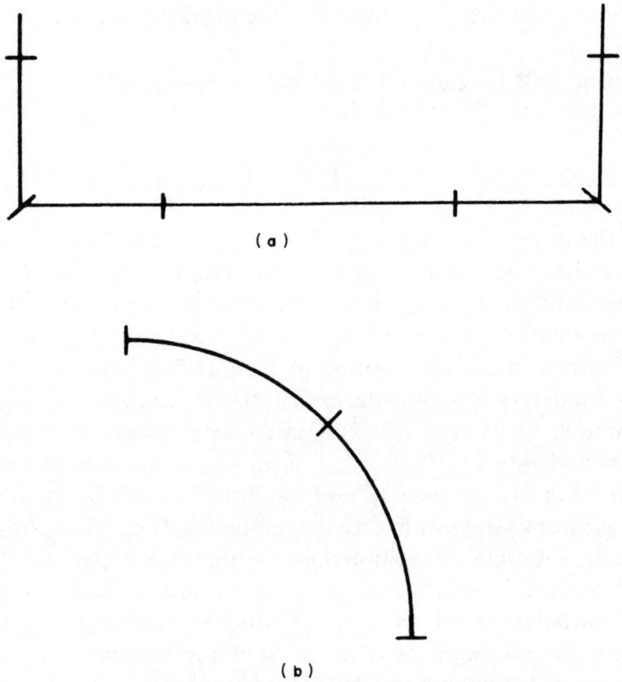

FIG. 16. Suggested boundary discretisations. (a) Boundary between two re-entrant corners; (b) 90° arc of circle.

midside node, compared with the exact value of 2·045. Infinite elements are a recent development and so rules for their use have not yet been formulated. However, some indication of their performance is given by the example of the calculation of displacements of an elastic halfspace, for which Young's modulus = 10 000 kN/m^2 and Poisson's ratio = 0·3, when subjected to a uniformly distributed load of 100 kN/m^2 over an area 2 m square. If one-quarter of the surface of the halfspace is represented by three finite and two infinite elements as shown in Fig. 17(b), then according to the three-dimensional analysis the maximum displacement is 0·021 19 m, compared with the exact value of 0·022 44 m.

The design of computer programs for analysis by the boundary integral equation method is a complex subject about which there is room to say very little here. Particular attention must be paid to the efficiency of routines for calculating coefficients of the simultaneous equations, and for calculating results at points inside subregions, because for most problems these

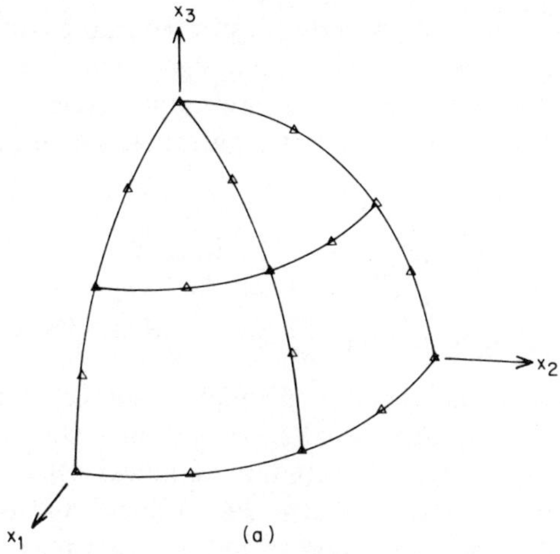

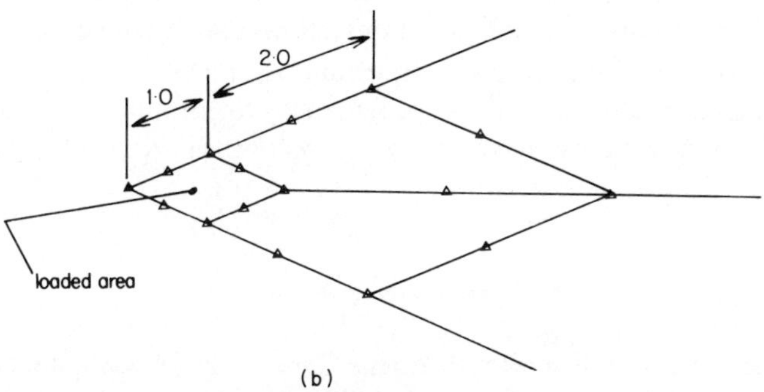

FIG. 17. Examples of discretisation of surfaces. Discretisation of surface of (a) spherical cavity; (b) halfspace.

operations are more costly than equation solution. In the current version of the program for three-dimensional analysis, equation solution time does not overtake matrix construction time until there are subregions with 1000 degrees of freedom. For a fuller discussion of computing strategies, see Lachat (1975) and Lachat & Watson (1975).

The formulation based upon Ergatoudis shape functions is now near the limit of its development, except for generalisations to non-linear analysis and the modelling of cracks. There is probably something to be gained by introducing singular shape functions at re-entrant corners, and in the three-dimensional analysis it would be desirable to replace the degenerate quadrilateral (Fig. 3(b)) by a properly constituted triangular element. It is likely that computational efficiency could significantly be improved by replacing the present shape functions by cubic splines, together with singular functions at re-entrant corners. The cubic spline functions are

$$\begin{aligned} N_0^{(1)}(\xi) &= \tfrac{1}{4}(2 + 3\xi - \xi^3) \\ N_1^{(1)}(\xi) &= -\tfrac{1}{4}(1 + \xi - \xi^2 - \xi^3) \\ N_0^{(2)}(\xi) &= \tfrac{1}{4}(2 - 3\xi + \xi^3) \\ N_1^{(2)}(\xi) &= \tfrac{1}{4}(1 - \xi - \xi^2 + \xi^3) \end{aligned} \tag{54}$$

There would be no midside nodes (a frequent source of error in data preparation), and a favourable distribution over the surface of collocation points. The tangent plane would be continuous across element boundaries, and the parameters in terms of which displacement and traction at an arbitrary point are expressed would include nodal values of tangential derivatives. As a consequence of this, the calculated surface strain and stress tensors would be continuous across element boundaries on smooth parts of the surface. Preliminary investigations show that a single cubic spline element can accurately represent the geometry of a 90° arc, and it is likely that this single element would also provide a representation of displacement or traction along the arc sufficiently accurate for the purpose of engineering analysis.

ACKNOWLEDGEMENTS

The author thanks the Science Research Council for financing much of the work presented here, and Imperial College for providing computing and other facilities.

REFERENCES

CRUSE, T. A. (1969). Numerical solutions in three dimensional elastostatics, *Int. J. Sol. Struct.*, **5,** 1259–74.

ERGATOUDIS, J. (1968). *Isoparametric finite elements in two and three dimensional analysis*, Ph.D. Thesis, University of Wales, Swansea.

JASWON, M. A. & PONTER, A. R. (1963). *Proc. Roy. Soc. London, Series A*, **273,** 237–46.

JENG, G. & WEXLER, A. (1977). Isoparametric finite element variational solution of integral equations for three dimensional fields, *Int. J. Num. Meth. Eng.*, **11,** 1455–74.

KUPRADZE, V. D. (1963). In: *Progress in Solid Mechanics*, Vol. III, Eds I. N. Sneddon and R. Hill, Wiley, London.

LACHAT, J. C. (1975). *A further development of the boundary integral technique for elastostatics*, Ph.D. Thesis, University of Southampton.

LACHAT, J. C. & WATSON, J. O. (1975). In: *Boundary Integral Equation Method: Computational Applications in Applied Mechanics*, Eds T. A. Cruse and F. J. Rizzo, AMD 11, ASME, New York.

LACHAT, J. C. & WATSON, J. O. (1976). Effective numerical treatment of boundary integral equations, *Int. J. Num. Meth. Eng.*, **10,** 991–1005.

LANGAN, D., O'FLYNN, M. & WELCH, A. K. (1972). *Proc. 1st Int. Conf. on Structural Mechanics in Reactor Technology* 1972, **4,** 369.

RIZZO, F. J. (1967). An integral equation approach to boundary value problems of classical elastostatics, *Quart. Appl. Math.*, **25,** 83–95.

STROUD, A. H. & SECREST, D. (1966). *Gaussian Quadrature Formulas*, Prentice-Hall, Englewood Cliffs, NJ.

SYMM, G. T. (1963). *Proc. Roy. Soc. London, Series A*, **275,** 33–46.

Chapter 4

TWO- AND THREE-DIMENSIONAL PROBLEMS OF ELASTO-PLASTICITY

P. K. BANERJEE,† D. N. CATHIE‡ and T. G. DAVIES†

† *Department of Civil and Structural Engineering, University College Cardiff, UK*

‡ *Department of Civil Engineering and Building Technology, University of Wales Institute of Science and Technology, Cardiff, UK*

SUMMARY

The direct formulation of the boundary element method has been developed for two- and three-dimensional problems of elasto-plasticity. A complete scheme of numerical analysis based on a new initial stress formulation is presented. The algorithm has been applied to a number of two- and three-dimensional problems to demonstrate that it can be used for efficient non-linear analyses of bulky solids.

4.1 INTRODUCTION

Although remarkable progress has been made in developing the boundary element method for analysing problems governed by linear differential equations, comparatively little progress has been made in applications to non-linear problems. We may surmise that the advanced state of development of non-linear finite element analyses has contributed to this situation.

However, it has been demonstrated by Cruse (1974, 1977, 1978), Lachat & Watson (1976), Banerjee (1976) and Banerjee & Butterfield (1977) that for linear problems, boundary element methods have distinct advantages over the finite element method for analysing bulky solids. Indeed, for a large class of three-dimensional problems the method is probably the only reliable means of getting adequately detailed results in the region where they are required, at a reasonable cost. A boundary element formulation for non-linear problems will inherit at least some of these advantages and the development of a procedure for general elasto-plastic analysis is therefore of considerable interest.

The earliest boundary element formulation for elasto-plasticity was due to Swedlow & Cruse (1971), although the basic ideas were well explored earlier by Reisner (1931) and Lin (1967, 1968). Riccardella (1973) implemented an algorithm based on Swedlow & Cruse's formulation, for two-dimensional problems. This attempt was not entirely successful, although it did demonstrate the potential of the method. Subsequently Rzasnicki & Mendelson (1975) and Mendelson & Albers (1975) implemented the same formulation and gave numerical solutions for some two-dimensional problems. These formulations were essentially based on an 'initial strain' approach, and hence are restricted to strain hardening materials. Kumar & Mukherjee (1977) used a similar initial strain formulation and presented solutions for simple problems of expanding cylinders and spheres. They simplified the formulation by using a special form of the constitutive relationships.

Banerjee & Mustoe (1978) described a completely general indirect formulation of the boundary element method using a 'modified body force' approach based on an initial stress formulation. Although early results show considerable promise, the body force approach still needs further development before it can be recommended for routine analysis of two-dimensional problems. One of its most attractive features is that the kernel functions involved in the volume integrals for interior stress calculations are numerically integrable.

In this chapter we describe a direct formulation of the boundary element method based on an initial stress distribution within the yielded zone. The formulation is applicable to strain hardening, elastic ideally plastic as well as strain softening materials.

4.2 GOVERNING DIFFERENTIAL EQUATIONS OF ELASTO-PLASTICITY

An elasto-plastic material must obey the equilibrium equations for the stress rate (increments):

$$\frac{\partial \dot{\sigma}_{ij}}{\partial x_j} = 0 \tag{1}$$

where the dot ($\dot{}$) indicates an increment.

The strain–displacement relations may be written as:

$$\dot{\varepsilon}_{ij} = \frac{1}{2}\left(\frac{\partial \dot{u}_i}{\partial x_j} + \frac{\partial \dot{u}_j}{\partial x_i}\right) \tag{2}$$

The total strain rate $\dot{\varepsilon}_{ij}$ may be decomposed into its elastic and plastic parts:

$$\dot{\varepsilon}_{ij} = \dot{\varepsilon}^{\mathrm{e}}_{ij} + \dot{\varepsilon}^{\mathrm{p}}_{ij} \tag{3}$$

The stresses are related to elastic strains via:

$$\dot{\sigma}_{ij} = \lambda \delta_{ij} \dot{\varepsilon}^{\mathrm{e}}_{\kappa\kappa} + 2\mu \dot{\varepsilon}^{\mathrm{e}}_{ij} \tag{4a}$$

where $\delta_{ij} = 1$ for $i = j$ and 0 for $i \neq j$ and λ, μ are elastic constants.

This equation may be written by using eqn. (3) as:

$$\dot{\sigma}_{ij} = \lambda \delta_{ij}(\dot{\varepsilon}_{\kappa\kappa} - \dot{\varepsilon}^{\mathrm{p}}_{\kappa\kappa}) + 2\mu(\dot{\varepsilon}_{ij} - \dot{\varepsilon}^{\mathrm{p}}_{ij}) \tag{4b}$$

By substituting eqn. (4b) in (1) and using the strain–displacement relations we obtain the governing differential equations for elasto-plastic flow (see, Lin, 1968; Swedlow & Cruse, 1971):

$$\mu \frac{\partial^2 \dot{u}_i}{\partial x_j^2} + (\lambda + \mu)\frac{\partial^2 \dot{u}_j}{\partial x_i \partial x_j} - \left(\lambda \delta_{ij} \frac{\partial \dot{\varepsilon}^{\mathrm{p}}_{\kappa\kappa}}{\partial x_j} + 2\mu \frac{\partial \dot{\varepsilon}^{\mathrm{p}}_{ij}}{\partial x_j}\right) = 0 \tag{5}$$

In the absence of any plastic strain the terms in the parentheses disappear and eqn. (5) is then the well known Navier equation for elasticity in incremental form. An elasto-plastic problem may therefore be posed as the solution of an elastic problem governed by the differential equation:

$$\mu \frac{\partial^2 \dot{u}_i}{\partial x_j^2} + (\lambda + \mu)\frac{\partial^2 \dot{u}_j}{\partial x_i \partial x_j} + \dot{f}_i = 0 \tag{6}$$

over the volume and subject to the boundary condition

$$\dot{t}_i = \dot{\sigma}_{ij} . n_j - \dot{t}^{\mathrm{o}}_i \tag{7}$$

over the surface. Where

$$\dot{f}_i = -\left(\lambda\delta_{ij}\frac{\partial\dot{\varepsilon}^{p}_{\kappa\kappa}}{\partial x_j} + 2\mu\frac{\partial\dot{\varepsilon}^{p}_{ij}}{\partial x_j}\right)$$

$$\dot{t}^{o}_i = -(\lambda\delta_{ij}\dot{\varepsilon}^{p}_{\kappa\kappa} + 2\mu\dot{\varepsilon}^{p}_{ij})\,.\,n_j$$

and n_j is the outward normal to the boundary.

We can see at once that eqn. (6) is identical to the incremental Navier equation with suitably modified body forces $\dot{f}_i$ generated by the irrecoverable component of the deformation field. Moreover, the non-linear terms occur only in the body force term of the inhomogeneous differential equation. Hence eqn. (6) is quasi-linear in character.

We may assume an *imaginary* elastic body whose body forces and the traction boundary conditions have been modified according to eqns. (6) and (7). The displacement field obtained from the solution of eqn. (6) would thus be correct for the real elasto-plastic body. The stresses corresponding to this displacement field would then be given by the elastic stress–strain relations in the elastic regions and elasto-plastic stress–strain relations in the elasto-plastic regions.

This procedure of solving an elasto-plastic problem by using a modified elastic problem is not new. Reisner (1931) suggested an essentially identical procedure by using intuitive reasoning and called it an initial stress approach (*eigenspannungen*). Zienkiewicz *et al.* (1969) developed an initial stress algorithm for the solution of elasto-plastic problems by the finite element method.

4.3 INCREMENTAL STRESS–STRAIN RELATIONS

The necessary requirements for deriving an incremental stress–strain relation based on the 'path independent in the small' theory of plasticity (incremental theory of plasticity) are:

(a) A yield function defining the limits of the elastic behaviour.
(b) A flow rule relating the irrecoverable plastic part of the strain rate to the stress state in the material.
(c) A hardening rule defining the subsequent yield surfaces resulting from continuous plastic deformation.

The nature of the yield surface and the hardening parameters are obviously dependent on the type of material. Therefore, we need to derive

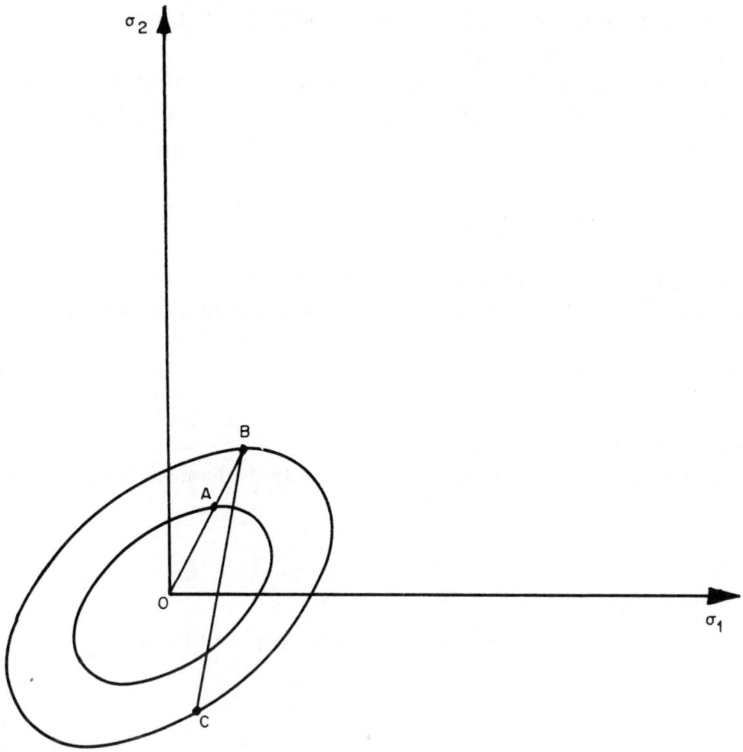

FIG. 1. Isotropic hardening.

the stress–strain relations for the general cases of isotropic and translational hardening.

4.3.1 Isotropic Hardening

In this theory it is assumed that during plastic flow the yield surface expands uniformly about the origin in the stress space maintaining its shape, centre and orientation as shown in Fig. 1. For example the path OA is elastic; at A the material is in the state of incipient yielding and upon further loading the yield surface will expand to B. The path AB represents isotropic strain hardening behaviour. If we unload, the behaviour will be elastic until the point C is reached when the material will start to yield again. If we assume that the continuously changing yield surfaces can be represented by a loading function of type:

$$F(\sigma_{ij}, \varepsilon_{ij}^{p}, h) = 0,$$

where h is a parameter, then for the plastic state $F = 0$, for the elastic state $F < 0$ and $F > 0$ is meaningless.

Since during plastic deformation the new stress state must lie on a newly developed yield surface characterised by a new strain hardening parameter $h + \dot{h}$, we have:

$$\mathrm{d}F = \frac{\partial F}{\partial \sigma_{ij}} \cdot \dot{\sigma}_{ij} + \frac{\partial F}{\partial \varepsilon_{ij}^{\mathrm{p}}} \cdot \dot{\varepsilon}_{ij}^{\mathrm{p}} + \frac{\partial F}{\partial h} \cdot \dot{h} = 0 \tag{8}$$

where h is a function of plastic strain.

By using Drucker's postulate for associated plastic flow we have:

$$\dot{\varepsilon}_{ij}^{\mathrm{p}} = \dot{\lambda} \frac{\partial F}{\partial \sigma_{ij}} \tag{9}$$

where $\dot{\lambda}$ is a non-negative scalar variable.

Since the hardening parameter h is a function of plastic strain we can write eqn. (8) as:

$$\frac{\partial F}{\partial \sigma_{ij}} \cdot \dot{\sigma}_{ij} + \frac{\partial F}{\partial \varepsilon_{ij}^{\mathrm{p}}} \cdot \dot{\varepsilon}_{ij}^{\mathrm{p}} + \frac{\partial F}{\partial h} \cdot \frac{\partial h}{\partial \varepsilon_{ij}^{\mathrm{p}}} \cdot \dot{\varepsilon}_{ij}^{\mathrm{p}} = 0 \tag{10}$$

Substituting (9) in (10) and solving for $\dot{\lambda}$ we get (Fung, 1965):

$$\dot{\lambda} = -\frac{\dfrac{\partial F}{\partial \sigma_{\kappa l}} \dot{\sigma}_{\kappa l}}{\left(\dfrac{\partial F}{\partial \varepsilon_{mn}^{\mathrm{p}}} + \dfrac{\partial F}{\partial h} \cdot \dfrac{\partial h}{\partial \varepsilon_{mn}^{\mathrm{p}}}\right) \dfrac{\partial F}{\partial \sigma_{mn}}}$$

Therefore the plastic strain can be obtained from:

$$\dot{\varepsilon}_{ij}^{\mathrm{p}} = G \cdot \frac{\partial F}{\partial \sigma_{ij}} \cdot \frac{\partial F}{\partial \sigma_{\kappa l}} \cdot \dot{\sigma}_{\kappa l} \tag{11}$$

where

$$G = -1 \Bigg/ \left(\frac{\partial F}{\partial \varepsilon_{mn}^{\mathrm{p}}} + \frac{\partial F}{\partial h} \cdot \frac{\partial h}{\partial \varepsilon_{mn}^{\mathrm{p}}}\right) \frac{\partial F}{\partial \sigma_{mn}}$$

Equation (11) together with the elastic component of the strain may be written as:

$$\dot{\varepsilon}_{ij} = \dot{\varepsilon}_{ij}^{\mathrm{e}} + \dot{\varepsilon}_{ij}^{\mathrm{p}} = C_{ij\kappa l}^{\mathrm{ep}} \dot{\sigma}_{\kappa l} \tag{12}$$

or

$$\dot{\sigma}_{ij} = D_{ij\kappa l}^{\mathrm{ep}} \dot{\varepsilon}_{\kappa l} \tag{13}$$

Equations (12) and (13) provide the necessary incremental stress–strain relations for a general case of isotropic hardening. Specifically if we use an isotropic strain-hardening Von Mises material we can write eqn. (13) explicitly as:

$$\dot{\sigma}_{ij} = 2\mu\left[\dot{\varepsilon}_{ij} + \frac{\nu}{1-2\nu}\delta_{ij}\dot{\varepsilon}_{\kappa\kappa} - \frac{3S_{ij}S_{\kappa l}}{2\sigma_0^2(1 + H/3\mu)}\cdot\dot{\varepsilon}_{\kappa l}\right] \tag{14}$$

where μ is the elastic shear modulus, σ_0 = the uni-axial yield stress $= \sqrt{(\frac{3}{2}\,.\,S_{ij}S_{ij})}$, S_{ij} = the deviatoric stress $= \sigma_{ij} - \frac{1}{3}\delta_{ij}\sigma_{\kappa\kappa}$, H = the plastic hardening modulus, the slope of the uni-axial plastic stress–strain curve.

4.3.2 Kinematic Hardening

The kinematic hardening theory of plasticity, which was originally developed by Prager (1955), assumes that during plastic deformation the loading surface (or the yield surface) translates as a rigid body in stress space. The basic aim of this theory was to model the Bauschinger effect apparent during the cyclic loading of metals.

Figure 2 shows typical kinematic hardening behaviour. The loading path OA is elastic. At A the yielding starts and the loading path AB represents the elasto-plastic kinematic loading behaviour. As a result of the translation of the yield surface during this loading, the centre of the yield surface moves to O′. Any unloading from B results in purely elastic behaviour until the loading path reaches C when the material yields once again—now at a lower yield stress.

If we assume that the translation of the centre of the yield surface can be represented by the tensor $\dot{\alpha}_{ij}$ then the loading function may be written as

$$F(\sigma_{ij}, \alpha_{ij}) = 0 \tag{15}$$

The consistency relation during the loading may be written as:

$$\mathrm{d}F = \frac{\partial F}{\partial \sigma_{ij}}\cdot\dot{\sigma}_{ij} + \frac{\partial F}{\partial \alpha_{mn}}\cdot\frac{\partial \alpha_{mn}}{\partial \varepsilon_{ij}^{\mathrm{p}}}\cdot\dot{\varepsilon}_{ij}^{\mathrm{p}} = 0 \tag{16}$$

where the translation α_{mn} is assumed to be a function of the plastic strain.

The kinematic hardening theory of Prager postulates that the increments of translation $\dot{\alpha}_{ij}$ of the loading surface in the 9-dimensional stress space occur in the direction of the outward normal to the surface at the instantaneous stress state, i.e.:

$$\dot{\alpha}_{ij} = C\dot{\varepsilon}_{ij}^{\mathrm{p}} \tag{17}$$

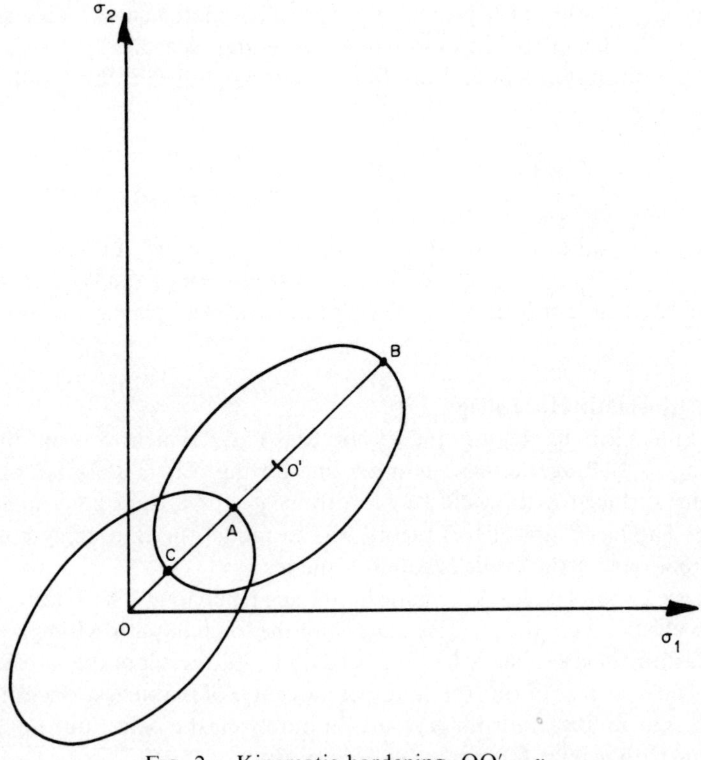

FIG. 2. Kinematic hardening. $OO' = \alpha_{ij}$.

where C is similar to the hardening parameter H in eqn. (14). Equations (16) and (17) can be used to formulate an incremental stress–strain relationship.

However when this theory is applied to various subspaces of stress several inconsistencies arise (Shields & Ziegler, 1958). To avoid these difficulties Ziegler (1959) proposed a modification to Prager's rule:

$$\dot{\alpha}_{ij} = \dot{\mu}_0(\sigma_{ij} - \alpha_{ij}) \tag{18}$$

where $\dot{\mu}_0$ is a positive scalar parameter.

Figure 3 illustrates the difference between the Prager and the Ziegler postulates. The increment of translation $\dot{\alpha}_{ij}$ in Ziegler's theory is colinear with the radius vector from the centre of the yield surface O′ to the stress point. The scalar function $\dot{\mu}_0$ can be determined by using the consistency condition that:

$$\mathrm{d}F(\bar{\sigma}_{ij}) = 0$$

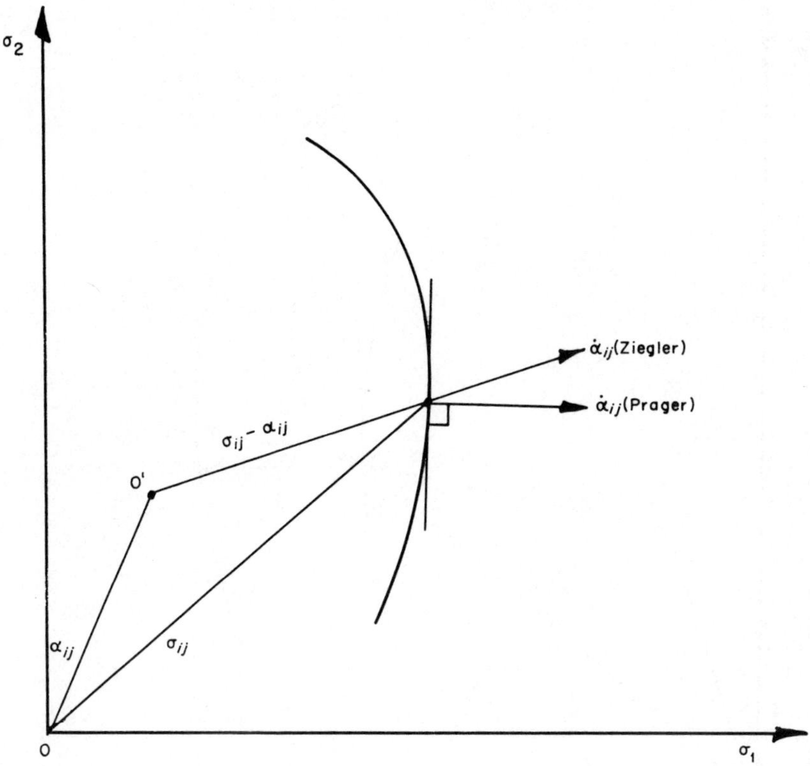

FIG. 3. Ziegler's and Prager's hardening rules.

where

$$\bar{\sigma}_{ij} = \sigma_{ij} - \alpha_{ij}$$

i.e.,

$$\frac{\partial F}{\partial \sigma_{ij}}(\dot{\sigma}_{ij} - \dot{\alpha}_{ij}) = 0$$

Substituting eqn. (18) in the above gives:

$$\dot{\mu}_0 = \frac{\left(\dfrac{\partial F}{\partial \sigma_{ij}}\right)\dot{\sigma}_{ij}}{(\sigma_{\kappa l} - \alpha_{\kappa l})\dfrac{\partial F}{\partial \sigma_{\kappa l}}} \tag{19}$$

In order to determine the plastic strain we have to determine $\dot{\lambda}$ in the flow rule (9). Ziegler assumed that the vector $C\dot{\varepsilon}_{ij}^{p}$ is the projection of $\dot{\sigma}_{ij}$ on the

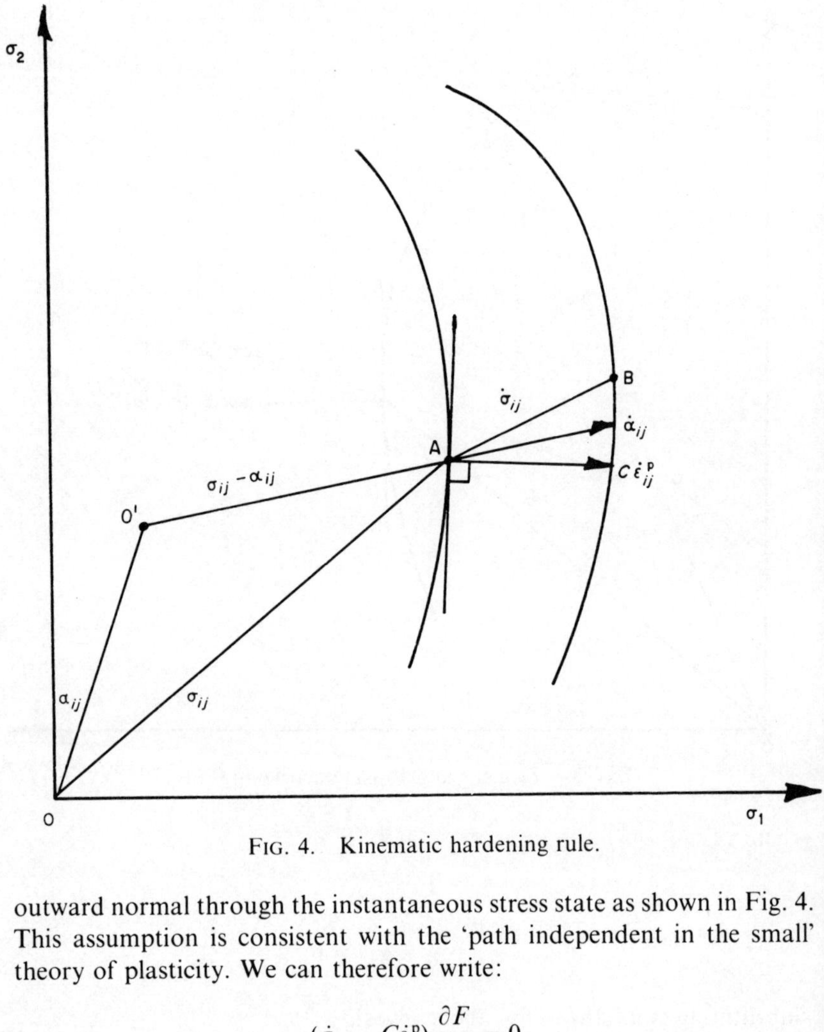

FIG. 4. Kinematic hardening rule.

outward normal through the instantaneous stress state as shown in Fig. 4. This assumption is consistent with the 'path independent in the small' theory of plasticity. We can therefore write:

$$(\dot{\sigma}_{ij} - C\dot{\varepsilon}^{p}_{ij})\frac{\partial F}{\partial \sigma_{ij}} = 0$$

Substituting eqn. (9) in the above we have:

$$\dot{\lambda} = \frac{1}{C}\frac{\left(\dfrac{\partial F}{\partial \sigma_{ij}}\right)\dot{\sigma}_{ij}}{\left(\dfrac{\partial F}{\partial \sigma_{\kappa l}}\right)\left(\dfrac{\partial F}{\partial \sigma_{\kappa l}}\right)}$$

Therefore the plastic strain increments are given by:

$$\dot{\varepsilon}_{ij}^{\mathrm{p}} = \frac{1}{C} \frac{\dfrac{\partial F}{\partial \sigma_{ij}} \cdot \dfrac{\partial F}{\partial \sigma_{mn}} \dot{\sigma}_{mn}}{\left(\dfrac{\partial F}{\partial \sigma_{\kappa l}}\right)\left(\dfrac{\partial F}{\partial \sigma_{\kappa l}}\right)} \tag{20}$$

The elastic and plastic strain components may be added to provide the complete stress–strain relations (eqns. (12) or (13)). Specifically, if we choose a Von Mises yield criterion, i.e.

$$F = \tfrac{1}{2}\bar{S}_{ij}\bar{S}_{ij} - \tfrac{1}{3}\bar{\sigma}_0^2 = 0$$

where

$$\bar{S}_{ij} = \bar{\sigma}_{ij} - \tfrac{1}{3}\delta_{ij}\bar{\sigma}_{\kappa\kappa},$$
$$\bar{\sigma}_{ij} = \sigma_{ij} - \alpha_{ij},$$

may be called a reduced stress state and

$$\bar{\sigma}_0 = \sqrt{(\tfrac{3}{2}\bar{S}_{ij}\bar{S}_{ij})},$$

the reduced uniaxial yield stress. We can write the explicit stress–strain relations as:

$$\dot{\sigma}_{ij} = 2\mu\left[\dot{\varepsilon}_{ij} + \frac{\nu}{1-2\nu}\delta_{ij}\dot{\varepsilon}_{\kappa\kappa} - \frac{3\bar{S}_{ij}\bar{S}_{\kappa l}}{2\bar{\sigma}_0^2(1 + C/2\mu)}\dot{\varepsilon}_{\kappa l}\right] \tag{21}$$

and

$$\dot{\alpha}_{ij} = C\dot{\varepsilon}_{ij}^{\mathrm{p}} = \frac{3\bar{S}_{ij}\bar{S}_{\kappa l}}{2\bar{\sigma}_0^2}\dot{\sigma}_{\kappa l}$$

It is of interest to note that since the normal to the Von Mises yield surface is colinear with the line joining the centre of the yield surface and the material point, the stress–strain relations derived from Prager's and Ziegler's hardening rules become identical.

By examining the uniaxial loading case it is easy to see that the relationship between C and H is $C = 2H/3$. Equations (14) and (21) can be used for any incremental stress–strain analysis.

4.4 INTEGRAL REPRESENTATIONS FOR THE BOUNDARY LOADING

The direct formulation may be developed by making use of the reciprocal theorem i.e.

$$\int_S \dot{t}_i^*(x)\dot{u}_i(x)\,\mathrm{d}S(x) + \int_V \dot{f}_i^*(z)\dot{u}_i(z)\,\mathrm{d}V(z) = \int_S \dot{t}_i(x)\dot{u}_i^*(x)\,\mathrm{d}S(x) + \int_V \dot{f}_i(z)\dot{u}_i^*(z)\,\mathrm{d}V \tag{22}$$

where ($\dot{t}_i^*, \dot{f}_i^*$ and $\dot{u}_i^*$) and ($\dot{t}_i, \dot{f}_i$ and $\dot{u}_i$) are two independent sets of tractions, body forces and displacements respectively. We can choose $\dot{t}_i, \dot{f}_i$ and $\dot{u}_i$ as those of the real problem and $\dot{t}_i^*, \dot{f}_i^*$ and $\dot{u}_i^*$ those due to a unit force in an infinite solid, i.e.

$$\dot{f}_i^*(z) = \delta_{ij}\delta(z,\xi)\dot{e}_j(\xi)$$

where $\delta(z,\xi)$ is the Dirac delta function.

Substituting the components of the unit fundamental solution in (22) for an interior point ξ within S we obtain:

$$\dot{u}_j(\xi) = \int_S [\dot{t}_i(x)U_{ij}(x,\xi) - T_{ij}(x,\xi)\dot{u}_i(x)]\,\mathrm{d}S(x) + \int_V \dot{f}_i(z)U_{ij}(z,\xi)\,\mathrm{d}V(z) \tag{23}$$

where

$$U_{ij}(x,\xi) = \frac{1}{[8\pi\mu(1-\nu)]}\left[-(3-4\nu)\delta_{ij}\log r + \frac{y_i y_j}{r^2}\right]$$

for 2-D problems (plane strain)

$$U_{ij}(x,\xi) = \frac{1}{[16\pi\mu(1-\nu)]}\left(\frac{1}{r}\right)\left[(3-4\nu)\delta_{ij} + \frac{y_i y_j}{r^2}\right]$$

for 3-D problems

$$T_{ij}(x,\xi) = \frac{1}{4\pi(1-\nu)}\left(\frac{1}{r}\right)\left[(1-2\nu)\left(n_j\frac{y_i}{r} - n_i\frac{y_j}{r}\right) + \left((1-2\nu)\delta_{ij} + \frac{2y_i y_j}{r^2}\right)\frac{y_\kappa n_\kappa}{r}\right]$$

for 2-D (plane strain problems)

$$T_{ij}(x,\xi) = -\frac{1}{8\pi(1-\nu)}\left(\frac{1}{r^2}\right)\left[(1-2\nu)\left(n_j\frac{y_i}{r} - n_i\frac{y_j}{r}\right)\right.$$
$$\left. + \left((1-2\nu)\delta_{ij} + \frac{3y_i y_j}{r^2}\right)\frac{y_\kappa n_\kappa}{r}\right] \quad \text{for 3-D problems}$$

with

$$y_i = (x_i - \xi_i) \qquad \text{and} \qquad r^2 = y_i y_i$$

The corresponding kernel function for plane stress may be obtained from those for the plane strain case given above by using an effective Poisson's ratio $\bar{\nu} = \nu/(1+\nu)$.

By taking the field point ξ onto the surface from equation (23) we can write

$$\tfrac{1}{2}\dot{u}_j(\xi) = \int_S [\dot{t}_i(x)U_{ij}(x,\xi) - T_{ij}(x,\xi)\dot{u}_i(x)]\,\mathrm{d}S(x) + \int_V \dot{f}(z)U_{ij}(z,\xi)\,\mathrm{d}V(z) \tag{24}$$

Obviously if we can calculate a suitable distribution of body forces in the interior of the body generated by the irrecoverable component of the deformation field we can use eqn. (24) as a basis for constructing the incremental solution algorithm similar to that indicated by Banerjee & Mustoe (1978). In this chapter however we shall describe a completely different approach in which we shall replace the body force integral by integrals involving a known distribution of initial stresses and initial strains.

4.5 DISPLACEMENTS DUE TO INITIAL STRESSES AND INITIAL STRAINS

Let us consider the virtual work equation:

$$\int_S \dot{t}^*(x)\dot{u}_i'(x)\,\mathrm{d}S + \int_V \dot{f}^*(x)\dot{u}_i'(x)\,\mathrm{d}V(x) = \int_V \dot{\sigma}_{ij}^*(x)\dot{\varepsilon}_{ij}'(x)\,\mathrm{d}V(x) \tag{25}$$

where ($\dot{t}_i^*, \dot{f}_i^*$ and $\dot{\sigma}_{ij}^*$) and ($\dot{\mu}_i'(x)$, $\dot{\varepsilon}_{ij}'(x)$) are two unrelated sets of (tractions, body forces and stresses) and (displacements and strains) respectively.

If we choose the (*) system as being that caused by a point body force system within an infinite solid and $\dot{u}_i'$ and $\dot{\varepsilon}_{ij}'$ as the real displacements and initial strains then it is easy to see that eqn. (25) gives:

$$\dot{u}_\kappa'(\xi) = \int_V T_{ij\kappa}(x,\xi)\dot{\varepsilon}_{ij}'(x)\,\mathrm{d}V \tag{26}$$

Where $T_{ij\kappa}$ are the stresses $\dot{\sigma}^*_{ij}(x)$ due to a unit force $\dot{e}_\kappa(\xi)$ vector.

$$T_{ij\kappa}(x, \xi) = -\frac{1}{[4\pi(1-\nu)]}\left(\frac{1}{r}\right) \times \left[(1-2\nu)\left(\delta_{i\kappa}\frac{y_j}{r} + \delta_{j\kappa}\frac{y_i}{r} - \delta_{ij}\frac{y_\kappa}{r}\right) + \frac{2y_iy_jy_\kappa}{r^3}\right]$$

for 2-D plane strain problems

$$T_{ij\kappa}(x, \xi) = -\frac{1}{[8\pi(1-\nu)]}\left(\frac{1}{r^2}\right) \times \left[(1-2\nu)\left(\delta_{i\kappa}\frac{y_j}{r} + \delta_{j\kappa}\frac{y_i}{r} - \delta_{ij}\frac{y_\kappa}{r}\right) + \frac{3y_iy_jy_\kappa}{r^3}\right]$$

for 3-D problems

Equation (26) is in fact the generalised version of the well known Maysel's expression for displacement caused by a steady state temperature field in an isotropic medium in which the initial strain is simply $\dot{\varepsilon}'_{ij} = \delta_{ij}\alpha\dot{T}$, where α is the coefficient of thermal expansion and $\dot{T}$ is the temperature increment.

By using the relations:

$$\int_V \dot{\sigma}^*_{ij}(x)\dot{\varepsilon}'_{ij}(x)\,\mathrm{d}V = \int_V C_{ij\kappa l}\dot{\varepsilon}^*_{\kappa l}(x)\dot{\varepsilon}'_{ij}(x)\,\mathrm{d}V = \int_V \dot{\varepsilon}^*_{\kappa l}(x)\dot{\sigma}'_{\kappa l}(x)\,\mathrm{d}V$$

We can write eqn. (26) as

$$\dot{u}'_\kappa(\xi) = \int_V B_{ij\kappa}(x, \xi)\dot{\sigma}'_{ij}(x)\,\mathrm{d}V \tag{27}$$

Where $B_{ij\kappa}$ are the strains $\dot{\varepsilon}^*_{ij}(x)$ due to a unit force $\dot{e}_\kappa(\xi)$ vector

$$B_{ij\kappa}(x, \xi) = -\frac{1}{[8\pi\mu(1-\nu)]}\left(\frac{1}{r}\right) \times \left[(1-2\nu)\left(\delta_{i\kappa}\frac{y_j}{r} + \delta_{j\kappa}\frac{y_i}{r}\right) - \delta_{ij}\frac{y_\kappa}{r} + \frac{2y_iy_jy_\kappa}{r^3}\right]$$

for 2-D plane strain problems

$$B_{ij\kappa}(x, \xi) = -\frac{1}{[16\pi\mu(1-\nu)]}\left(\frac{1}{r^2}\right) \times \left[(1-2\nu)\left(\delta_{i\kappa}\frac{y_j}{r} + \delta_{j\kappa}\frac{y_i}{r}\right) - \delta_{ij}\frac{y_\kappa}{r} + \frac{3y_iy_jy_\kappa}{r^3}\right]$$

for 3-D problems

Equation (27) now provides the displacements at a point ξ due to a known distribution of initial stresses $\dot{\sigma}'_{ij}$ at x. These stresses may be present in a body as a result of plastic deformations, temperature gradients or any other non-linear events.

Swedlow & Cruse (1971) presented an initial strain formulation in which the volume integral in eqn. (26) was used in place of the body force integral of eqns. (23) and (24). Except for strain hardening materials it is generally difficult to calculate the magnitude of the initial strains. Indeed for this reason in the finite element method, the initial stress method has virtually replaced the initial strain method. In the present work we shall therefore be using eqn. (27). Thus we can write the boundary integral as:

$$\tfrac{1}{2}\dot{u}_j(\xi) = \int_S [\dot{t}_i(x)U_{ij}(x,\xi) - T_{ij}(x,\xi)\dot{u}_i(x)]\,\mathrm{d}S(x) + \int_V B_{i\kappa j}(x,\xi)\dot{\sigma}'_{i\kappa}(x)\,\mathrm{d}V \tag{28}$$

If physical body forces are present in the body and result in the plastic deformation, the body force integral must also be retained in (28). The solution must also trace the entire history of this body force development.

4.6 DISCRETISATION OF THE BOUNDARY AND VOLUME INTEGRALS

If the initial strains are known, eqn. (28) provides an exact formulation of the problem for any general well posed set of boundary conditions. The numerical errors arise solely due to subsequent discretisation of eqn. (28). These errors can be minimised by implementing an advanced numerical solution scheme such as that outlined in Chapter 3 of this volume.

In this chapter however, we shall use a much simpler implementation by using a piece-wise constant approximation for $\dot{t}_i, \dot{u}_i$ and $\dot{\sigma}'_{i\kappa}$. For our purposes we represent boundaries by straight-line segments for two-dimensional problems, and by triangular or quadrilateral elements for three-dimensional problems. The interior region which is expected to yield as a result of the loading is then divided into a suitable number of triangular or quadrilateral cells for two-dimensional problems, and tetrahedra or brick-like cells in three-dimensions. Although such a discretisation has the appearance of that used in the finite element method, the cells are used here simply to evaluate the volume integral in eqn. (28) as a piece-wise summation.

We can therefore write eqn. (28) for a field point ξ on the pth boundary element as:

$$\tfrac{1}{2}\dot{u}_j(\xi^p) = \sum_{q=1}^{n}\left[\dot{t}_i(x^q)\int_{\Delta S_q} U_{ij}(x^q, \xi^p)\,\mathrm{d}S - \dot{u}_i(x^q)\int_{\Delta S_q} T_{ij}(x^q, \xi^p)\,\mathrm{d}S\right] + \sum_{l=1}^{m}\dot{\sigma}'_{i\kappa}(x^l)\int_{\Delta V_l} B_{i\kappa j}(x^l, \xi^p)\,\mathrm{d}V \qquad (29)$$

where n is the total number of boundary elements and m is the total number of interior cells.

By taking the field point ξ over each boundary element, i.e., $p = 1, 2 \ldots n$, and absorbing the term $\frac{1}{2}\dot{u}_j$ within the diagonal blocks of T_{ij} (for $p = q$) we have

$$\mathbf{T}\dot{\mathbf{u}} - \mathbf{U}\dot{\mathbf{t}} = \mathbf{B}\dot{\boldsymbol{\sigma}}' \qquad (30)$$

which can be solved to obtain the unknown boundary data, provided that the initial stress increments are known. The determination of these initial stresses is discussed in the next section.

In order to construct the system of eqn. (30) it is necessary to evaluate the various integrals in eqn. (29) accurately. For elements and cells where the load point and field point do not coincide all the integrals can be evaluated by using the standard Gaussian quadrature formulae, the order of which can be varied depending on the distance between the 'loaded' element (or cell) and the field point.

When the field point coincides with the loaded boundary element the surface integrals can be calculated exactly by splitting the boundary element through the field point and constructing a local polar co-ordinate system as originally discussed by Cruse (1969). When the field point ξ is anywhere within a cell the volume integrals involved can be calculated by splitting the cell (joining the field point and the vertices of the cell) into a number of triangles for two-dimensional problems or tetrahedra for three-dimensional problems. The integrals over these sub-volume cells may be evaluated by using Gaussian quadrature formulae.

Having obtained the unknown components of boundary displacements and boundary tractions the displacements at any interior point can be obtained from:

$$\dot{u}_j(\xi) = \int_S [\dot{t}_i(x)U_{ij}(x, \xi) - T_{ij}(x, \xi)\dot{u}_i(x)]\,\mathrm{d}S(x) + \int_V B_{i\kappa j}(x, \xi)\dot{\sigma}'_{i\kappa}(x)\,\mathrm{d}V \qquad (31)$$

which can be written in the same manner as eqn. (29) for the discretised problem.

The strains corresponding to these displacements can be calculated by using a finite difference approximation to eqn. (2), i.e.

$$\dot{\varepsilon} = \mathbf{M}\dot{\mathbf{u}}_{\mathrm{n}} \tag{32}$$

where u_n are nodal values of displacements.

The stresses corresponding to these strains may be calculated from:

$$\dot{\boldsymbol{\sigma}}^{\mathrm{e}} = \mathbf{D}^{\mathrm{e}}\,.\,\dot{\varepsilon} \text{ in the elastic regions}$$

or

$$\dot{\boldsymbol{\sigma}}^{\mathrm{ep}} = \mathbf{D}^{\mathrm{ep}}\,.\,\dot{\varepsilon} \text{ in the elasto-plastic regions} \tag{33}$$

where $\mathbf{D}^{\mathrm{e}}$ is the elasticity matrix, $\mathbf{D}^{\mathrm{ep}}$ is the elasto-plastic matrix calculated by using a suitably up-dated stress history.

4.7 THE INCREMENTAL COMPUTATION

The above analysis requires the complete knowledge of the initial stress distribution $\sigma'_{i\kappa}$ within the yielded region, which is not known *a priori* for a particular increment. These must be obtained from an interative process as decribed below. For this purpose it is convenient to write eqn. (30) as:

$$\mathbf{A}\dot{\mathbf{x}} = \dot{\mathbf{b}} + \mathbf{B}\dot{\boldsymbol{\sigma}}' \tag{34}$$

where $\dot{\mathbf{x}}$ is the vector of unknown boundary tractions and boundary displacements, and $\dot{\mathbf{b}}$ is the matrix formed by multiplying the appropriate columns of **U** and **T** and the given increments of the boundary tractions and boundary displacements.

The steps of the incremental solution may then be described as follows:

(i) Apply a load increment $\dot{\mathbf{b}}$, assuming $\dot{\boldsymbol{\sigma}}'$ as zero. Calculate $\dot{\mathbf{x}}$, the strains $\dot{\varepsilon}$ and the elastic stress increments $\dot{\boldsymbol{\sigma}}$. Scale the solution so that the most highly stressed cell is at the point of yielding. Store the current values of stresses $\boldsymbol{\sigma}_1$.

(ii) Apply a small load increment. Calculate the stress increments in all the cells by using the elastic stress–strain relations, i.e., $\dot{\boldsymbol{\sigma}}^{\mathrm{e}} = \mathbf{D}^{\mathrm{e}}\dot{\varepsilon}$. Calculate the equivalent stresses by using $\boldsymbol{\sigma}_2 = \boldsymbol{\sigma}_1 + \dot{\boldsymbol{\sigma}}^{\mathrm{e}}$ as the stress history and compile a list of yielded cells. Calculate the correct stresses in the elasto-plastic cells by using the elasto-plastic stress–strain relations $\dot{\boldsymbol{\sigma}}^{\mathrm{ep}} = \mathbf{D}^{\mathrm{ep}}\,.\,\dot{\varepsilon}$, using the elastic strain increments as a first approximation. The initial stresses generated are $\dot{\boldsymbol{\sigma}}' = \dot{\boldsymbol{\sigma}}^{e} - \dot{\boldsymbol{\sigma}}^{ep}$ as a first approximation. Modify the stress history for the yielded cells to $\boldsymbol{\sigma}_2 = \boldsymbol{\sigma}_1 + \dot{\boldsymbol{\sigma}}^{ep}$ and make $\boldsymbol{\sigma}_1 = \boldsymbol{\sigma}_2$.

(iii) Assuming $\dot{\mathbf{b}} = 0$ and with the generated initial stresses $\dot{\boldsymbol{\sigma}}'$ calculate a new $\dot{\mathbf{x}}$ vector by using eqn. (34) and calculate the nodal displacement increments $\dot{\mathbf{u}}_n$, the strain increment $\dot{\boldsymbol{\varepsilon}}$ and the stress increments by using the elastic stress–strain relations $\dot{\boldsymbol{\sigma}}^{e} = \mathbf{D}^{e}.\dot{\boldsymbol{\varepsilon}}$. Calculate the equivalent stresses by using the history $\boldsymbol{\sigma}_2 = \boldsymbol{\sigma}_1 + \dot{\boldsymbol{\sigma}}^{e}$ and compile a list of yielded cells. For the elasto-plastic cells calculate the correct stresses $\dot{\boldsymbol{\sigma}}^{ep} = \mathbf{D}^{ep}.\dot{\boldsymbol{\varepsilon}}$. The initial stresses generated are $\dot{\boldsymbol{\sigma}}' = \dot{\boldsymbol{\sigma}}^{e} - \dot{\boldsymbol{\sigma}}^{ep}$. Modify the stress history for the yielded cells to $\boldsymbol{\sigma}_2 = \boldsymbol{\sigma}_1 + \dot{\boldsymbol{\sigma}}^{ep}$ and make $\boldsymbol{\sigma}_1 = \boldsymbol{\sigma}_2$.

(iv) Check if the initial stresses $\dot{\boldsymbol{\sigma}}'$ are less than an acceptable norm, if so go to the step (ii), if not go back to the step (iii). If the number of iterations exceeds 50 it can be assumed that collapse has been achieved.

Within the framework of the algorithm described above several improvements are possible. For example a certain degree of forward extrapolation may result in quicker convergence. The accuracy can be improved if a check is carried out on the sign of the dissipated plastic work rate to ensure that all the plastic cells are in fact plastic. The incremental algorithm outlined above is essentially similar to that described by Zienkiewicz (1971).

4.8 EXAMPLES OF SOLVED PROBLEMS

The algorithm described above has been programmed for ICL 2980 and CDC 7600 computers. The current version of the two-dimensional program is completely general, but the three-dimensional program has been written for restricted geometries, primarily to simplify the input data. Both the isotropic strain hardening model and the kinematic strain hardening model have been implemented. The examples presented below are intended to demonstrate the utility of the proposed algorithm.

4.8.1 Expansion of a Thick Cylinder under Internal Pressure

Two solutions to this problem were obtained using a coarse and a fine boundary and cell discretisation (see Figs. 5(a) and 5(b)). Although the problem is radially symmetric, one quadrant was used in the analysis. The material was assumed to be elastic ideally plastic (i.e., the hardening parameter $H = 0$).

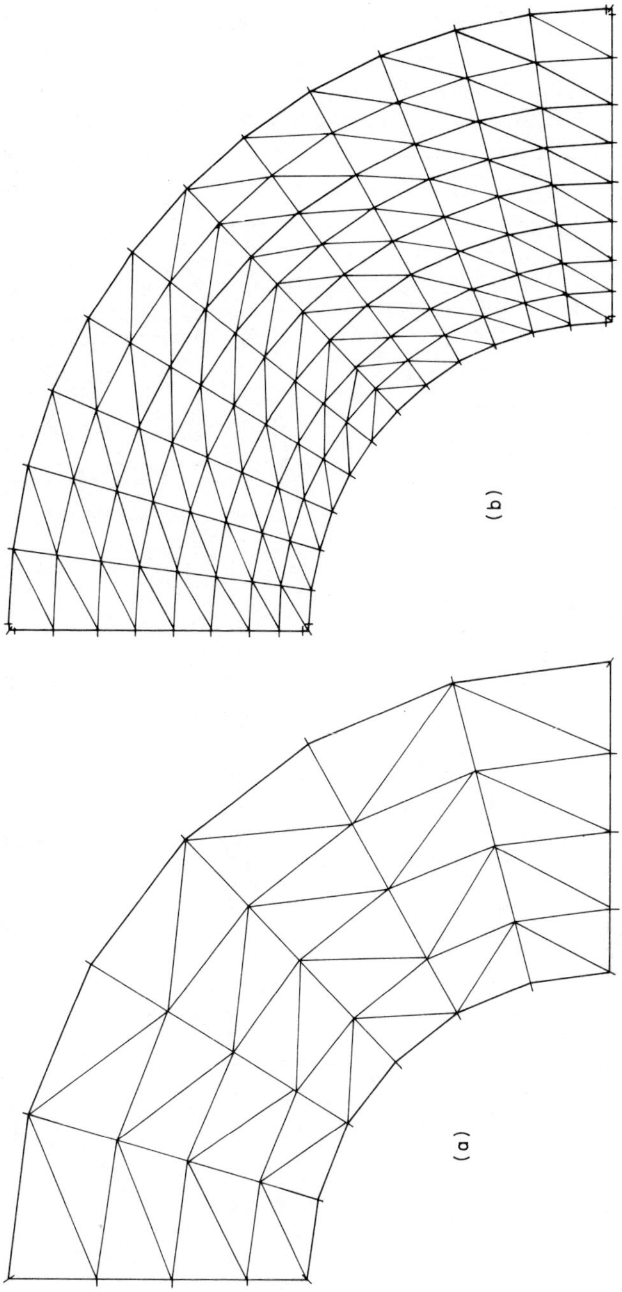

FIG. 5. Boundary discretisation and cell geometry for thick cylinder solution. (a) Coarse mesh, (b) fine mesh.

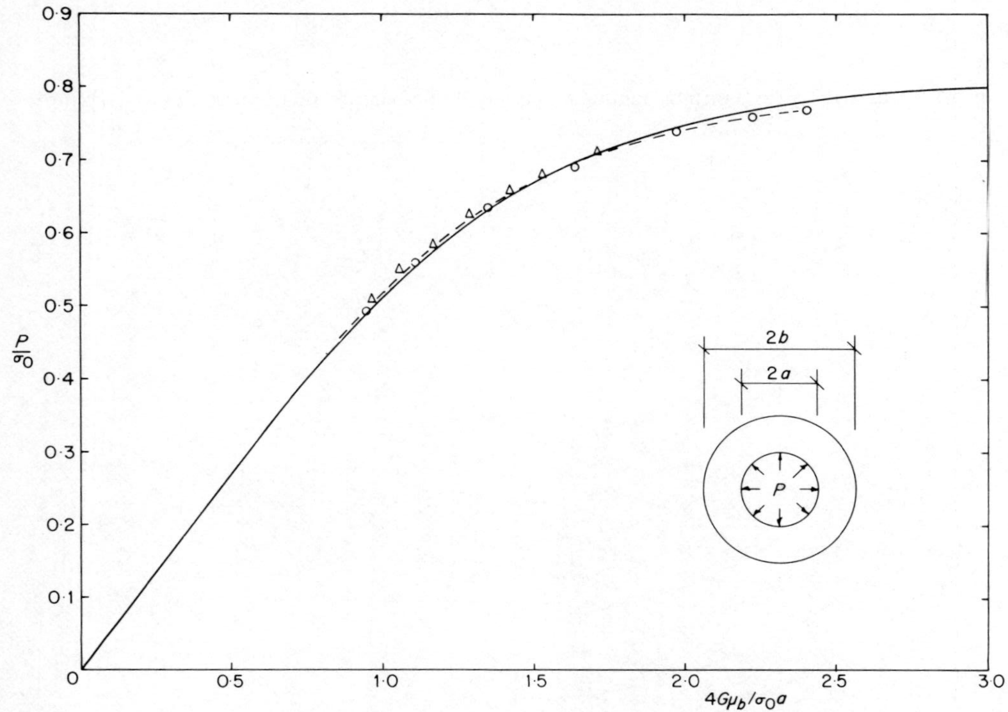

FIG. 6. Expansion of a thick cylinder under internal pressure—comparison of analytical and numerical results. Variable load increment sizes were used, not all results were plotted. —— Analytical; --△-- boundary element (coarse mesh); --○-- boundary element (fine mesh); $\mu = 1000$; $v = 0{\cdot}3$; $\sigma_0 = 600$; $H = 0{\cdot}0$; $a = 10$; $b/a = 2$ (plane strain).

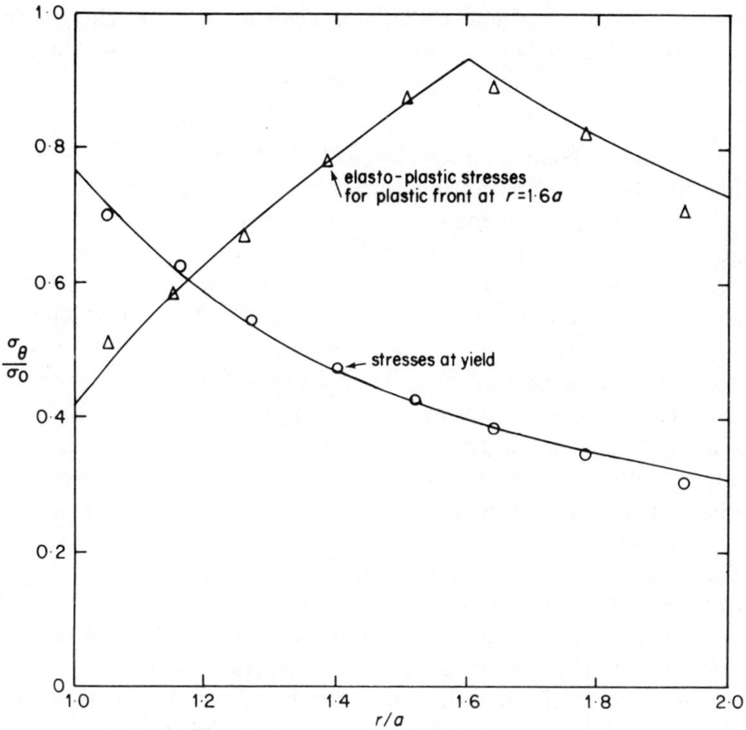

FIG. 7. Circumferential stress distribution in a thick cylinder—analytical and boundary element results. —— Analytical; –△– boundary element; –○– boundary element.

The computed results for the expansion of the outer surface of the cylinder with a gradually increasing internal pressure together with the analytical results from Hodge & White (1950) are shown in Fig. 6. Small differences between the two numerical solutions at first yield are to be expected since these are strongly influenced by the size of the cells used near the inner boundary. However, the comparison between analytical and computed results is still very good. These solutions for two different discretisation schemes, establish the stability and convergence of the solution method.

Typical results for circumferential stresses for the analysis using the finer discretisation at first yield and at an instant when the elasto-plastic boundary is at $r = 1{\cdot}6a$ are shown in Fig. 7. The computed results are in reasonably good agreement with the analytical solution. The solutions obtained by using load increments of 5 % and 7·5 % of the load at first yield were essentially identical at a given load level.

4.8.2 A Flexible Footing on an Elasto-Plastic Half Space

Whilst for the example shown earlier the computational cost was about 50 % higher (which can be reduced by further programming effort) than the corresponding solutions obtained by using the finite element method, the present method is particularly attractive for problems where the elastic region extends to infinity and the yielded region is confined to the vicinity of the loaded area. The infinitely distant elastic boundary is automatically included without discretisation, a feature unique to the present method.

In order to solve the problem of an elasto-plastic half space loaded by a gradually increasing surface pressure (plane strain problem), symmetry about the centreline was exploited. This quarter-space was then represented by 40 boundary elements and 64 triangular cells. The cells were confined to the vicinity of the loaded strip.

The problem was solved for strain hardening ($H > 0$), elastic ideally plastic ($H = 0$) and strain softening ($H < 0$, restricted) cases. The load–displacement plots are presented in Fig. 8 for the three cases. It is of interest to note that for the elastic ideally plastic case the load at which convergence could not be achieved is about 3 % above the exact collapse load of 5·14. The solution for the strain softening case is of course not valid for unrestricted choice of the strain softening parameter $H < 0$, since, during the incremental solution, the possibility of non-uniqueness of the solution may arise. Furthermore the loading and unloading criteria of the incremental theory of plasticity do not allow for strain softening. Nevertheless the solution for the particular case presented here appears reasonable for the following reasons.

(a) The stresses were always on the current yield surface of reduced size.

(b) Decreasing the magnitude of the load increments resulted in almost identical loads, stresses and displacements at collapse.

Obviously a great deal of work is needed in this important area before current incremental theories of plasticity can be used for strain softening materials.

Finally, the computational cost for this problem was about 10 % lower than that of a finite element analysis of comparable accuracy.

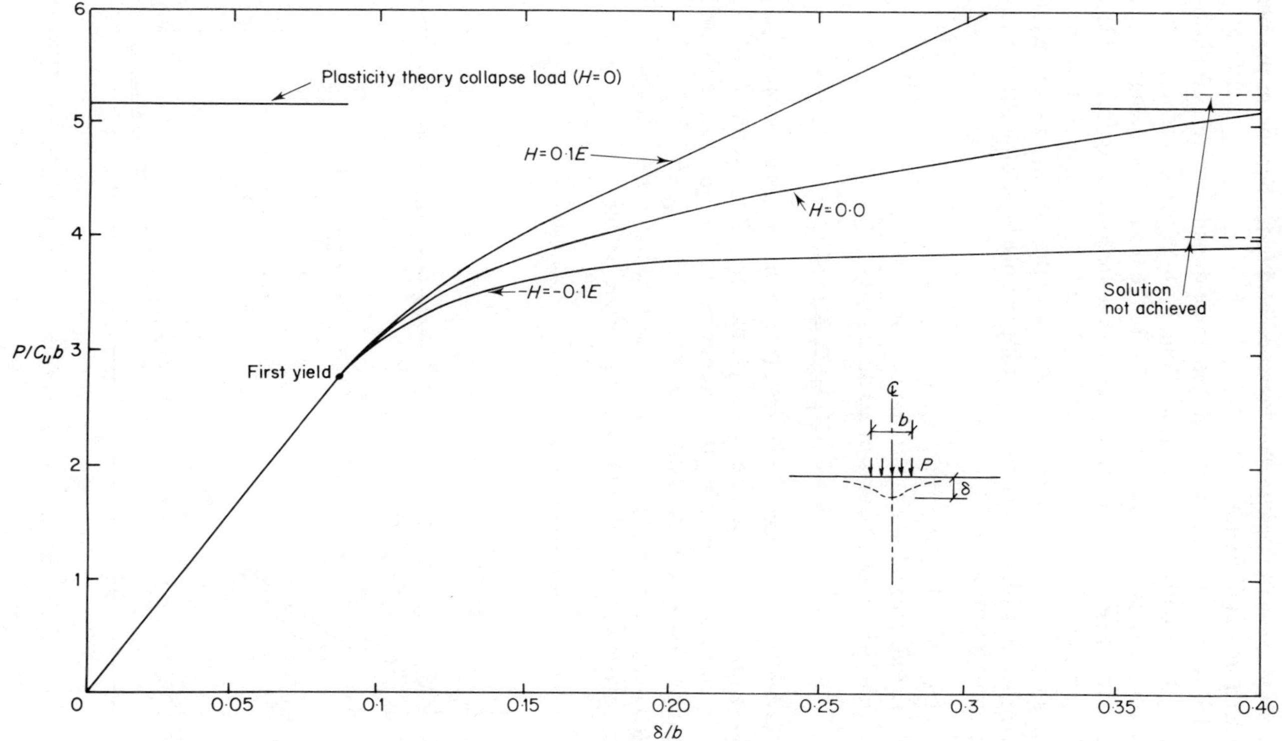

FIG. 8. Load–displacement behaviour of a flexible strip footing on semi-infinite elasto-plastic half space. $E = 10\,000\,\mathrm{kN/m^2}$; $\nu = 0{\cdot}25$; $\sigma_0 = 173\,\mathrm{kN/m^2}$; $b = 1{\cdot}0\,\mathrm{m}$; $C_u = \sigma_0/\sqrt{3}$.

4.8.3 Displacements Due to the Presence of Plastic Strains in a Cube Embedded in a Half Space

In order to test the accuracy of the three-dimensional elasto-plastic analysis, the recently solved problem (Chiu, 1978) of the displacement field due to plastic strains in a cube embedded in the interior of an elastic half space was analysed. Chiu presented a very accurate solution to this problem by using an integral transform method, and plotted the result of the vertical displacement of a surface point vertically above the cube against the depth of the cube below the surface for three different plastic strain distributions in the cube. These are:

(i) the vertical plastic strain $\dot{\varepsilon}_{11}^{p}$ only,
(ii) the horizontal plastic strain $\dot{\varepsilon}_{22}^{p}$ only,
(iii) uniform plastic strains in all directions

$$\dot{\varepsilon}_{11}^{p} = \dot{\varepsilon}_{22}^{p} = \dot{\varepsilon}_{33}^{p}$$

In order to reproduce Chiu's results, the traction free surface was discretised by using 34 square boundary elements. The symmetry of the problem was utilised by using the unknowns for one quadrant only and assuming the same element pattern existed in other quadrants.

If the plastic strains are known accurately the initial stresses generated are simply given by $\dot{\boldsymbol{\sigma}}' = \mathbf{D}^{e} . \dot{\boldsymbol{\varepsilon}}^{p}$. Numerical results for the vertical

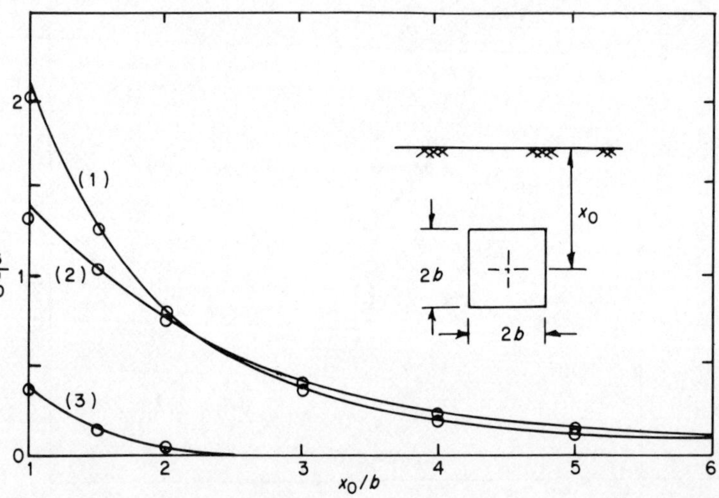

FIG. 9. Vertical surface displacement due to initial strains in a cube. —— Chiu; ○ numerical; (1) $\dot{\varepsilon}_{11}' = \dot{\varepsilon}_{22}' = \dot{\varepsilon}_{33}' = \dot{\varepsilon}_0'$; (2) $\dot{\varepsilon}_{33}' = \dot{\varepsilon}_0'$; (3) $\dot{\varepsilon}_{11}' = \dot{\varepsilon}_0'$.

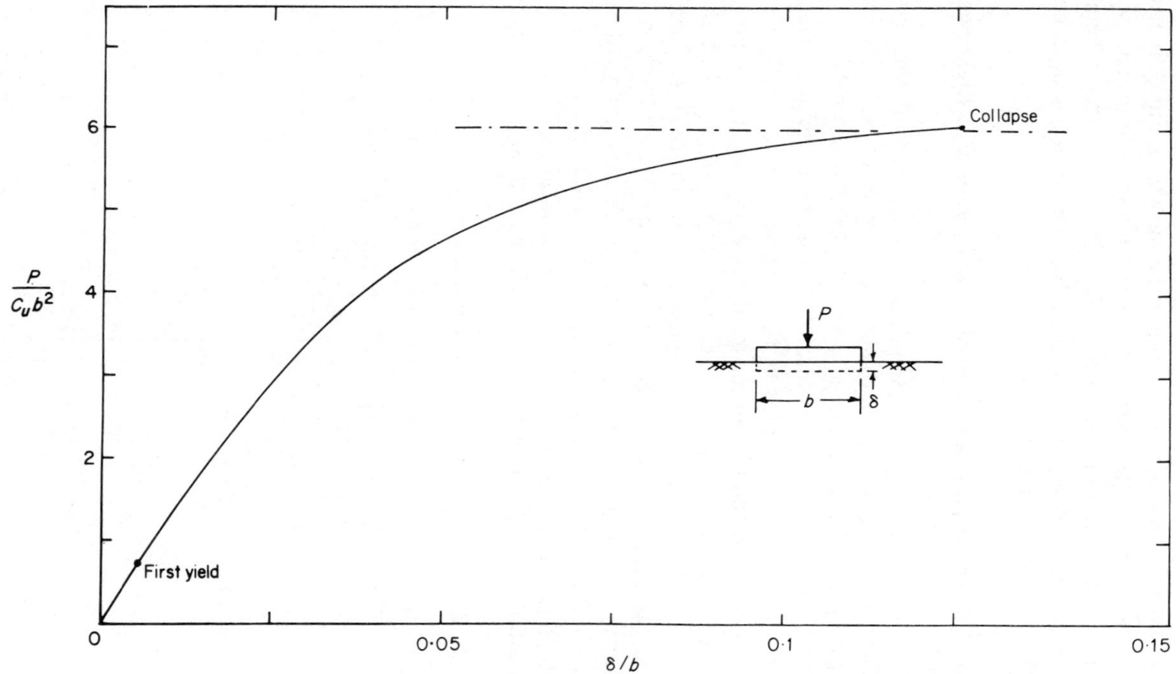

FIG. 10. Load–displacement behaviour of a rigid square footing. $C_u = \sigma_0/\sqrt{3}$; $\mu/C_u = 1000$.

displacements of the point on the free surface vertically above the cube, due to the initial stresses corresponding to the three values of plastic strains, are plotted against the depth of the cube below the surface, in Fig. 9. It can be seen that these are in excellent agreement with Chiu's results except for the case when the cube is very close to the surface. This is due to the difficulties involved in evaluating the weakly singular volume integral numerically. Nevertheless the maximum error is only about 4%.

4.8.4 The Problem of an Indentation of a Half Space by a Rigid Square Punch

Figure 10 shows the load–displacement curve for a square footing resting on the surface of a semi-infinite elastic ideally plastic solid. The exact solution to this problem is not known, however, the collapse load for a rigid circular footing resting on the surface of a semi-infinite solid is given by $P = 6\pi r^2 C_u$ where r is the radius of the footing and $C_u = \sigma_0/\sqrt{3}$. The collapse load for a circular footing having the same contact area is shown in Fig. 10 for comparison. The result of the present analysis gives the collapse load slightly higher than $6C_u b^2$. However, based on their engineering experience, the authors would expect the actual collapse load to be a little higher than this result. It is also of interest to note that the first yield under the corner of the footing occurs at a very early stage of the loading. The surface discretisation used for this problem is identical to that used in the previous example, in addition 60 cubical cells were used to calculate the

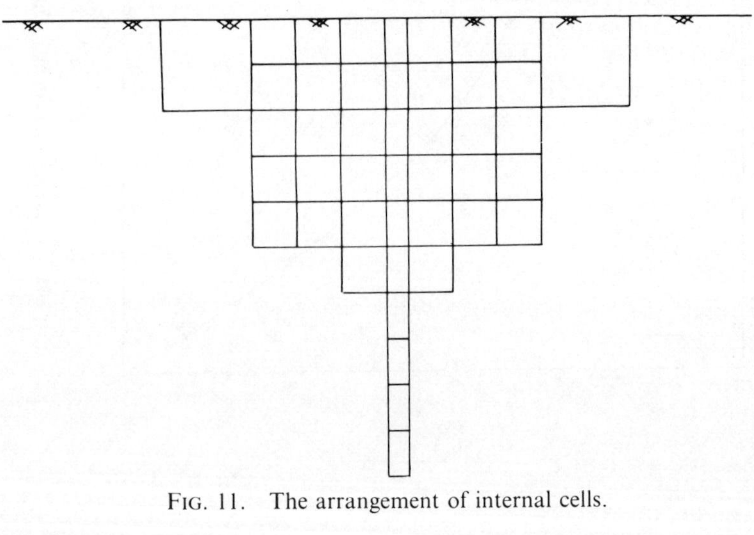

FIG. 11. The arrangement of internal cells.

contributions of the generated initial stresses. The computation time was about 6 minutes by using a C.D.C. 7600 computer.

4.8.5 The Problem of a Laterally Loaded Pile Subjected to Cyclic Loading

While all the previous examples were solved by using the general method of analysis with (i) Kelvin's kernel function (U_{ij}) and (ii) the isotropic strain hardening model, this particular problem was analysed by using Mindlin's solution and the kinematic hardening model. The traction free surface of the half space is automatically satisfied by Mindlin's solution (Mindlin, 1936) therefore only the discretisation of the pile–soil interface is necessary.

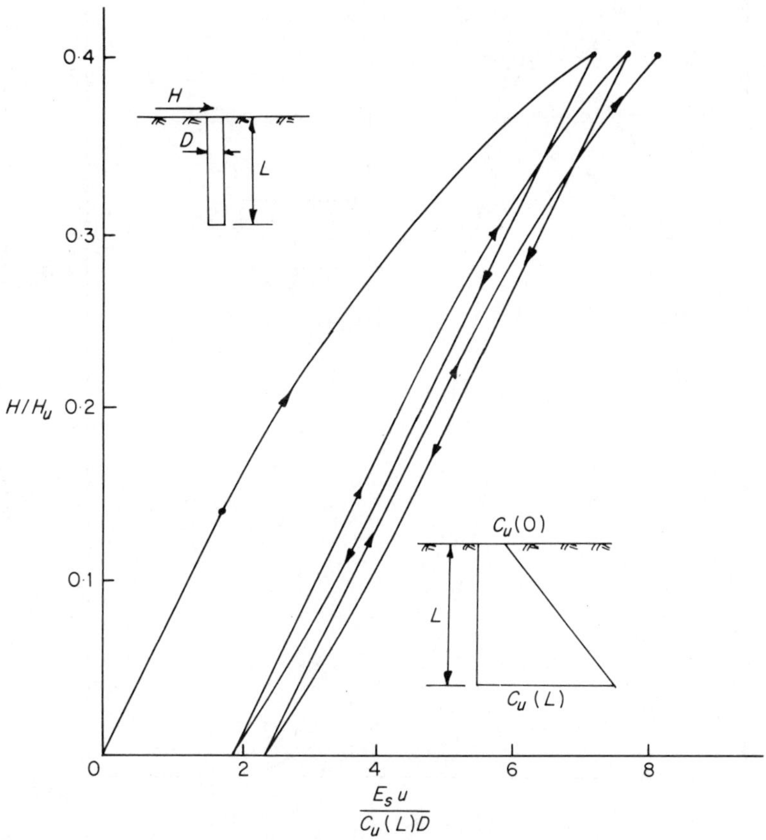

FIG. 12. Loading and unloading behaviour of a laterally loaded pile. u = Lateral deflection; H_u = ultimate load; $K_R = E_p I_p/(E_s \,.\, L^4) = 3 \times 10^{-5}$; $L/D = 20$; $\nu = 0{\cdot}3$; $E_s/C_u(L) = 100$; $C_u(0)/C_u(L) = 0{\cdot}2$.

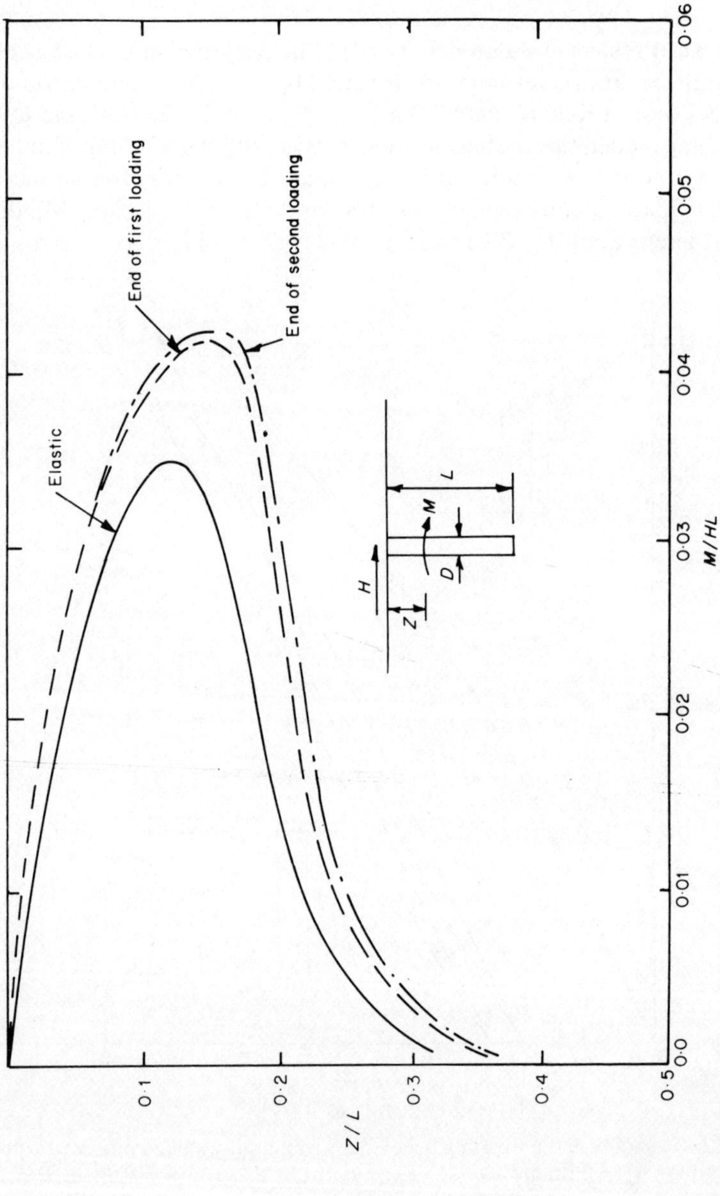

FIG. 13. The bending moment distribution. $K_R = 3 \times 10^{-5}$; $L/D = 20$.

The pile was represented as a linear structural element (see, Banerjee & Driscoll, 1976; Banerjee, 1978; Banerjee & Davies, 1978). Figure 11 shows the typical cell pattern used around the embedded pile. It is also of interest to note that these cells are clustered around the expected elasto-plastic zone. In this problem although E_s was assumed to be constant the initial yield stress $C_u = \sigma_0/\sqrt{3}$ was varied with the depth and the kinematic hardening parameter taken as $C = 0{\cdot}001 \times E_s$.

Figure 12 shows the loading, unloading and reloading curves for the pile, and Fig. 13 shows the initial elastic bending moment distribution and the subsequent ones at the end of each reloading. Clearly, an analysis of this type is needed to investigate any possible lack of shakedown load for cyclically loaded structures. We shall explore these problems in greater detail in some forthcoming publications.

4.9 CONCLUSIONS

A new boundary element formulation based on the initial stress method has been described which eliminates many difficulties that were present in implementing earlier formulations.

A complete algorithm for obtaining numerical solutions of two- and three-dimensional engineering plasticity problems is described.

A series of examples solved by using the present algorithm have been presented to demonstrate the elegance and the versatility of the present method for routine analysis of practical problems.

REFERENCES

Banerjee, P. K. (1976). Integral equation methods for the analysis of piece-wise non-homogeneous elastic solids of arbitrary shape, *Int. J. Mech. Sci.*, **18,** 293–301.

Banerjee, P. K. (1978), 'Analysis of axially and laterally loaded pile groups'. In: *Developments in Soil Mechanics*, Ed. C. R. Scott, Applied Science Publishers, London, Chapt. 9, pp. 317–40.

Banerjee, P. K. & Butterfield, R. (1977). 'Boundary element methods in geomechanics'. In: *Finite Elements in Geomechanics*, Ed. G. Gudehus, Wiley, New York, Chapt. 16, pp. 519–70.

Banerjee, P. K. & Driscoll, R. M. C. (1976). Three-dimensional analysis of raked pile groups, *Proc. Inst. Civil Engineers, Part 2 research and theory*, **61,** 653–70.

BANERJEE, P. K. & DAVIES, T. G. (1978). Behaviour of axially and laterally loaded single piles embedded in non-homogeneous soils, *Geotechnique*, **28** (1), 309–26.

BANERJEE, P. K. & MUSTOE, G. G. W. (1978). Boundary element methods in two-dimensional problems of elasto-plasticity, *Proc. Int. Conf. on Recent Developments in Boundary Element Methods*, Southampton University, pp. 283–300.

CHIU, Y. (1978). On the stress field and surface deformation in a half space with a cuboidal zone in which initial strains are uniform. *J. Appl. Mech, Trans. ASME*, **45,** 302–6.

CRUSE, T. A. (1969). Numerical solutions in three-dimensional elasto-statics, *Int. J. Sol. and Struct.*, **5,** 1259–74.

CRUSE, T. A. (1974). An improved boundary-integral equation method for three-dimensional elastic stress analysis, *Int. J. Comp. and Struct.*, **4,** 741–57.

CRUSE, T. A. & MEYERS, G. J. (1977). Three-dimensional fracture mechanics analysis, *J. Structural Division, ASCE* **103,** (ST2), 309–20.

CRUSE, T. A. & WILSON, R. B. (1978). Advanced applications of boundary-integral equation methods, *Nuclear Engineering and Design*, **46,** 223–34.

FUNG, Y. C. (1965). *Foundations of Solid Mechanics*, Prentice-Hall, New York.

HODGE, P. G. & WHITE, G. N. (1950). A quantitative comparison of flow and deformation theory of plasticity, *J. Appl. Mech.*, **17,** 180–5.

KUMAR, V. & MUKHERJEE, S. (1977). A boundary integral formulation for time dependent in elastic deformation of metals. *Int. J. Mech. Sci*, **19,** 713–24.

LACHAT, J. C. & WATSON, J. O. (1976). Effective numerical treatment of boundary integral equations: a formulation for three-dimensional elasto-statics, *Int. J. Num. Meth. Eng.*, **10,** 991–1005.

LIN, T. Y. (1967). Reciprocal theorem for displacements in inelastic bodies, *J. Composite Matter*, **1,** 144–51.

LIN, T. Y. (1968). *Theory of Inelastic Structures*, Wiley, New York.

MENDELSON, A. & ALBERS, L. U. (1975). Application of boundary integral equation to elasto-plastic problems, *Proc. ASME Conf. on Boundary Integral Equation Methods*, New York.

MINDLIN, R. D. (1936). Force at a point in the interior of a semi-infinite solid, *Physics*, **7,** 195–202.

PRAGER, W. (1955). Theory of plasticity: a survey of recent achievement, (James Clayton Lecture), *Proc. Inst. Mech. Engineers*, **169,** 493–8.

REISNER, H. (1931). Initial stresses and sources of initial stresses, (in German), *Z. Agnew, Math. Mech.*, **11,** 1–8.

RICCARDELLA, P. (1973). *An implementation of the boundary integral technique for plane problems of elasticity and elasto-plasticity*, Ph.D. Thesis, Carnegie Mellon University, Pittsburg.

RZASNICKI, W. & MENDELSON, A. (1975). *Application of boundary integral equation method to elasto-plastic analysis of V-notched beams*, NASA technical report, TMX-71472.

SHIELDS, R. T. & ZIEGLER, H. (1958). On Prager's hardening rule, *Z.A.M.P.*, **9a,** 260.

SWEDLOW, J. L. & CRUSE, T. A. (1971). Formulation of boundary integral equations for three-dimensional elasto-plastic flow, *Int. J. Sol. and Struct.*, **7,** 1673–81.

ZIEGLER, H. (1959). A modification of Prager's hardening rule, *Quart. Appl. Math.* **17,** (1), 55–65.

ZIENKIEWICZ, O. C., VALLIAPPAN, S. & KING, I. P. (1969). Elasto-plastic solutions of engineering problems 'initial stress', finite element approach, *Int. J. Num. Meth. Eng.*, **1,** 75–100.

ZIENKIEWICZ, O. C. (1971). *Finite Element Method in Engineering Science*, McGraw-Hill, New York.

Chapter 5

TWO- AND THREE-DIMENSIONAL PROBLEMS OF FRACTURE MECHANICS

T. A. CRUSE

Pratt & Whitney Aircraft Group, East Hartford, USA

SUMMARY

This chapter describes various numerical and analytical approaches that have been devised for the application of the boundary integral equation formulation to fracture mechanics problems. Because the method is intrinsically more capable of resolving high stress gradients than is the finite element method, great numerical success has been achieved, including for the very difficult three-dimensional crack problem.

The basic singular relations for the elasticity solution near a crack tip are first reviewed in order to define basic terms. Direct formulation of a boundary integral equation, dislocation model for the general crack problems is introduced. This formulation leads to the required crack surface, traction integral equation which is analytically solved for the two-dimensional problem.

A second approach, useful in two-dimensional problems, is the Green's function for cracked bodies. This formulation is reviewed and its accuracy established. Three-dimensional problems require direct numerical analysis. In this chapter, some new results are shown using a special crack tip modelling procedure based on isoparametric shape function formulations. These new results show great promise in further resolving the surface crack fracture mechanics problem.

5.1 INTRODUCTION

The life of a structure subjected to repeated moderate-to-high load levels is often governed by the cyclic growth of crack-like defects. It has been established that parameters deduced from elastic fracture mechanics analysis can correlate crack growth rate in specimens and structures which undergo nominally elastic loading. These parameters, referred to as stress intensity factors, incorporate stress levels, geometry and crack size in a systematic manner and may be evaluated from the stress analysis of cracked structures.

The boundary integral equation (BIE) method has been adopted for elastic stress analysis of two- and three-dimensional bodies. Inherent modelling accuracy and efficiency have also led to the use of BIE techniques for elastic fracture mechanics analysis. The purpose of this chapter is to illustrate some of the numerical and modelling problems and results for BIE fracture mechanics analysis.

Basic aspects of elastic fracture mechanics analysis are reviewed in the Section 5.2, including definition of stress intensity factors and reference to some techniques for numerical evaluations. Section 5.3 reviews the BIE method of elastic stress analysis and its extensions to fracture mechanics analysis. Included in this discussion are numerical results for two- and three-dimensional cracks as well as the development of a special purpose BIE method for fracture mechanics problems. Finally, some current limitations and areas of future research are discussed.

5.2 ELASTIC FRACTURE MECHANICS ANALYSIS

5.2.1 Basis of the Analytical Model

The physical presence of a sharp crack in a structure generally reduces the static and fatigue strength of the structure because stresses and strains are highly magnified at the crack tip. Analytical fracture mechanics seeks to determine the characteristics of this local magnification or intensification. Stress and strain fields in crack tip regions generally involve all three spatial variables to some degree, often necessitating the use of three-dimensional analysis.

Elastic fracture mechanics analysis idealises the physical crack problem in three major ways: the crack plane is generally taken to be flat; the crack is assumed to be sufficiently large that local material microstructure can be modelled as a continuum; and, inelastic crack tip effects such as plasticity are restricted to a sufficiently small volume that they may be neglected. The validity of these assumptions must be verified on the basis of the actual behaviour of the cracked structure. In some problems, such as planar structures containing through cracks, the stress and strain fields ahead of the crack may be approximated as two-dimensional.

The mathematical construct seeks to determine the stresses and strains in an infinite body containing a finite-size crack; the body is taken to be uniformly loaded at infinity. In two dimensions the crack is taken to be a straight-line slit of length $2a$; in three dimensions the crack is considered herein to be of elliptical plan-form and to have an infinitesimal width. The elliptical shape is given by a major axis of length $2a$ and a minor axis of length $2b$. These geometries are shown in Fig. 1.

5.2.2 Elastic Stress Intensity Factors

The displacement of any point in a body has three degrees of freedom. Elastic modelling of crack tip behaviour therefore makes use of three degrees of freedom or modes of motion for near crack tip points. The three accepted modes are portrayed in Fig. 2 as the opening (mode I), sliding (mode II), and tearing (mode III) modes. As will be shown below, the use of these three modes allows for a decoupling of stresses, strains and displacements near the crack tip for isotropic materials.

Consider the two-dimensional elastic region near a crack tip as shown in Fig. 3. Williams (1952) has shown that the stresses for traction-free crack surfaces may be given as infinite series in r, the distance from the crack tip

$$\begin{aligned} \sigma_x &= -a_1 r^{-1/2}\cos\frac{\theta}{2}\left[1-\sin\frac{\theta}{2}\sin\frac{3\theta}{2}\right] \\ &\quad + b_1 r^{-1/2}\sin\frac{\theta}{2}\left[2+\cos\frac{\theta}{2}\cos\frac{3\theta}{2}\right] + 0(r^{\circ}) \\ \sigma_y &= -a_1 r^{-1/2}\cos\frac{\theta}{2}\left[1+\sin\frac{\theta}{2}\sin\frac{3\theta}{2}\right] - b_1 r^{-1/2}\sin\frac{\theta}{2}\cos\frac{\theta}{2}\cos\frac{3\theta}{2} + 0(r^{\circ}) \\ \tau_{xy} &= a_1 r^{-1/2}\sin\frac{\theta}{2}\cos\frac{\theta}{2}\cos\frac{3\theta}{2} - b_1 r^{-1/2}\cos\frac{\theta}{2}\left[1-\sin\frac{\theta}{2}\sin\frac{3\theta}{2}\right] + 0(r^{\circ}) \end{aligned} \tag{1}$$

In eqn. (1), a_1 refers to loading symmetric to the $\theta = 0$ axis (mode I); b_1 refers to loading antisymmetric with respect to the $\theta = 0$ axis (mode II). The term $0(r^{\circ})$ refers to terms proportional to non-negative powers of r.

The coefficients a_1, b_1 in (1) define the strength of the stress singularity in the vicinity of the crack tip. That is, over a sufficiently small distance from

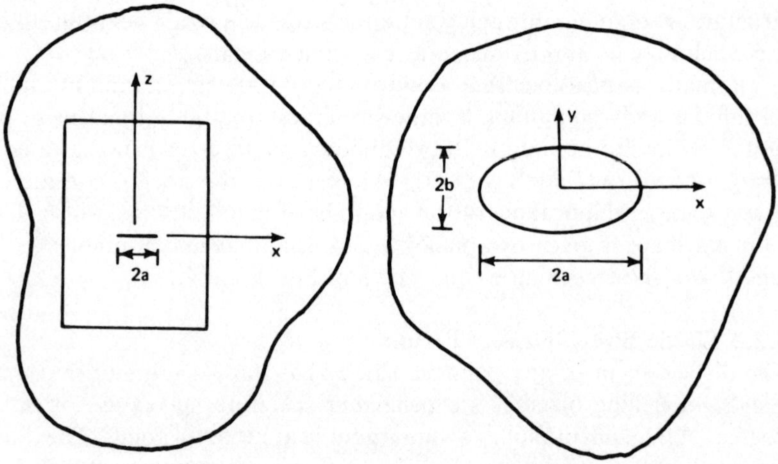

FIG. 1. Definition of crack geometries.

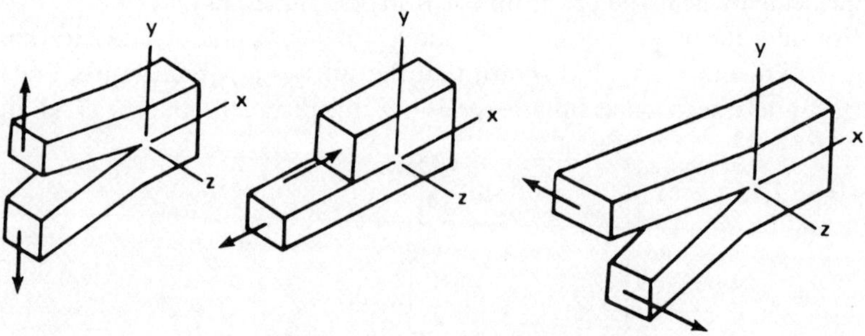

FIG. 2. Crack tip loading and displacement modes.

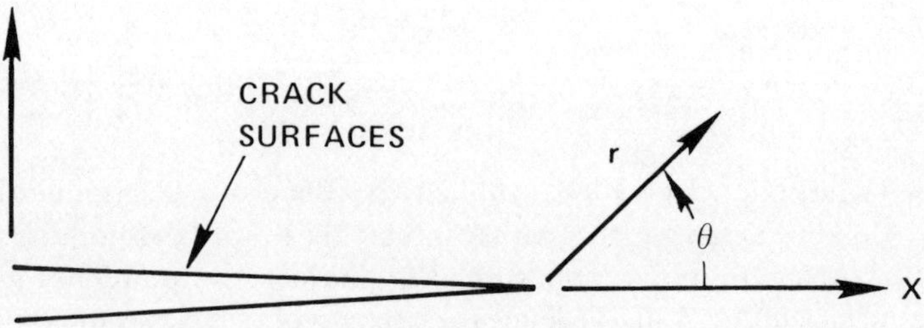

FIG. 3. Two-dimensional elastic region near a crack tip.

$r = 0$ the stresses are fully characterised by a_1 and b_1. Common usage is to replace a_1 and b_1 by the expressions

$$a_1 = -K_{\mathrm{I}}/\sqrt{2\pi}$$
$$b_1 = -K_{\mathrm{II}}/\sqrt{2\pi} \tag{2}$$

where K_{I} and K_{II} are the mode I and II stress intensity factors. For two-dimensional plane strain problems, these two stress intensity factors completely describe the near crack tip displacements

$$u_x = \frac{K_{\mathrm{I}}}{G}\left(\frac{r}{2\pi}\right)^{1/2} \cos\frac{\theta}{2}\left[1 - 2\nu + \sin^2\frac{\theta}{2}\right] + \frac{K_{\mathrm{II}}}{G}\left(\frac{r}{2\pi}\right)^{1/2} \sin\frac{\theta}{2}\left[2(1-\nu) + \cos^2\frac{\theta}{2}\right]$$

$$u_y = \frac{K_{\mathrm{I}}}{G}\left(\frac{r}{2\pi}\right)^{1/2} \sin\frac{\theta}{2}\left[2(1-\nu) - \cos^2\frac{\theta}{2}\right] + \frac{K_{\mathrm{II}}}{G}\left(\frac{r}{2\pi}\right)^{1/2} \cos\frac{\theta}{2}\left[-1 + 2\nu + \sin^2\frac{\theta}{2}\right] \tag{3}$$

where ν is Poisson's ratio and G is the shear modulus.

As will be described further in the next section, K_{I} and K_{II} depend on the geometry of the body, the loading, and the length of the crack. Thus, K_{I} and K_{II} are to be determined by elastic stress analysis of a given geometry. The third mode of deformation is associated with out-of-plane motion of the crack faces and is given by a crack tip stress singularity of strength K_{III}.

5.2.3 Evaluation of Some Stress Intensity Factors

The literature is replete with elastic solutions of cracked, two-dimensional bodies. An example compilation of these stress intensity factors is given by Tada *et al.* (1973). Three-dimensional fracture mechanics problems are beyond closed-form solution except for certain infinite body problems with idealised crack shapes. A few examples of two- and three-dimensional stress intensity factors are cited below.

The simplest two-dimensional problem to consider is an infinite elastic plate under remote tension containing a crack inclined at an angle from the load axis. The solution for this problem is given by Paris & Sih (1965) as

$$K_{\mathrm{I}} = \sigma \sin^2 \beta (\pi a)^{1/2}$$
$$K_{\mathrm{II}} = \sigma \sin\beta \cos\beta (\pi a)^{1/2} \tag{4}$$

For finite boundaries, advanced numerical techniques such as reported in this chapter must be used to model the cracked geometry; in isotropic materials the effect of finite geometry is included in a correction term

$$K_{\rm I} = \sigma\sqrt{a}\,Y\,(\text{geometry}) \tag{5}$$

Tada *et al.* (1973) compile many such finite body solutions for various crack shapes and loading conditions.

One of the few three-dimensional problems for which stress intensity factor solutions exist is the buried elliptical crack as solved by Green & Sneddon (1950). The simplest problem, and one which is numerically modelled below, is for remote tension applied transverse to the crack plane. The values of $K_{\rm II}$ and $K_{\rm III}$ are zero; the opening mode stress intensity factor is given by

$$K_{\rm I} = \frac{\sigma\sqrt{\pi a}}{E(k)}\left[\sin^2\theta + \frac{a^2}{b^2}\cos^2\theta\right]^{1/4} \qquad (a < b) \tag{6}$$

where $E(k)$ is the elliptic integral for $k^2 = 1 - a^2/b^2$; θ is the elliptical angle measured from the x-axis. Additional terms in (6) are defined in Fig. 1. More complex load conditions have been treated by Kassir & Sih (1966).

5.2.4 Strain Energy Release Rate

Consider a remotely loaded, cracked elastic body containing a through-crack of length $2a$ as shown in Fig. 1. The elastic strain energy (per unit area for plane strain) can be shown to decrease for an infinitesimal static extension of the crack as shown by Paris & Sih (1965). The rate of strain energy change is given by

$$\mathrm{G} = \frac{\mathrm{d}U}{\mathrm{d}a} = \frac{1-\nu^2}{E}(K_{\rm I}^2 + K_{\rm II}^2) + \frac{1}{2G}K_{\rm III}^2 \tag{7}$$

The value of the strain energy release rate, G, may be approximated by comparing the elastic strain energy for two geometries with slightly different crack lengths (U_2, U_1) as follows

$$\mathrm{G} \cong \Delta U/\Delta a = (U_2 - U_1)/(a_2 - a_1) \tag{8}$$

For three-dimensional problems Δa in eqn. (8) is to be interpreted as the change in the area of one crack face. Equation (8) forms the basis of fracture mechanics compliance testing to determine G for a given test specimen; some of the numerical data presented below approximates G in the same manner as eqn. (8).

5.3 BOUNDARY INTEGRAL EQUATION FRACTURE MECHANICS ANALYSIS

5.3.1 Dislocation Model for Buried Cracks

Consider the geometry, illustrated in Fig. 4, of a solid containing an open cavity approximating a crack. Let S, Γ^+, and Γ^- represent the regular, upper and lower crack surfaces as shown. Then the Somigliana identity for the displacement at an interior point $p(x)$ and boundary point $Q(x)$ is given by

$$u_i(p) = -\int_{S+\Gamma^+ +\Gamma^-} T_{ij}(p, Q)u_j(Q)\,\mathrm{d}S + \int_{S+\Gamma^+ +\Gamma^-} U_{ij}(p, Q)t_j(Q)\,\mathrm{d}S \qquad (9)$$

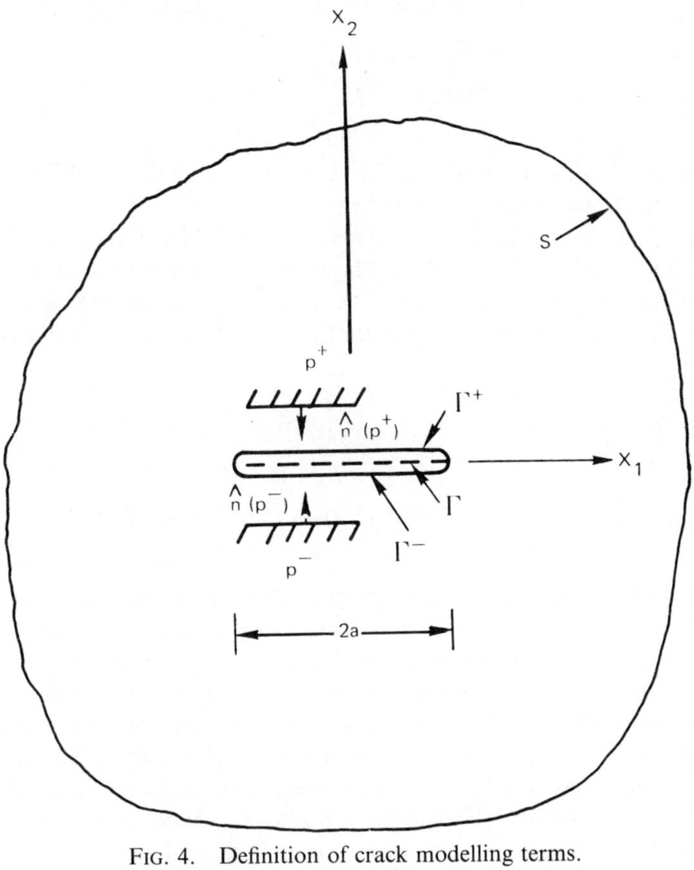

FIG. 4. Definition of crack modelling terms.

The kernels in eqn. (9) are defined in Rizzo (1967) and in Cruse (1969) where it is seen that

$$\begin{aligned} T_{ij}(p, Q^+) &= -T_{ij}(p, Q^-) \\ U_{ij}(p, Q^+) &= +U_{ij}(p, Q^-) \end{aligned} \tag{10}$$

except near the small crack tip region; this sign change is due to the opposite normal directions of the two crack surfaces. Letting the two crack surfaces collapse to the plane Γ (with the same normal as Γ^+) eqn. (9) is seen to give

$$\begin{aligned} u_i(p) = &-\int_S T_{ij}(p, Q)u_j(Q)\,\mathrm{d}S + \int_S U_{ij}(p, Q)t_j(Q)\,\mathrm{d}S \\ &-\int_\Gamma T_{ij}(p, Q)\Delta u_j(Q)\,\mathrm{d}S + \int_\Gamma U_{ij}(p, Q)\Sigma t_j(Q)\,\mathrm{d}S \end{aligned} \tag{11}$$

where

$$\begin{aligned} \Delta u_j(Q) &= u_j(Q^+) - u_j(Q^-) \\ \Sigma t_j(Q) &= t_j(Q^+) + t_j(Q^-) \end{aligned} \tag{12}$$

In general, we will consider problems where the crack is loaded by equal and opposite tractions such that $\Sigma t_j(Q) = 0$.

Equation (11) is not a function solely of boundary data and the usual operation of taking $p(x) \rightarrow P(x)$, a boundary point, is necessary to obtain the BIE. Letting $P(x)$ be a point on the plane Γ (not at the edge of Γ where $\Delta u_j = 0$) and taking the usual term for the jump in the boundary integral we obtain

$$\begin{aligned} u_i(P) - \Delta u_i(P)/2 &+ \int_S T_{ij}(P, Q)u_j(Q)\,\mathrm{d}S \\ &+ \int_{\langle\Gamma\rangle} T_{ij}(P, Q)\Delta u_j(Q)\,\mathrm{d}S = \int_S U_{ij}(P, Q)t_j(Q)\,\mathrm{d}S \end{aligned} \tag{13}$$

where the brackets on the second integral denote a principal value integral. The first two terms in eqn. (13) are easily seen to be equal to $\Sigma u_i(P)/2$.

Equation (13) has two critical deficiencies as a mathematical model for crack geometries. First consider a problem where there are no tractions on the outer boundary and only crack surface loading; since any set of equal and opposite boundary tractions gives the same BIE as eqn. (13) the result is non-unique. Second, while a single surface Γ is now being treated, two variables are unknown: Δu_j and Σu_j. Thus, eqn. (13) under-determines the mathematical basis for a solution of the flat crack modelling problem.

Two deficiencies identified by eqn. (13) for flat crack modelling identify the necessary (but not unique) correction. A second integral identity for the boundary data which is independent of eqn. (13) must be devised; in addition, the new BIE must uniquely depend on the state of loading of the crack surface.

Development of an additional BIE for the boundary data on the flat crack surface is based on a careful extension of the equation for the internal stresses. Internal stresses are given by Cruse (1969) as

$$\sigma_{ij}(p) = \int_S D_{kij}(p, Q) t_k(Q)\, \mathrm{d}S - \int_S S_{kij}(p, Q) u_k(Q)\, \mathrm{d}S$$
$$+ \int_\Gamma D_{kij}(p, Q) \Sigma t_k(Q)\, \mathrm{d}S - \int_\Gamma S_{kij}(p, Q) \Delta u_k(Q)\, \mathrm{d}S \qquad (14)$$

In eqn. (14) the integrals over the crack surface are determined as in eqn. (11). Further, it is assumed that the crack is loaded by equal and opposite tractions. If there is no crack, the first two integrals in eqn. (14) are the usual terms for internal stress calculation and are identified for convenience as

$$\sigma^*_{ij}(p) = \int_S D_{kij}(p, Q) t_k(Q)\, \mathrm{d}s - \int_S S_{kij}(p, Q) u_k(Q)\, \mathrm{d}S \qquad (15)$$

Let $\sigma^\circ_{ij}(p)$ be the stress due to the presence of the crack surface in eqn. (14) such that

$$\sigma_{ij}(p) = \sigma^*_{ij}(p) + \sigma^\circ_{ij}(p) \qquad (16)$$

In the general problem of analysing a cracked body, $\sigma^*_{ij}(p)$ is not truly independent of the crack due to the coupling of the outer boundary and crack boundary tractions and displacements through eqn. (13). However, for an indefinitely large body unloaded on the external surface the internal stress is given solely by

$$\sigma^\circ_{ij}(p) = -\int_\Gamma S_{kij}(p, Q) \Delta u_k(Q)\, \mathrm{d}S \qquad (17)$$

Cruse (1975), and Guidera & Lardner (1975) both showed the reduction of eqn. (15) to integrable form for the three-dimensional crack problem. Following the three-dimensional analyses, the two dimensional traction BIE for the symmetric (mode I) crack problem is given by

$$\sigma^\circ_{22}(\xi_1) = \frac{\mu}{2\pi} \frac{1}{1 - v} \int_{-a}^{a} \frac{1}{x_1 - \xi_1} \Delta u_{2,1}(x_1)\, \mathrm{d}x_1 \qquad (18)$$

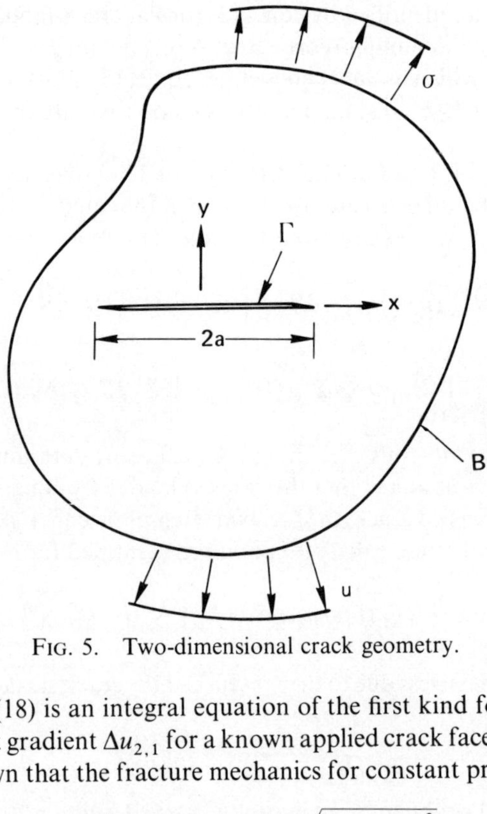

FIG. 5. Two-dimensional crack geometry.

Equation (18) is an integral equation of the first kind for the unknown displacement gradient $\Delta u_{2,1}$ for a known applied crack face pressure σ_{22}^{o}. It is easily shown that the fracture mechanics for constant pressure, given by

$$\mu \Delta u_2 = 2(1 - \nu) pa \sqrt{1 - (x/a)^2} \tag{19}$$

satisfies eqn. (18). The fracture mechanics problem for finite bodies is much more complex and must be solved numerically, as discussed by Cruse (1975).

5.3.2 Green's Function Formulation for Two Dimensions

The present discussion is limited to the solution of the two-dimensional elastic fracture mechanics problem of a finite plate containing a single, traction-free crack as shown in Fig. 5. In general, the surface B may intersect the crack surface Γ with the requirement that loading on B at Γ be compatible with the traction-free condition on Γ. The crack surface Γ is taken to be centred at $(x, y) = 0$, with crack tips at $x = \pm a$.

The elastic Green's function for the cracked plate problem with a point load at (x_0, y_0) was given by Snyder & Cruse (1975). Utilising the reported

Green's function, the Somigliana identity for the cracked plate is given by:

$$u_j(z_0) = -\int_B [T^{\circ}_{ji}(z_Q, z_0) + T^{*}_{ji}(z_Q, z_0)]u_i(z_Q)\,\mathrm{d}s(z_Q) + \int_B [U^{\circ}_{ji}(z_Q, z_0) + U^{*}_{ji}(z_Q, z_0)]t_i(z_Q)\,\mathrm{d}s(z_Q) \quad (20)$$

The displacement vector $u_j(z_0)$ in eqn. (20) satisfies the Navier equations of elasticity at all points $z_0 = x + \mu y$ ($\mu = \alpha + i\beta$); tractions derived from eqn. (20) satisfy the traction-free condition for $z_0 \in \Gamma$. The singular kernel functions T°_{ji}, U°_{ji} are the usual Kelvin traction and displacement functions for unit point loads at z_0; the kernel functions T^{*}_{ji}, U^{*}_{ji} are required by the presence of the traction-free crack. In what follows, the kernel functions will be combined and the superscript deleted for simplicity.

The kernel functions reported by Snyder & Cruse (1975) are real functions written in terms of the complex variable $z_k = x + \mu_k y$; μ_k are the distinct complex roots of the anisotropic characteristic equation:

$$\beta_{11}\mu^4 - 2\beta_{16}\mu^3 + (2\beta_{12} + \beta_{66})\mu^2 - 2\beta_{26}\mu + \beta_{22} = 0 \quad (21)$$

where β_{ij} are the elastic compliances. For notational convenience below, the μ_k notation will be replaced by μ. For isotropic materials, solutions may be obtained by taking $\beta_{11} = \beta_{22} = 1/E, \beta_{16} = \beta_{26} = 0, \beta_{12} = -\nu\beta_{11}$, and $\beta_{66} = (1 + \varepsilon)/G$, where E, ν, G are the usual isotropic elastic constants and ε is a small number.

The displacement gradient at z_0 is computed by differentiation of eqn. (20) as follows:

$$\frac{\partial u_j}{\partial(x_0, y_0)} = \frac{\partial u_j}{\partial z_0}\frac{\partial z_0}{\partial(x_0, y_0)} = \frac{\partial z_0}{\partial(x_0, y_0)}\left\{-\int_B \frac{\partial T_{ji}(z_Q, z_0)}{\partial z_0}u_i(z_Q)\,\mathrm{d}s + \int_B \frac{\partial U_{ji}(z_Q, z_0)}{\partial z_0}t_i(z_Q)\,\mathrm{d}s\right\} \quad (22)$$

Stresses are then computed directly from eqn. (22) using the symmetric part of the displacement gradient together with Hooke's law. It is well established in linear elastic fracture mechanics that the stresses become singular at $x = \pm a$; the stress intensity factor is defined as follows, using the notation from Fig. 2:

$$\begin{aligned} K_{\mathrm{I}}/\sqrt{2\pi} &= \lim_{r\to 0}\sqrt{r}\,\sigma_{y|\theta=0} \\ K_{\mathrm{II}}/\sqrt{2\pi} &= \lim_{r\to 0}\sqrt{r}\,\tau_{xy|\theta=0} \end{aligned} \quad (23)$$

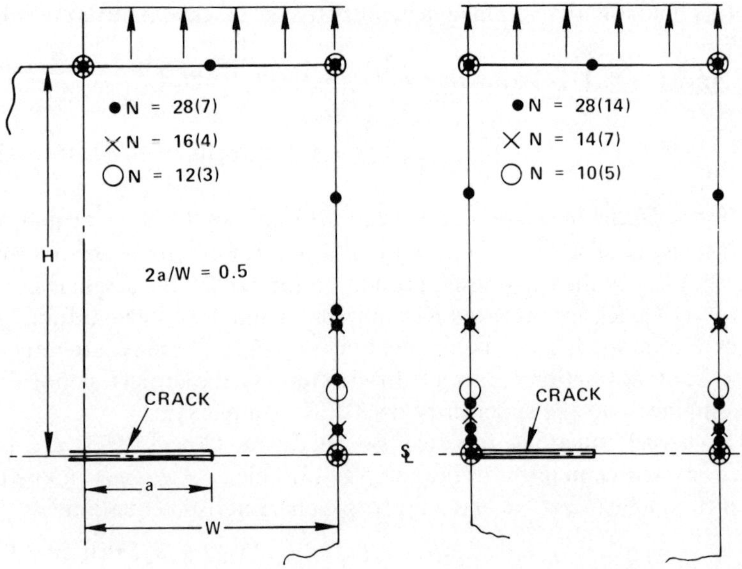

FIG. 6. BIE maps for fracture mechanics modelling.

Snyder & Cruse (1975) showed that the two-dimensional stress intensity factors ($K_{\rm I}$, $K_{\rm II}$) can be written by eqn. (23) and the stresses by eqn. (22) to obtain:

$$(K_{\rm I}, K_{\rm II}) = -\int_B R_i^{\rm I,II}(z_Q)u_i(Q)\,{\rm d}s + \int_B L_i^{\rm I,II}(z_Q)t_i(Q)\,{\rm d}s \qquad (24)$$

for each crack tip. Equation (24) constitutes a set of path independent integrals for the elastic stress intensity factors and was derived in an equivalent form at a later date by Stern *et al.* (1976). It should be emphasised that the boundary data (u_i, t_i) may be obtained by any suitable numerical analysis; further, B may be any boundary enclosing a crack tip.

The linear variation BIE model of the Green's function method described by Cruse (1978) is as easily used as the earlier piece-wise constant model discussed by Snyder & Cruse (1973). Two fracture mechanics test problems have been selected for BIE modelling: a centre-cracked and an edge cracked plate. The boundary models used for the two problems are shown in Fig. 6 for the symmetric part of the geometry. Since the BIE computer program includes a symmetry capability the number of nodes (N) is indicated in Fig. 6 for the total boundary and, parenthetically, the symmetric part of the boundary. The numerical results for the mode I stress intensity factor are given in Table 1.

The data in Table 1 are now compared to the collocation results tabulated by Tada *et al.* (1973); Tada's data are cited as being within 0·5 % of the collocation results over the range of crack length including the value used in this study. In general, while no effort has been made in this study to optimise the map, the agreement with Tada's values is quite good. It is seen

TABLE 1
FRACTURE MECHANICS MODELLING RESULTS

	N	$K_I/\sigma\sqrt{\pi a}$	*c.p.u. (sec.)*	*% Difference*
Centre crack	12 (3)	1·169	1·5	1·6
	16 (4)	1·180	2·4	0·6
	28 (7)	1·196	6·4	−0·7
Edge crack	10 (5)	2·703	1·7	4·6
	14 (7)	2·697	3·2	4·9
	28 (14)	2·791	11·8	1·3

that the most difficult problem for BIE analysis is the edge crack problem. This stems from the significant rotation and deformation of the boundary for this problem. Some numerical analysis of the edge crack problem using the piece-wise constant model showed errors on the order of 30–50 %. Thus linear variation is seen as the minimal modelling level for such problems. Further numerical results including an evaluation of the path independent integral in eqn. (24) are given by Cruse & Wilson (1978*a*).

5.3.3 Three-Dimensional Fracture Mechanics Analysis

Several different strategies have been explored in recent years for BIE modelling of cracked three-dimensional structures as shown in Fig. 7. The more general approach is to model the exterior of the cracked body and the crack surface, but not the unbroken region ahead of the crack, in which the traction distribution is singular. This modelling technique is applicable without symmetry restriction on either geometry or loading. The first approach along this line was the idealisation of the crack as an open notch. Useful results were obtained by Cruse (1974) with this technique, but it poses a major dilemma. The accuracy of the results suffers if the notch is too thick, that is if the (coincident) upper and lower surfaces of the mathematical crack are modelled too far apart. On the other hand, the equation system becomes badly conditioned if the surfaces are too close to

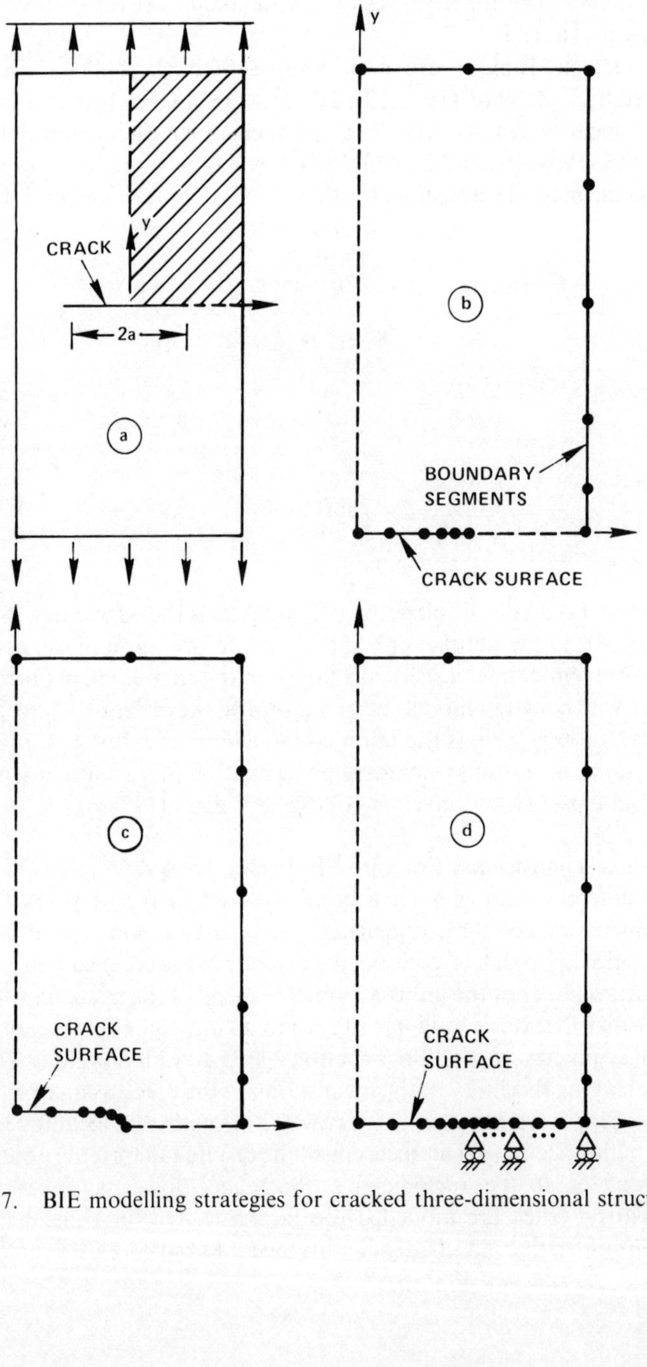

FIG. 7. BIE modelling strategies for cracked three-dimensional structures.

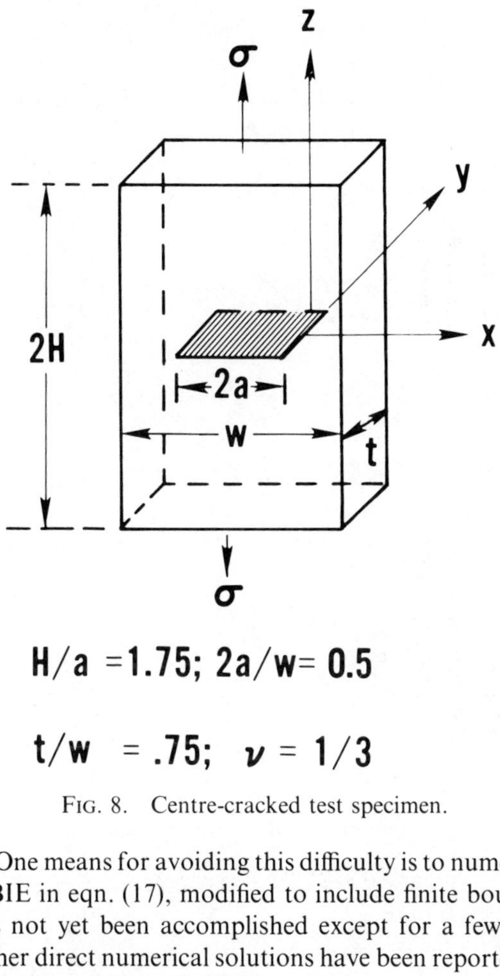

FIG. 8. Centre-cracked test specimen.

one another. One means for avoiding this difficulty is to numerically model the traction BIE in eqn. (17), modified to include finite boundaries. This approach has not yet been accomplished except for a few simple crack problems. Other direct numerical solutions have been reported by Cruse & Van Buren (1971) and Cruse & Meyers (1977).

Significant improvements in modelling three-dimensional cracks, using the approach shown in Fig. 7(d), have been accomplished using the isoparametric models developed by Lachat & Watson (1976). Near crack tip displacement variation is modelled by placement of the row of modes adjacent to the crack tip at the one-quarter point, in the same fashion as in finite element analysis described by Barsoum (1976). The shape functions for quadratic, isoparametric boundary elements are able to reproduce local crack opening behaviour functionally equivalent to the two-dimensional result. The details of this approach are given in Cruse & Wilson (1978*b*).

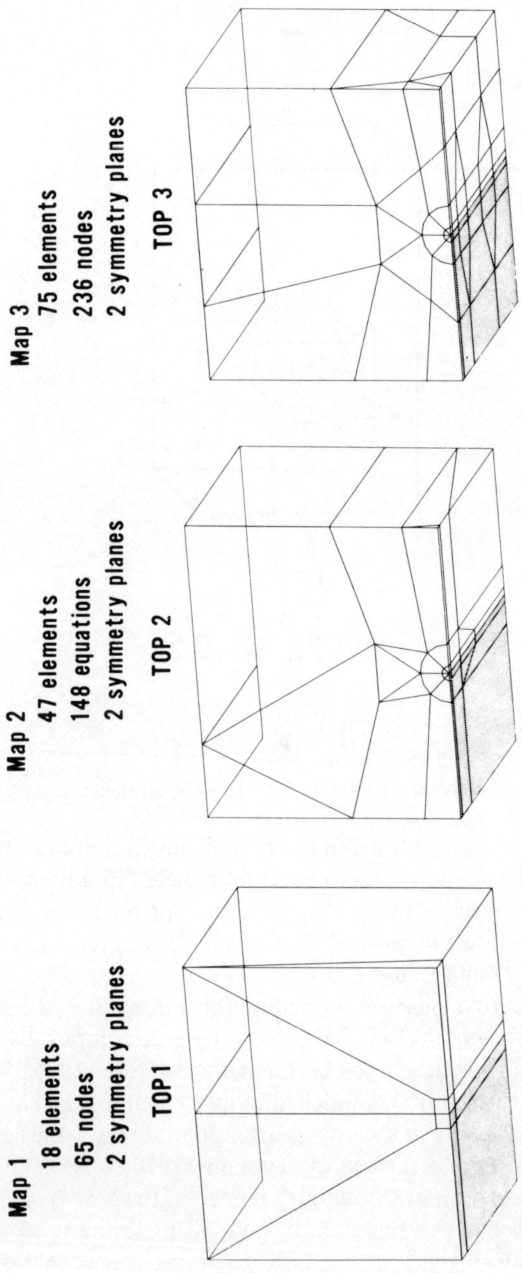

FIG. 9. BIE maps of one-eighth of centre-cracked test specimen.

This same report also describes a singularity element for the tractions ahead of the crack tip, which contains a leading term singularity in the tractions. This behaviour also closely models the two-dimensional result. A number of example problems have been solved using this approach, with very satisfactory results.

An example problem analysed was that of the centre-cracked test specimen shown in Fig. 8. One-eighth of the geometry was modelled, using two unmodelled symmetry planes but modelling the crack plane. The BIE maps used in the analysis are shown in Fig. 9. Both two-dimensional (plane strain) and three-dimensional analyses were carried out by changing boundary conditions on the plane $y = t/2$. Singularity finite element results are available for the three-dimensional problem and were generated by Raju & Newman (1977). Attention was given to several different issues, in particular the effect of the use of crack tip and traction singularity elements, evaluation of K_I without extrapolation, and exploration of the region near the crack tip/free surface intersection in the three-dimensional problem.

The specimen was first analysed in plane strain using Map 1 and a variety of choices of element type and boundary data variation. Figure 10 gives the stress component in the direction of the applied load for the various cases considered. Interior stresses ($\theta = 90°$) were considered, as it was found that they were more accurate than the traction values in the crack plane. This result is attributed to the averaging effect of the boundary integral representation for interior stresses. Further, it was intended to use stress, rather than displacement, as the basis for K_1 calculations, avoiding any plane stress or plane strain assumptions in the three-dimensional problem.

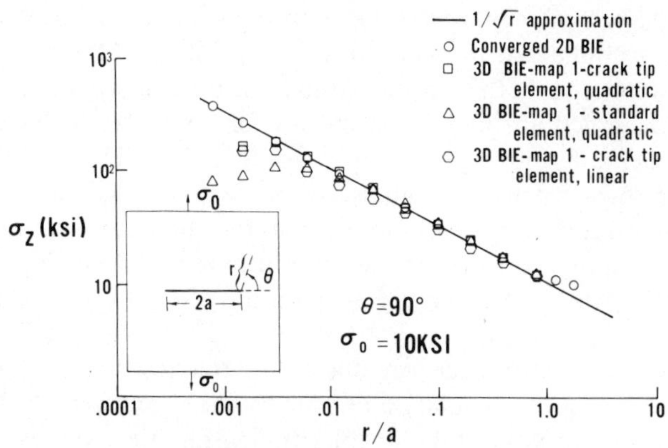

FIG. 10. Effect of modified crack tip element on interior solution.

Several comments can be made about the results shown in Fig. 10. The use of quadratic variation and crack tip elements gave the best results. The divergence of the finite geometry plane strain solution from the $1/\sqrt{r}$ approximation sets an outer limit (in terms of r/a) to the data to be used in calculating K_I. Lastly, the boundary layer effect of finite surface element size on the three-dimensional BIE interior stress calculation sets an inner limit to r/a.

Based on these observations, it was decided to evaluate K_I using the relationship

$$K_I = \sqrt{2\pi r}\left\{\cos\frac{\theta}{2}\left[1 + \sin\frac{\theta}{2}\sin\frac{3\theta}{2}\right]\right\}^{-1}\sigma_z(r,\theta) \tag{25}$$

with $\theta = 90°$, $0{\cdot}01 < r/a < 0{\cdot}1$. Either point estimates of K_I or an average value of K_I over the interval (in r/a) can be used. In the present study, point estimates at $r/a = 0{\cdot}025$ and $0{\cdot}05$ were taken because a major objective in the three-dimensional problem was to examine stress behaviour as a function of r/a near the free surface.

It must be emphasised that K_I values were not extrapolated back to $r/a = 0$. Such extrapolation was felt to be inappropriate, at least in the BIE analysis, because of the risk of misinterpreting the numerical boundary layer as a real variation of K_I with r/a. In addition, an extrapolation technique would be especially dangerous near the crack tip/free surface intersection where the nature of the stress singularity is not known. Even ignoring numerical boundary layer effects, the basic information needed to guide meaningful extrapolation does not yet exist.

After the preliminary work done with the relatively crude Map 1, the specimen was analysed in plane strain using Map 2. Modified crack tip elements were used, both with and without the traction singularity. Table 2 contains the crack opening displacements for both BIE analyses at three different locations through the specimen thickness. The results from a converged two-dimensional analysis are included as well. There is significant improvement in results near the crack tip due to inclusion of the traction singularity. Further, the results are uniform through the specimen thickness, indicating that the BIE map on the plane $y = t/2$ is adequately refined.

The improvement in accuracy due to the traction singularity element carries over to the K_I calculation as well. Table 3 compares the K_I values obtained with and without the singularity element at $r/a = 0{\cdot}05$. The two-dimensional result given is believed accurate to within $0{\cdot}25\,\%$. It is clear that

TABLE 2

$Ew/\sigma a$: CRACK OPENING DISPLACEMENTS (PLANE STRAIN)

r/a	*Plane strain*[a]	*3-D Map 2 Crack tip element with traction singularity* $y/t = 0$	$y/t = 0{\cdot}25$	$y/t = 0{\cdot}5$	*3-D Map 2 Crack tip element without traction singularity* $y/t = 0$	$y/t = 0{\cdot}25$	$y/t = 0{\cdot}5$
0·0125	0·399	0·398	—	0·398	0·381	—	0·380
0·05	0·795	0·791	0·792	0·790	0·784	0·785	0·785
0·125	1·243	1·238	—	1·236	1·234	—	1·233
0·2	1·552	1·547	1·547	1·545	1·543	1·544	1·541
0·6	2·444	2·435	—	2·433	2·432	—	2·429
1·0	2·693	2·688	2·688	2·685	2·684	2·684	2·682

[a] These BIE results are from the program described by Cruse (1978).

the use of the traction singularity element improves both the accuracy and the consistency between the stress and displacement fields.

Solution of the three-dimensional problem was carried out using Map 2; later, the more refined Map 3 was also used with no significant change in results. Figure 11 gives non-dimensional crack opening displacements for the three-dimensional problem at both the centre and free surface of the specimen. Results are given for the three-dimensional BIE analysis, the singular finite element analysis of Raju & Newman (1977), and for comparison, for the two-dimensional BIE solution. The results of the three-dimensional BIE analysis show displacements significantly larger than the finite element results at both the centre and free surface of the specimen. K_I at the centre of the specimen is within 1% of the plane strain value.

The effect of the free surface is shown in Fig. 12. Stress is plotted as a function of position through the specimen thickness. The BIE results are

TABLE 3

$K_I/\sigma\sqrt{\pi a}$ FOR CENTRE-CRACKED PANEL[a] (PLANE STRAIN)

Analysis	*From interior stress* ($\theta = 90°$)	*From crack opening displacements*
2-D BIE	1·421[b]	1·421[b]
3-D BIE: with traction singularity	1·420	1·416
3-D BIE: without traction singularity	1·436	1·395

[a] Evaluated at $r/a = 0{\cdot}05$.
[b] Evaluated using path independent integral, Snyder & Cruse (1975).

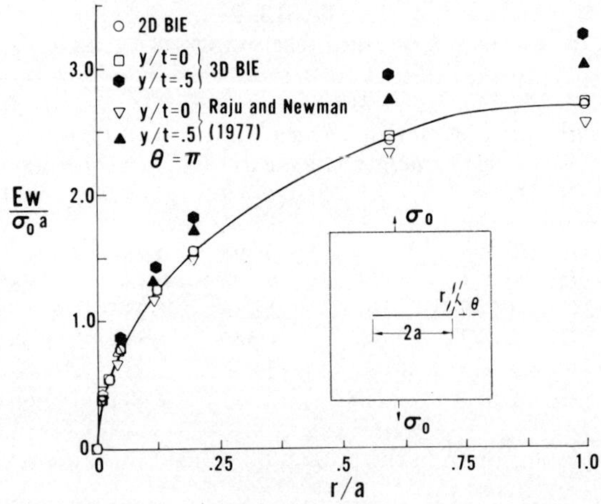

FIG. 11. Crack opening displacement in the three-dimensional problem.

given at $r/a = 0{\cdot}025$ and $0{\cdot}05$. The finite element results are based on extrapolation. The BIE and finite element results agree qualitatively, showing an increase in stress, peaking at $y/t \approx 0{\cdot}42$, followed by a rapid drop off in a boundary layer near the free surface. The BIE stress data (at both r/a locations) rise to a higher peak ($+2{\cdot}5\%$) closer to the free surface than the finite element data. The BIE stresses then drop faster to lower values at the free surface, relative to the singular finite element data.

In addition, comparison of the BIE results at the two different r/a values indicates a possible change in the nature of the stress singularity at the free

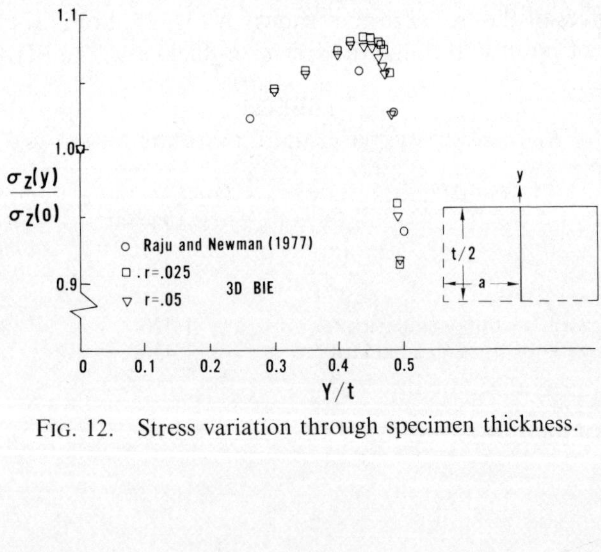

FIG. 12. Stress variation through specimen thickness.

surface. The behaviour of stress near the peak stress location is consistent with a singularity of higher order than $1/\sqrt{r}$. It is reasonable to suppose, however, that behaviour near the crack tip–free surface intersection depends on position relative to both the crack and the free surface. It may well be inappropriate to interpret the data near the free surface simply in terms of distance from the crack tip.

Finally, it is possible that the nature of behaviour near a free surface–crack tip intersection cannot be resolved by numerical modelling. In the work discussed here, it has been verified that the behaviour shown was not induced by the use of the traction singularity element. Nevertheless, both the BIE and finite element methods assume specific interpolation functions, and these assumptions limit the solution behaviour which can be exhibited without excessive mesh refinement. Pending the analytical resolution of the nature of the singularity at the crack tip–free surface intersection, the most promising approach is the implementation of the flat crack BIE described in Section 5.3.1. This approach would completely eliminate the traction modelling problem and allow needed mesh refinement to be concentrated on the more accurate definition of displacements on the free surface and crack near their intersection.

5.4 CONCLUSIONS

The Green's function method has been found to be an accurate, yet efficient BIE approach to the two-dimensional fracture mechanics problem. Linear variation modelling of the boundary data produces significant improvements in modelling accuracy for problems with significant bending.

Indeterminancy is shown to be a basic limitation to BIE fracture mechanics modelling when the Green's function for a given crack geometry is not accessible. An additional BIE for the crack face loading was shown to contain the essential features of the analytical fracture mechanics problem for two-dimensional cracks. The new BIE is equivalent to evaluating the stresses at any interior point for the flat crack problem. Evaluation of an expansion of the new BIE for near crack tip locations reproduces the singular stress term for the two-dimensional case.

Expansion of these fracture mechanics models to three dimensions is a significant challenge. To date, no closed-form Green's function expression has been generated for even the simple class of circular cracks. However, the flat crack BIE has been developed for three-dimensional cracks, although it has yet to be numerically exploited except for simple cases. Finally, the possibility of using the near crack front expansion of this stress equation to deduce three-dimensional crack singularity expansions exists and should be further explored.

ACKNOWLEDGEMENTS

Significant contributions were made to this effort by Dr R. B. Wilson; support for parts of the reported research were sponsored by the Air Force Office of Scientific Research, Contract F44620-74-C-0060.

REFERENCES

Barsoum, R. S. (1976). On the use of isoparametric finite elements in linear fracture mechanics, *Int. J. Num. Meth. Eng.*, **10,** 25–37.

Cruse, T. A. (1969). Numerical solutions in three dimensional elastostatics, *Int. J. of Sol. Struct.*, **5,** 1259–74.

Cruse, T. A. (1974). An improved boundary-integral equation method for three dimensional elastic stress analysis, *Computers and Structures*, **4,** 741–54.

Cruse, T. A. (1975). *Boundary-integral equation method for three dimensional elastic fracture mechanics analysis*, AFOSR-TR-75-0813, Accession No. ADA 011660.

Cruse, T. A. (1978). Two dimensional BIE fracture mechanics analysis, *Applied Mathematical Modeling*, **2,** 287–93.

Cruse, T. A. & Meyers, G. J. (1977). Three dimensional fracture mechanics analysis, *J. Structural Division, ASCE*, **103** (ST2), 309–20.

Cruse, T. A. & Van Buren, W. (1971). Three-dimensional elastic stress analysis of a fracture specimen with an edge crack, *Int. J. Fract. Mech.*, **7,** 1–15.

Cruse, T. A. & Wilson, R. B. (1978*a*). Advanced applications of boundary-integral equation methods, *Nuclear Engineering and Design*, **46,** 223–34.

Cruse, T. A. & Wilson, R. B. (1978*b*). *Boundary-integral equation method for elastic fracture mechanics analysis*, AFOSR-TR-78-0355, Accession No. ADA 051992.

Green, A. E. & Sneddon, I. N. (1950). The stress distribution in the neighbourhood of a flat elliptical crack in an elastic solid, *Proceedings of the Cambridge Philosophical Society*, **47,** 159–63.

Guidera, J. T. & Lardner, R. D. (1975). Penny-shaped cracks, *Journal of Elasticity*, **5,** 59–73.

Kassir, M. K. & Sih, G. C. (1966). Three dimensional stress distribution around an elliptical crack under arbitrary loadings, *J. Appl. Mech.*, **33,** 601–11.

Lachat, J. C. & Watson, J. O. (1976). Effective numerical treatment of boundary integral equations: A formulation for three-dimensional elastostatics, *Int. J. Num. Meth. Eng.*, **10,** 991–1005.

PARIS, P. C. & SIH, G. C. (1965). Stress analysis of cracks, *Fracture Toughness Testing*, American Society for Testing and Materials, STP 381, Philadelphia, 67–81.

RAJU, I. S. & NEWMAN, J. C., JR. (1977). *Three dimensional finite element analysis of finite thickness fracture specimens*, NASA TN D-8414.

RIZZO, F. J. (1967). An integral equation approach to boundary value problems of classical elastostatics, *Quart. Appl. Math.*, **25,** 83–95.

SNYDER, M. D. & CRUSE, T. A. (1973). *Crack tip stress intensity factors in finite anisotropic plates*, Air Force Materials Laboratory Technical Report AFML-TR-73-209, Accession No. AD-777 588.

SNYDER, M. D. & CRUSE, T. A. (1975). Boundary-integral equation analysis of cracked anisotropic plates, *Int. J. Fract.*, **11,** 315–28.

STERN, M., BECKER, E. B. & DUNHAM, R. S. (1976). A contour integral computation of mixed-mode stress intensity factors, *Int. J. Fract.*, **12,** 359–68.

TADA, H., PARIS, P. C. & IRWIN, G. R. (1973). *The Stress Analysis of Cracks Handbook*, Del Research Corporation, Hellertown, Pennsylvania.

WILLIAMS, M. L. (1952). Stress singularities resulting from various boundary conditions in angular corners of plates in extension, *J. Appl. Mech.*, **74,** 526–8.

Chapter 6

BOUNDARY INTEGRAL EQUATION METHODS APPLIED TO WAVE PROBLEMS

R. P. SHAW

Department of Engineering Science, Aerospace Engineering and Nuclear Engineering, State University of New York at Buffalo, USA

SUMMARY

An introduction is given to the application of boundary integral equation methods to the areas of wave propagation, radiation and scattering. Such applications were among the first, for this method, which made use of the power of modern digital computers; however, there is a long history of the use of integral methods in this field. The emphasis here is on fundamental concepts, coupled with illustrations of the application of this method. The breadth of the applications of this approach precludes any attempt at overall completeness, and many details are left to an extensive, but undoubtedly incomplete, bibliography. There is however sufficient material to formulate the solutions of most problems, with some of the inevitable pitfalls indicated.

6.1 INTRODUCTION

In view of the role of integral equations in the development of wave theory, it is not strange that some of the first applications of Boundary Integral Equation (BIE) methods were to examples in wave propagation and scattering. The use of Green's functions in all branches of mathematical physics is pervasive enough that individual references are scarcely required; one readily available reference which includes a wide range of applications of Green's functions is Morse & Feshbach (1953). A review of Green's function techniques used by the mathematical physics community at that time, as exemplified by this text, would however reveal that this use was primarily either to obtain analytical (exact) solutions to a fairly restricted class of problems (e.g., those simple geometries and boundary conditions for which a particular Green's function could be constructed so as to reduce Green's theorem to a simple quadrature) or to obtain asymptotic expansion solutions. A third form of application lay in the development of various integral identities, existence and uniqueness proofs, etc.; these have played a major role in the development of general theory.

The use of boundary integral equations to solve wave problems numerically is a fairly recent development, with the first papers using computer solutions based on the boundary integral equation method for transient acoustic wave scattering appearing in the early 1960s, e.g., Friedman & Shaw (1962), Shaw & Friedman (1962), followed by a similar treatment of time harmonic elastic and acoustic wave scattering, Banaugh & Goldsmith (1963*a*, *b*). Prior to this time, there was of course a great deal of work done on these forms of equations, e.g., Kupradze (1935), but these papers appear to mark the shift to computer solutions. At this same time, Hess (1962*a*, *b*) was developing computer solutions to potential flow problems and Jaswon (1963) and Symm (1963) for other potential problems. Since that time, the BIE method has been employed (in some cases, rediscovered) as a numerical solution method by researchers in many areas of wave research. Before going into details of individual cases, it is useful to separate these problems into sub-groups by recognising some distinctive characteristics. One separation is that of transient versus time harmonic behaviour; both are of great physical importance in many areas of physics and engineering but require quite different formulations and solution methods. Another division is that between exterior and interior problems; for the former, solutions are sought in an infinite domain exterior to some prescribed surface, while the latter requires solution in a bounded region interior to such a surface. (This distinction blurs somewhat for surfaces of infinite extent.) Appropriate boundary conditions are specified on the surface in either case with the exterior problem requiring some further specification of 'behaviour at infinity'. A further grouping may be made on the basis of the inhomogeneity or forcing function. A problem driven by some prescribed boundary condition, e.g., a given displacement or velocity field on a specified surface, would be considered a mechanical radiation problem. A problem wherein an obstacle with a prescribed boundary condition (usually homogeneous) interacts with some incident wave field generated by sources elsewhere (frequently considered as plane waves coming from 'infinity') would be called a scattering problem. Of interest as well is the unforced or eigenvalue problem. Still further distinction may be made between two and three spatial dimensional problems; this distinction being straightforward in time harmonic problems but presenting a serious difficulty in transient problems. Other categorisation may be made on the basis of the type of boundary conditions prescribed (Dirichlet, Neumann, Robin, etc.). Finally, and possibly most important physically, is the particular medium through which the waves propagate, e.g., acoustic, elastic, electromagnetic, water (especially free surface gravity waves), and thus the physical significance of the dependent variables and their boundary and initial conditions. Fortunately, the boundary integral equation approach is applicable to all of these 'variations on a theme'. Its primary limitations are the requirement of a known fundamental solution for the particular governing equation and linearity of the equation, but not necessarily of the boundary conditions. Emphasis here will be given to mechanical rather than electromagnetic wave problems; integral equation approaches to the latter are discussed at length by Mittra (1973).

The bulk of the research based on boundary integral equation methods applied to wave problems has been directed towards numerical solution, using large scale digital computers, and this will form the primary theme in this chapter. Since efficient algorithms exist for the solution of systems of algebraic equations, the most common procedure is to reduce the governing boundary integral equation to such a system by approximating the boundary values of the dependent variable by suitable trial functions over selected boundary elements. The nature of the boundary integral equation approach is to reduce the number of approximating dependent variables, e.g., trial function coefficients, by only requiring boundary values of the original dependent variable as opposed to other solution schemes which require a solution everywhere in the domain of the problem. Such a procedure is particularly useful in exterior problems where other numerical solution methods require some 'finite' closing surface at which their approximating mesh may be terminated, e.g., see Vastano & Reid (1967) and Shaw (1974) for a direct comparison of exterior solutions by finite differences and BIE respectively. The price paid for this reduction is a full coefficient matrix as opposed to the sparse matrices found in corresponding finite difference or finite element methods. Some overall reviews of this approach are available; Schenck (1967) for time harmonic acoustic scattering and radiation, Shaw (1970) for transient and time harmonic acoustic scattering and radiation, Chertock (1971) and Kleinman & Roach

(1974) again for Helmholtz problems and Mittra (1973) for the electromagnetic case. In addition, the works by Kupradze (1935, 1963) discuss integral equation formulations for elastodynamics, while Pao & Mow (1973) give a section on numerical solution for elastic scattering problems using boundary integral equations. It is clear, however, the boundary integral equation method is *more* than a numerical technique; it is a reformulation completely equivalent to any original partial differential equation representation. As such, it lends itself to other forms of solution, i.e., analytical and approximate as well as numerical. Analytical solutions are widely covered in classical mathematical physics, e.g., Morse & Feshbach (1953), and will not be pursued further here. The distinction between approximate and numerical solutions is one of semantics since, clearly, numerical solutions are also approximate in some sense. What is meant here is that approximate solutions will be taken as global or functional solutions to a system of equations which represent some form of expansion of the original equation in terms of a physical parameter which characterises the problem. Numerical solutions on the other hand will be solutions to the original equation with some piece-wise and local approximate representation of the dependent variable. Thus in the former, it is the governing equation which is 'approximated' while in the latter it is the dependent variable. Typical of such approximate techniques are the high frequency asymptotic expansions of classical optics, e.g., Northover (1971). Also of interest are expansions in terms of geometrical parameters, for example the offset distance in shells of slightly non-concentric circular or other simple boundaries, Shaw & Tai (1974), Shaw (1978*a*) or the plate thickness in the theory of thin elastic plates, Shaw (1978*b*). Examples of the latter approach will be discussed in Section 6.4. Solution schemes based on Galerkin or related moment methods can be employed as well in the solution of integral equations, e.g., Delves & Walsh (1974) and Fox (1962) for general theory, and Sharma (1967) for a specific example. Indeed the piece-wise numerical approximation mentioned above could be thought of as a special case of a Galerkin approach with particularly simple trial functions.

6.2 FORMULATION

Wave propagation problems arise in so many areas of mathematical physics and engineering sciences that it is important to concentrate on their similarities, which are primarily mathematical, rather than their differences, which are primarily physical. In some cases, the physical differences require a difference in formulation and these will be treated as they arise. In general however, wave propagation problems may be classified as hyperbolic (time dependent) or elliptic (time harmonic) differential systems. The most common governing equations of these forms are the wave equation

$$\nabla^2\psi(\bar{r}, t) = c^{-2}\,\partial^2\psi(\bar{r}, t)/\partial t^2 \tag{1}$$

and the Helmholtz equation (with a time harmonic behaviour required) i.e., $\psi(\bar{r}, t) = \phi(\bar{r})\exp[-i\omega t]$

$$\nabla^2\phi(\bar{r}) + k^2\phi(\bar{r}) = 0 \tag{2}$$

respectively with c the phase speed and $k = \omega/c$ the wave number.

The integral equation formulation for the wave equation, developed in 1882 by Kirchhoff (see Kirchhoff, 1883), has been widely interpreted as a mathematical formulation of an even earlier physical statement, Huygens principle, e.g., see Baker & Copson (1939). While the form of Kirchhoff's integral formulation can be derived directly by an application of Green's theorem in four-dimensional space time, e.g., Friedman & Shaw (1962), it follows most readily, albeit in a more restricted sense, from Helmholtz's integral formulation; this shall be the approach here.

Consider the Helmholtz equation with an inhomogeneous term, applied to a domain D with a boundary B upon which some appropriate boundary conditions on ϕ (e.g., Dirichlet, Neumann, Robin) are specified. The fundamental or point source or infinite space Green's function solution, G, (these terms shall be used here interchangeably) satisfies (using ∇_0^2 to represent the Laplacian in $\bar{r}_0$ co-ordinates)

$$\nabla_0^2 G(\bar{r}; \bar{r}_0) + k^2 G(\bar{r}; \bar{r}_0) = -4\pi\delta(\bar{r} - \bar{r}_0) \tag{3}$$

and may be found as

$$G(\bar{r}; \bar{r}_0) = G(|\bar{r} - \bar{r}_0|) = \exp[ikR]/R \tag{4}$$

where $R \equiv |\bar{r} - \bar{r}_0|$. The sign in the exponential term is chosen to ensure outgoing waves from the source with a time behaviour of the form $\exp(-i\omega t)$. The unknown ϕ satisfies

$$\nabla_0^2\phi(\bar{r}_0) + k^2\phi(\bar{r}_0) = -4\pi f(\bar{r}_0) \tag{5}$$

Equation (5) multiplied by G minus eqn. (4) multiplied by ϕ may be integrated (in $\bar{r}_0$ co-ordinates) over the domain of interest D and the resulting integral by Green's theorem reduced to an integral over the surface B which forms the boundary of D. If any portion of D is infinite such as to require a boundary at infinity, appropriate limiting conditions must be included for this portion of B (this is not difficult for most wave problems but can require care in some cases, particularly in some water wave applications. This results in the inhomogeneous Helmholtz integral equation, e.g., Morse & Feshbach (1953), Section 7.2

$$\varepsilon\phi(\bar{r}) = \int_D f(\bar{r}_0)G(R)\,\mathrm{d}V_0 + (1/4\pi)\int_B [G(R)\nabla_0\phi(\bar{r}_0) - \phi(\bar{r}_0)\nabla_0 G(R)]\,\mathrm{d}\bar{A}_0 \tag{6}$$

with $\mathrm{d}\bar{A}_0$ and $\mathrm{d}V_0$ a directed area element and a volume element respectively. The first integral represents the contribution of the inhomogeneous or forcing function given in the domain D; the second represents the contribution of the boundary values with the area element $\mathrm{d}\bar{A}_0$ positive away (outward) from the domain D. The gradient operator ∇_0 is expressed in $\bar{r}_0$ co-ordinates which will be referred to as source or integration point co-ordinates; $\phi(\bar{r})$ is evaluated at $\bar{r}$ which will be called the field point. The parameter ε is zero if the field point is exterior to the domain D (i.e., the delta function $\delta(\bar{r} - \bar{r}_0)$ is zero everywhere within D) and is one if the field point lies within the domain D. Of particular interest to boundary integral equation methods is the case where the field point lies on the boundary B. For this case, the boundary integral has a singular contribution at the point $R = 0$ which may be evaluated by distorting the actual boundary to exclude the field point from the domain of integration and taking a limiting process as the distorted boundary returns to the original form. For a plane boundary surface, this results in a value of 1/2 for ε; for a field point at a corner, the appropriate choice for ε is $(1/4\pi)$ (solid angle of the domain excluded) (e.g., Banaugh & Goldsmith (1963*a*), Appendix C). Furthermore, the boundary integral must be taken as a principal value integral, i.e., the singular point $R = 0$ is to be excluded from the integration (since its effect has already been included in the modification of ε to some value other than zero or one). Equation (6) may be written as

$$\begin{aligned}\varepsilon\phi(\bar{r}) = &\int_D \{f(\bar{r}_0)\exp[ikR]/R\}\,\mathrm{d}V_0 \\ &+ (1/4\pi)\int_B \{(1/R)\exp[ikR]\,\partial\phi(\bar{r}_0)/\partial n_0 \\ &- \phi(\bar{r}_0)\exp[ikR].[ik/R - 1/R^2]\,\partial R/\partial n_0\}\,\mathrm{d}A_0\end{aligned} \tag{7}$$

where n_0 is the outward normal (from D). If the time harmonic dependence is reintroduced, this equation becomes

$$\varepsilon\phi(\bar{r})\exp[-i\omega t] = \int_D \{f(\bar{r}_0)\exp[-i\omega(t - R/c)]/R\}\,\mathrm{d}V_0$$

$$+(1/4\pi)\int_B \{(1/R)\,\partial\phi(\bar{r}_0)/\partial n_0 - \phi(\bar{r}_0)[i\omega/cR - 1/R^2]\,\partial R/\partial n_0\}$$

$$.\exp[-i\omega(t - R/c)]\,\mathrm{d}A_0 \qquad (8)$$

These terms may be thought of as representing the inversion of a Fourier transform in time of the time dependent functions $\psi(\bar{r}, t)$ and $F(\bar{r}, t)$ which satisfy:

$$\nabla^2\psi(\bar{r}, t) - (1/c^2)\,\partial^2\psi(\bar{r}, t)/\partial t^2 = -4\pi F(\bar{r}, t) \qquad (9)$$

or upon transformation

$$\nabla^2\tilde{\psi}(\bar{r};\omega) + k^2\tilde{\psi}(\bar{r};\omega) = -4\pi\tilde{F}(\bar{r};\omega) \qquad (10)$$

with $\tilde{\psi}(\bar{r};\omega)$ identified with ϕ and $\bar{F}(\bar{r};\omega)$ with f, i.e.,

$$\tilde{\psi}(\bar{r};\omega) = \int_{-\infty}^{\infty} \psi(\bar{r}, t)\exp[i\omega t]\,\mathrm{d}t = \phi(\bar{r})$$

$$\psi(\bar{r}, t) = (1/2\pi)\int_{-\infty}^{\infty} \phi(\bar{r})\exp[-i\omega t]\,\mathrm{d}\omega$$

Then

$$\varepsilon\psi(\bar{r}, t) = \int_D (1/2\pi)\int_{-\infty}^{\infty} \{f(\bar{r}_0)/R\}\exp[-i\omega(t - R/c)]\,\mathrm{d}\omega\,\mathrm{d}V_0$$

$$+ (1/4\pi)\int_B (1/2\pi)\int_{-\infty}^{\infty} \{(1/R)\,\partial\phi(\bar{r}_0)/\partial n_0$$

$$-\phi(\bar{r}_0)[i\omega/cR - 1/R^2]\,\partial R/\partial n_0\}.\exp[-i\omega(t - R/c)]\,\mathrm{d}\omega\,\mathrm{d}A_0 \qquad (11)$$

These integrals may be readily interpreted as time retarded values, i.e., by means of the shifting theorem in Fourier transform theory. f, ϕ and $\partial\phi/\partial n_0$ do not contain ω explicitly and the single appearance of ω within the integral may be interpreted as the transform of a time derivative. Thus, using a retarded time $t_0 = t - R/c$, this inversion leads to

$$\varepsilon\psi(\bar{r}, t) = \int_D \{F(\bar{r}_0, t_0)/R\}\,\mathrm{d}V_0 + (1/4\pi)\int_B \{(1/R)\,\partial\psi(\bar{r}_0, t_0)/\partial n_0$$

$$+ [(1/cR)\,\partial\psi(\bar{r}_0, t_0)/\partial t_0 + \psi(\bar{r}_0, t_0)/R^2]\,\partial R/\partial n_0\}\,\mathrm{d}A_0 \qquad (12)$$

which is Kirchhoff's 'retarded time' integral equation, equivalent to the original wave equation. The concept of a retarded time arises naturally in the sense that an event occurring at a source point $\bar{r}_0$ at time t_0 is felt at a field point $\bar{r}$ only at time t where $t = t_0 + R/c$. As mentioned above, the restrictions on this derivation, e.g., that $\psi(\bar{r}, t)$ have Fourier transform zero initial values, limits of integration over B which do not depend on t, etc., may be removed by a direct derivation, e.g., Morse & Feshbach (1953) using singular solutions or Friedman & Shaw (1962) using a four-dimensional Green's theorem.

These two integral equations, (6) and (12) form the basis of the boundary integral equation method. The three-dimensional Helmholtz integral equation may be reduced to a two-dimensional form called the Weber integral equation without difficulty;

$$\varepsilon\phi(\bar{r}) = (i/4)\int_B [H_0^{(1)}(kR)\,\partial\phi(\bar{r}_0)/\partial n_0 - \phi(\bar{r}_0)\partial[H_0^{(1)}(kR)]/\partial n_0]\,\mathrm{d}s_0 + (i/4)\int_D H_0^{(1)}(kR)f(\bar{r}_0)\,\mathrm{d}A_0 \qquad (13)$$

where B is now the one-dimensional boundary curve of the two-dimensional domain D, and $\mathrm{d}s_0$, $\mathrm{d}A_0$ are a line element and an area element respectively.

The Kirchhoff retarded time integral equation cannot be reduced simply for two-dimensional problems. This is due to the inherent difference in the physical nature of a transient point source solution in two and in three dimensions, e.g., Hadamard (1952). The two-dimensional point source solution requires an additional integration over time which means that two-dimensional transient problems may just as well be left as three-dimensional problems for cylindrical surfaces (the two-dimensional domain extended as a cylinder into the third (z) direction) with no direct z dependence of the dependent variables. Due to the appearance of R within the retarded time, t_0, there will be a variation of $\phi(\bar{r}_0, t_0)$ in the z_0 direction; the integration over z_0 then plays a role equivalent to the time integration for the two-dimensional point source, e.g., Friedman & Shaw (1962).

The forms of these two formulations indicate a significant difference in their physical nature and thus in their numerical solution. The time harmonic case, i.e., the Helmholtz or Weber integral equations, leads to a Fredholm equation which upon approximation by suitable trial functions and boundary elements leads to a system of simultaneous algebraic equations (linear for linear boundary conditions). The transient case leads to a Volterra integral equation; this may not appear obvious from the form of the Kirchhoff retarded time equation, but taking account of the general initial value nature of such problems (and a corresponding null field prior to such values, i.e., for $t < 0$) there will be portions of the boundary surface sufficiently far away from the field point so as to have negative retarded time values, i.e., they are evaluated at times when the field there has zero values and thus do not contribute, at least for some interval of time. This in turn implies that the limits of integration in the integral depend on time, increasing as time increases, i.e., a Volterra integral equation. Upon suitable numerical approximation, these equations lead to successive, rather than simultaneous, algebraic equations, thereby considerably simplifying the solution procedure. For finite bodies, there will come a time when all surface values are non-zero, but the retardation still leads to successive equations. This presents a new difficulty however. The derivation of the original Kirchhoff equation was valid only for continuous dependent variables. A discontinuous problem, e.g., scattering of a step wave, requires additional terms. Fortunately, these can be evaluated in closed form, e.g., a general discussion in Shaw (1968*b*) with examples in Friedman & Shaw (1962) and Shaw & Friedman (1962).

An interesting sidelight developed in these last two references is the formulation of transient problems in terms of a total rather than a scattered field. The Kirchhoff retarded time equation as given above when applied to an exterior scattering problem must be written in terms of a dependent variable which decays appropriately at large distances from the scatterer to allow a closing surface 'at infinity' in Green's theorem. This clearly is the scattered (total minus incident) field. The required boundary conditions could then be written on the scattered field. However, if such boundary conditions were originally homogeneous for the total field, they would now be inhomogeneous for the scattered field. This difficulty can be avoided by writing an integral equation on the incident field as well. This may be developed in several ways. The most direct is to consider the boundary integral equation applied to the scattered field at a boundary point. The time harmonic case is used as an example; the approach is equally valid for the transient problem. The inhomogeneity, $f(\bar{r})$, is dropped for convenience.

$$(1/2)\phi_S(\bar{r}_B) = (1/4\pi)\int_B \left\{\frac{\partial\phi_S(\bar{r}_0)}{\partial n_0}\cdot\frac{\exp[ikR]}{R} - \phi_S(\bar{r}_0)\frac{\partial}{\partial n_0}\left[\frac{\exp[ikR]}{R}\right]\right\}\mathrm{d}A \tag{14}$$

Next, consider a formulation for the incident wave, $\phi_W(\bar{r})$. This is the field which would exist in the absence of the scattering body defined by the surface B. An exterior boundary integral equation on ϕ_W is not available since ϕ_W does not in general vanish at large distances from the scatterer, and thus the integral over a closing surface 'at infinity' in Green's theorem may not vanish. However, an interior formulation is available; $\phi_W(\bar{r})$ may be expressed at some interior point to B, i.e., inside the scattering surface, by considering the surface to be absent but by forming the integral equation over those values of ϕ_W and $\partial\phi_W/\partial n$ which exist at those points where the scattering surface was. This leads to a similar boundary integral equation as on ϕ_S as the interior point approaches the location of B, but with a change of sign for the normal direction. Keeping n_0 the same as that in eqn. (9) leads to

$$(1/2)\phi_W(\bar{r}_B) = (-1/4\pi) \times \int_B \left\{ \frac{\partial\phi_W(\bar{r}_0)}{\partial n_0} \cdot \frac{\exp[ikR]}{R} - \phi_W(\bar{r}_0)\frac{\partial}{\partial n_0}\left[\frac{\exp[ikR]}{R}\right]\right\} \mathrm{d}A_0 \quad (15)$$

The total field, $\phi_T = \phi_W + \phi_S$, may be found by subtracting these equations, yielding

$$(1/2)\phi_T(\bar{r}_B) = \phi_W(\bar{r}_B) + (1/4\pi) \times \int_B \left\{ \frac{\partial\phi_T(\bar{r}_0)}{\partial n_0} \cdot \frac{\exp[ikR]}{R} - \phi_T(\bar{r}_0)\frac{\partial}{\partial n_0}\left[\frac{\exp[ikR]}{R}\right]\right\} \mathrm{d}A_0 \quad (16)$$

An alternative, and in some ways more useful, derivation may be described in the form of a 'saltus' problem. The exterior and interior wave problems for the same geometry are obviously closely related. If the domain in question is exterior to the boundary B, then ε is one for a field point exterior to B and zero for a field point interior to B and vice-versa for the interior problem, with a change in sign for the normal, n_0, which is always directed outward from the domain in question. The kernel functions, G and $\partial G/\partial n$, in the integral equation formulation may be interpreted as layers of monopole and dipole distributions of strength $\partial\phi/\partial n$ and ϕ respectively, which replace the boundary surface by an equivalent singular distribution in an infinite fluid domain including the interior of the body for the exterior case and vice-versa, i.e., a 'saltus' problem. These surface singularity distributions cause discontinuities in $\partial\phi/\partial n$ and ϕ of strengths $\partial\phi/\partial n$ and ϕ, i.e., they carry the value of ϕ and $\partial\phi/\partial n$ from their appropriate values on one side of the surface B to zero values on the other side. This property of distributions of monopoles and/or dipoles is well known and discussed in many references, e.g., Morse & Feshbach (1953). Now reconsider the equation on ϕ_S, at some field point $\bar{r}$ in the fluid writing ϕ_S and $\partial\phi_S/\partial n_0$ within the integral as the actual discontinuities in these values, i.e., $[\phi_S]$ and $[\partial\phi_S/\partial n]$ where [ ] is the value on the 'fluid' side of B minus the value on the 'empty' side of B (that side excluded from the Green's theorem integration). ϕ_S is replaced by $\phi_T - \phi_W$ and $[\phi_S]$, $[\partial\phi_S/\partial n_0]$ by $[\phi_T]$, $[\partial\phi_T/\partial n_0]$ since ϕ_W is continuous everywhere in the fluid, even across B.

$$\phi_T(\bar{r}) - \phi_W(\bar{r}) = (1/4\pi)\int_B \left\{\left[\frac{\partial\phi_T(\bar{r}_0)}{\partial n_0}\right] \times \frac{\exp[ikR]}{R} - [\phi_T(\bar{r}_0)]\frac{\partial}{\partial n_0}\left[\frac{\exp[ikR]}{R}\right]\right\} dA_0 \tag{17}$$

Now as the field point $\bar{r}$ approaches B, the integrals contribute a term $1/2[\phi_T(\bar{r})]$ leading again to eqn. (16), since ϕ_T inside B is zero identically, by this formulation.

The introduction of the concept of singularity distributions which replace the original problem by an equivalent 'saltus' problem leads naturally in a discussion of the 'indirect' boundary integral formulations; the form above is called the direct approach since it deals with the actual dependent variables ϕ and $\partial\phi/\partial n$ 'directly'. The indirect approach is based on the observation that a distribution of monopoles or dipoles alone or in fact any combination of these will be sufficient to solve these problems; however, only the direct approach has the strengths of these distributions tied directly to the actual dependent variables of the problem. Excellent discussions of these alternative formulations are given by Chertock (1971) and by Kleinman & Roach (1974) in terms of the present formulations; many other classical discussions are also available, e.g., Lamb (1932).

Consider a monopole surface distribution of strength $\sigma(\bar{r}_0)$. The field at a point $\bar{r}$ is given by

$$\phi(\bar{r}) = (1/4\pi)\int_B \left\{\sigma(\bar{r}_0)\cdot\frac{\exp[ikR]}{R}\right\} dA_0 \tag{18}$$

where $\sigma(\bar{r}_0)$ must be $\partial\phi/\partial n_0 - \partial\tilde{\phi}/\partial n_0$ where $\partial\phi/\partial n_0$ is the outward normal derivative of the exterior (interior) field under consideration and $\partial\tilde{\phi}/\partial n_0$ is the normal derivative (in the same direction) of a complementary field; that which satisfied the same boundary values for ϕ for the interior (exterior) problem. For a problem with Dirichlet boundary conditions, this forms a 'boundary' integral equation on the monopole distribution strength $\sigma(\bar{r}_0)$ in terms of the prescribed $\phi(\bar{r}_0)$. There is no singular contributions as $\bar{r}$

approaches the surface B. Neumann boundary conditions could be treated by differentiating the equation on $\phi(\bar{r})$ in an n direction before approaching the surface; this leads to an alternate form

$$\partial\phi(\bar{r}_0)/\partial n = \int_B \left\{\sigma(\bar{r}_0)\frac{\partial}{\partial n}\left[\frac{\exp[ikR]}{R}\right]\right\} dA_0 - (1/2)\sigma(\bar{r}_B) \qquad (19)$$

Similar results could be obtained for the dipole layer

$$\phi(\bar{r}) = \int_B \left\{\mu(\bar{r}_0)\frac{\partial}{\partial n_0}\left[\frac{\exp[ikR]}{R}\right]\right\} dA_0 \qquad (20)$$

or any combination of monopole and dipole distributions. These alternate formulations have some advantages, as is most apparent in the solution of exterior time harmonic scattering at wave numbers near to the eigenvalues of the corresponding interior problem. The Fredholm alternative indicates no unique solution to the integral equation at these values, even though the exterior problem does have a unique and well behaved solution for all wavenumbers. This is due to the close coupling between boundary integral equation formulations of exterior and their complementary interior problems. This coupling has presented substantial difficulty in the application of BIE methods to time harmonic exterior scattering problems as described for example by Chertock (1971), and Kleinman & Roach (1974) which describe monopole and/or dipole combinations to circumvent this problem and has led to several 'modified' procedures, e.g., those of Schenck (1968), which uses additional equations for field points within the scatterer, Ursell (1973) which uses a modified wave source solution which satisfies a dissipative boundary condition on some surface interior to the scatterer. Of course, the BIE method is quite suitable for the determination of these eigenvalues for the interior problem, e.g., Tai & Shaw (1974), DeMey (1976), Hutchinson (1978); this is discussed further in Section 6.3.

Finally, an alternative formulation to that usual boundary integral equation needs to be given. If $\partial\phi/\partial n$ is specified on B for eqns. (6) or (16), for example, the resulting integral equation is one of the second kind and numerical procedures may be used without too much difficulty. If however, ϕ is specified, i.e., a Dirichlet problem, the result is an integral equation of the first kind on the dependent variable, $\partial\phi/\partial n$. While this may be solved numerically, e.g., Shaw & English (1972) for transient scattering by a pressure release sphere, it may be useful to convert the original form into an equation of the second kind before numerical approximation. This is done by taking the field point off the surface B, differentiating with respect to the normal direction n and then allowing the field point, at which $\partial\phi/\partial n$ is now defined, to approach B;

$$(1/2)\frac{\partial\phi(\bar{r}_B)}{\partial n} = \frac{1}{4\pi}\int_B \left\{\frac{\partial\phi(\bar{r}_0)}{\partial n_0}\cdot\frac{\partial}{\partial n} \times \left[\frac{\exp[ikR]}{R}\right] - \phi(\bar{r}_0)\frac{\partial^2}{\partial n\,\partial n_0}\left[\frac{\exp[ikR]}{R}\right]\right\} \mathrm{d}A_0 \qquad (21)$$

and an analogous equation for the transient case where such an approach was used by Shaw & Friedman (1962). Further discussion on the difficulty of numerical solution of Fredholm integral equations of the first kind is given for example in Delves & Walsh (1974).

6.3 NUMERICAL SOLUTION

Due to the large number of examples available in the literature, as given in part in the References and the Bibliography, and due to the several review articles available, it does not seem appropriate to give details of either numerical procedures or actual programs. In the sense that this is still a developing field, any of the numerical techniques and/or computer programs are subject to significant improvement. What does appear appropriate here is a description of the forms of numerical solution techniques and those pitfalls which have been encountered and, where this has been found possible, circumvented.

Consider first the class of interior time-harmonic problems for a finite region bounded by B where some suitable boundary conditions are given. Here the homogeneous and inhomogeneous problems must be distinguished. The former has as its goal the determination of the eigenvalues and eigenmodes for that region with those particular boundary conditions, e.g., Tai & Shaw (1974), DeMey (1976), Gregoire *et al.* (1975) and Hutchinson (1978). As an example, consider the two-dimensional resonance of an acoustic domain with a rigid boundary, i.e., a Neumann boundary condition. The governing homogeneous boundary integral equation is

$$(1/2)\phi(\bar{r}_B) = (i/4)\int_B \phi(\bar{r}_0)\frac{\partial}{\partial n_0}[H_0^{(1)}(kR)]\,\mathrm{d}s_0 \qquad (22)$$

In the simplest approach, ϕ is approximated by a piece-wise constant over N segments of B, preferably equal in size, and the field point, $\bar{r}_B$, is placed at the midpoint of each segment with $\phi(\bar{r}_B)$ for $\bar{r}_B = \bar{r}_K$ defined as the constant for that segment, ϕ_K. Then

$$\tfrac{1}{2}\phi_K = \sum_{L=1}^{N} \phi_L \alpha_{KL}; \qquad K = 1, 2, \ldots, N \tag{23}$$

with

$$\alpha_{KL} = (i/4) \,⨍_{B_L} \frac{\partial}{\partial n_0} [H_0^{(1)}(k|\bar{r}_0 - \bar{r}_K|)]\, \mathrm{d}s_0 \tag{24}$$

The principle value occurs only for that coefficient where the interval L includes the field point K, i.e., L equals K. All other coefficients are calculated as ordinary integrals. The integral for α_{KK} excludes the point $\bar{r}_0$ equal to $\bar{r}_K$; an examination of the form of the integrand shows it to be $-kH_1^{(1)}(kR)\partial R/\partial n_0$. As a first approximation, consider α_{KK} to be evaluated by any numerical scheme, e.g., a trapezoidal rule, in which the (curved) segment B_L is replaced by a (large) number of straight subsegments, e.g., the intervals used in the trapezoidal rule. That straight segment containing the field point, i.e., R equals zero, also has $\partial R/\partial n_0$ equal to zero everywhere. Since the point R equals zero is excluded from the integration (this is a principle value integral), the contribution of this subelement to the coefficient α_{KK} is zero. Of course, for a simple curved geometry, the integral might be evaluated in closed form as a principle value integral directly, but this is unlikely for an arbitrary shape. It is useful from a numerical view to take the subsegments in the integration over B_L such that the actual field point falls at the midpoint of one of these subsegments. Details for the calculation of these 'influence coefficients' (which are also used in forced problems) can be found in almost any of the references, e.g., Banaugh & Goldsmith (1963*a*, *b*), for a somewhat more accurate evaluation.

The resulting system of linear algebraic equations must have a zero determinant for the coefficient matrix, i.e.,

$$|(1/2)\delta_{KL} - \alpha_{KL}| = 0$$

where α_{KL} are functions of the parameter k, to have a non-trivial solution. Those values of k satisfying this equation are then the eigenvalues of the original problem.

Some care must be taken with the form of the Green's function used. Some authors, DeMey (1977) and Hutchinson (1978), have suggested that a simplification could be achieved by using $Y_0(kR)$ in place of $H_0^{(1)}(kR)$ as the fundamental solution. This possesses the appropriate singular behaviour and satisfies the Helmholtz equation and furthermore leads to real equations (real α_{KL}) on the (real) eigenvalues, k. Unfortunately, this approach is incorrect since the $Y_0(kR)$ fundamental solution contains both outgoing *and* incoming waves whereas the $H_0^{(1)}(kR)$ alone possesses the appropriate behaviour of outgoing waves at large R. This can be readily seen in a simple example, a circular boundary, to which the solutions are well known. In this particular case, described in detail in the report version of Tai & Shaw (1974), the Neumann boundary integral equation becomes

$$(1/2)\phi(\theta) = (ika/4) \int_0^{2\pi} \phi(\theta + 2\beta) H_1^{(1)}(2ka \sin \beta) \sin \beta \, \mathrm{d}\beta \tag{25}$$

where $\beta = (\theta - \theta_0)/2$. An expansion of $\phi(\theta)$ as a Fourier series and a corresponding modal separation gives

$$\phi(\theta) = \sum_{m=0}^{\infty} [A_m \sin(m\theta) + B_m \cos(m\theta)]$$

with

$$A_m = \left(\frac{ika}{2}\right) \int_0^{2\pi} A_m \cos(2m\beta) H_1^{(1)}(2ka \sin \beta) \sin \beta \, \mathrm{d}\beta$$

and

$$B_m = \left(\frac{ika}{2}\right) \int_0^{2\pi} B_m \cos(2m\beta) H_1^{(1)}(2ka \sin \beta) \sin \beta \, \mathrm{d}\beta$$

These equations are identical; clearly either A_m or B_m could vanish, but not both for a non-trivial solution. Thus, in real and imaginary parts,

$$1 = -ka \int_0^{2\pi} \cos(2m\beta) Y_1(2ka \sin \beta \, \mathrm{d}\beta = 1 + \pi ka Y_m(ka) J'_m(ka) \tag{26a}$$

$$0 = ka \int_0^{2\pi} \cos(2m\beta) J_1(2ka \sin \beta) \sin \beta \, \mathrm{d}\beta = \pi ka J_m(ka) J'_m(ka) \tag{26b}$$

The only common root is $J'_m(ka)$ equal to zero and these are of course the correct solutions (apart from ka itself zero). Note however that the use of the $Y_0(kR)$ fundamental solution would have led to only the first equation which has an additional set of spurious roots, $Y_m(ka)$ equal to zero. Thus it is necessary to keep the original fundamental solution and pay the price of the complex determinant calculation in order to obtain the appropriate eigenvalues.

The corresponding eigenmodes can be calculated from the original set of algebraic equations by choosing a reference value for one of the values ϕ_K, e.g., ϕ_1 equal 1. As long as this is not a node, the resulting set of N equations on $N-1$ unknowns may be solved for the eigenmode on the boundary; interior values may be obtained from the original Helmholtz or Weber integral equation with the field point interior to B, i.e.,

$$\phi(\bar{r}) = (i/4)\int_B \phi(\bar{r}_0)\frac{\partial}{\partial n_0}[H_0^{(1)}(kR)]\,\mathrm{d}s_0$$

where this is a simple quadrature since $\phi(\bar{r}_0)$ on B is now known from the previous calculation.

Finally some comment should be made on the use of boundary integral equation methods for higher modes. To accurately represent a rapidly varying function, many boundary segments would be required leading to a large system of algebraic equations and thus a high order determinant. This numerical approach then appears unwieldy; however it is for these higher modes (large values of ka) that the asymptotic solutions to the boundary integral equation may play a significant role.

If Dirichlet boundary conditions are given, the resulting boundary integral equation is one of the first kind. These are inherently less stable for numerical solution than those of the second kind, e.g., Delves & Walsh (1974), and should probably be converted over to equations of the second kind, as described in Section 6.2, before proceeding with numerical solution.

The interior inhomogeneous time harmonic problem has not been as widely studied as the corresponding exterior problem, which has great application in terms of wave scattering and radiation in an infinite domain. Indeed, this is probably the most widely studied case for the boundary integral equation method as applied to wave problems. The major difficulty with this group of problems is not the actual numerical solution procedure but rather the inherent difficulty in obtaining solutions to the exterior boundary integral equations at wave numbers which match eigenvalues for a corresponding (but altered boundary condition) interior problem, e.g.,

Exterior Dirichlet → Interior Neumann eigenvalues

Exterior Neumann → Interior Dirichlet eigenvalues

This is not a physical difficulty inherent in the exterior problem since there are no 'exterior' eigenvalues for these acoustic problems, although such questions do arise for long period water waves over variable topography, e.g., Meyer (1971). Rather this is a difficulty inherent in the boundary integral equation approach, as discussed by a number of authors, with a corresponding number of methods of circumventing the problem, as described in the previous section. To illustrate this point, consider the example of a two-dimensional circle of radius, a, radiating into an infinite acoustic domain with a prescribed time harmonic surface velocity, $v_n = -(\partial\phi/\partial n) = v_0$. For simplicity take the axisymmetric case, i.e., v_0 is constant. Then eqn. (6) becomes

$$\varepsilon\phi(\bar{r}) = (i/4)\int_0^{2\pi}\left\{-v_0 \,.\, H_0^{(1)}(kR) - \phi(\theta_0)\frac{\partial}{\partial n_0}[H_0^{(1)}(kR)]\right\}a\,\mathrm{d}\theta_0 \qquad (27)$$

where ϕ is a constant ϕ_0 on the surface, $r = a$, i.e.,

$$\varepsilon\phi(\bar{r}) = -\frac{iav_0}{4}\int_0^{2\pi} H_0^{(1)}(kR)\,\mathrm{d}\theta_0 - \frac{ia\phi_0}{4}\int_0^{2\pi}\frac{\partial}{\partial n_0}[H_0^{(1)}(kR)]\,\mathrm{d}\theta_0$$

The field point is *not* placed on the surface since the expansion of the fundamental solution, as given in Morse & Feshbach (1953), Section 7.2, requires a different value for $|\bar{r}|$ and $|\bar{r}_0|$.

$$R = [r^2 + r_0^2 - 2rr_0\cos(\theta - \theta_0)]^{1/2}$$

$$\partial R/\partial n_0 = -\partial R/\partial r_0 = -(r_0 - r\cos(\theta - \theta_0))/R$$

$$H_0^{(1)}(kR) = \sum_{m=0}^{\infty}\varepsilon_m\cos(m(\theta - \theta_0))\begin{cases}J_m(kr)H_m(kr_0); & r < r_0\\ J_m(kr_0)H_m(kr); & r > r_0\end{cases}$$

where $\varepsilon_m = 1$ if $m = 0$ and 2 if $m \neq 0$.

$$\frac{\partial H_0^{(1)}(kR)}{\partial r_0} = k\sum_{m=0}^{\infty}\varepsilon_m\cos(m(\theta - \theta_0))\begin{cases}J_m(kr)H'_m(kr_0); & r < r_0\\ J'_m(kr_0)H_m(kr); & r > r_0\end{cases}$$

For simplicity, take $\theta = 0$ and $r = a^-$; since $r_0 = a$, this places r infinitesimally less than r_0 and also places the field point exterior to the infinite fluid, i.e., interior to the circular boundary. Then $\varepsilon = 0$ and

$$0 = -\frac{iav_0}{4}J_0(ka^-)H_0(ka) - \frac{ia\phi_0}{4}kJ_0(ka^-)H'_0(ka)$$

or

$$\phi_0 = -\frac{v_0}{k}\cdot\frac{H_0(ka)}{H'_0(ka)} \qquad (28)$$

which agrees with the well known separation of variables solution *if* $J_0(ka^-)$ is not zero, i.e., if k is not an eigenvalue to the corresponding Dirichlet problem.

Apart from this very real but treatable difficulty, the two- and three-dimensional time harmonic exterior problems are no more difficult than the interior counterparts; this is in strong contrast to the alternative numerical methods of finite differences and finite elements.

Transient problems form the next class for discussion. As mentioned above, the two-dimensional transient fundamental solution contains a time integration and it has been found convenient to treat two-dimensional problems in terms of an extension into a three-dimensional form with infinite cylinders of constant cross-section replacing the original two-dimensional body, e.g., Friedman & Shaw (1962). There have been a number of two- and three-dimensional problems considered; these have been exclusively exterior problems (see the References and the Bibliography). The main advantage of time dependent boundary integral equation solutions is that they lead to uncoupled or weakly coupled systems of algebraic equations when Φ is approximated by some trial function, e.g., a piece-wise constant, over the boundary elements. Furthermore, solutions for discontinuous forcing functions, e.g., shock wave or step pressure loading, can be treated with the contribution from the discontinuity evaluated exactly, e.g., Friedman & Shaw (1962), Shaw (1968*a*). While integral equation formulations of the second kind are preferable, e.g., Shaw & Friedman (1962), some solutions have been carried out using integral equation formulations of the first kind, e.g., Shaw & English (1972).

As a simple illustration, consider the two-dimensional scattering of a step plane pressure pulse by an infinite right angle wedge with a rigid boundary condition; this forms the 'front' portion of the problem described by Friedman & Shaw (1962). The problem is treated as three-dimensional with z the co-ordinate axis in the third direction. The two surfaces of the wedge are taken as the xz and yz planes and the governing equation for a field point at (x, o, o) is then

$$(1/2)\phi(x,o,o,t) = \phi_W(x,o,o,t) + \frac{1}{4\pi}\int_B \left\{ \frac{1}{cR}\frac{\partial\phi(\bar{r}_0, t_0)}{\partial t_0} + \frac{\phi(\bar{r}_0, t_0)}{R^2} \right\} \frac{\partial R}{\partial n_0}\, dA_0 \tag{29}$$

Here S represents both surfaces but since $\partial R/\partial n_0$ vanishes along the $x_0 z_0$ surface (the singular contribution from the point R equal zero has already been accounted for, i.e., the principle value excludes this point) only integrations on the $y_0 z_0$ plane remain. Taking a symmetric incident wave for simplicity, the solution on the $y_0 z_0$ surface is equal to that on the $x_0 z_0$ surface and the equation becomes

$$\tfrac{1}{2}\phi(x,o,o,t) = \phi_{\mathrm{W}}(x,o,o,t) + \frac{1}{4\pi}\int_0^{y_{0u}}\int_{z_{0L}}^{z_{0u}}\left\{\frac{1}{cR}\frac{\partial\phi(\bar{r}_0,t_0)}{\partial t_0} + \frac{\phi(\bar{r}_0,t_0)}{R^2}\right\}\frac{\partial R}{\partial n_0}\,\mathrm{d}z_0\,\mathrm{d}y_0 \tag{30}$$

The limits of integration are determined by the condition

$$[ct - (\sqrt{2}/2)y_0]^2 = x^2 + y_0^2 + z_0^2 \tag{31}$$

which yields

$$y_{0u} = ct/(1 + \sqrt{2}/2) \tag{32a}$$

$$\begin{aligned} z_{0u} = -z_{0L} &= [(ct - (\sqrt{2}/2)y_0)^2 - y_0^2]^{1/2} \\ &= 0 \qquad \text{at } y_{0u} \end{aligned} \tag{32b}$$

Thus, eqn. (30) has time dependent limits of integration. If an incident pressure field, $p_w = \rho_0\,\partial\phi/\partial t$, is prescribed as a unit step function, $H[ct - (x + y)\sqrt{2}/2]$, this equation on the velocity potential ϕ may be differentiated, using Leibnitz' rule to interchange the spatial integration with the temporal differentiation

$$\begin{aligned} \tfrac{1}{2}p(x,o,o,t) &= p_w(x,o,o,t) \\ &+ \frac{\rho_0}{4\pi}\int_0^{y_{0u}} 2\frac{\partial z_{0u}}{\partial t}\left[\frac{\phi(\bar{r}_0,t_0)}{R^2} + \frac{1}{cR}\frac{\partial\phi(\bar{r}_0,t_0)}{\partial t_0}\right]\frac{\partial R}{\partial n_0}\,\mathrm{d}y_0 z_0 = z_{0u} \\ &+ \frac{1}{4\pi}\int_0^{y_{0u}} 2\int_0^{z_{0u}}\left[\frac{p(0,y_0,z_0,t_0)}{R^2} + \frac{1}{cR}\frac{\partial p(0,y_0,0,t_0)}{\partial t_0}\right]\mathrm{d}z_0\,\mathrm{d}y_0 \end{aligned} \tag{33}$$

The first integral is evaluated at z_{0u} which is immediately behind the wave front; ϕ and $\partial\phi/\partial t$ are known here as the direct reflected solution; ϕ is zero and $\partial\phi/\partial t$ is $2/\rho_0$. Thus this term may be integrated as a simple quadrature, yielding a function $A(x,t)$.

The numerical solution of this boundary integral equation may be accomplished by taking p to have a constant value, $p_{K,L}$, for $(K-1)\Delta Y < Y_0 < K\Delta Y$ and for $(L-1)\Delta t < t < L\Delta t$. The time derivative may be replaced by a backwards time difference,

$$\begin{aligned} \frac{\partial p(0,y_0,z_0,t_0)}{\partial t_0} &= \frac{p(0,y_{0K},z_0,t_{0L}) - p(0,y_{0K},z_0,t_{0L}-\Delta t)}{\Delta t} \\ &= \frac{p_{K,L} - p_{K,L-1}}{\Delta t} \end{aligned} \tag{34}$$

The location of the field point, x_J, enters through the term R in the retarded time t_0. Define

$$\beta_{KLJ} = (1/2\pi) \int_{B_{KL}} (1/R^2)\, \partial R/\partial n_0 \, dy_0 \, dz_0 \tag{35a}$$

$$\gamma_{KLJ} = (1/2\pi) \int_{B_{KL}} (1/cR)\, \partial R/\partial n_0 \, dy_0 \, dz_0 \tag{35b}$$

where B_{KL} is the surface element defined by $(K-1)\Delta Y < Y_0 < K\Delta Y$ and $(L-1)\Delta t < c < L\Delta t$, such that $(M-L)\Delta t < t_0 < (M-L+1)\Delta t$ there. It is simpler to use the condition $ct = R$ in place of eqn. (31); this would have source points contributing to the integral before an appropriate time interval has passed allowing these points to be stimulated by the incident wave and then send a signal to the field point. As long as the strength of these contributions is zero, this integral in theory does not change. In practice, there is a small effect due to the finite discretisation of the integral but this effect is negligible. It would not have been appropriate however to make this modification before the wave front discontinuity contribution, $A(x, t)$, was calculated.

The boundary integral equation is then replaced by

$$(1/2)p_{J,M} = p_W(x_J, t_M) + A(x_J, t_M) + \sum_{K=1}^{N}\sum_{L=1}^{M}\left\{\beta_{KLJ}p_{K,M-L+1} + \gamma_{KLJ}\left[\frac{p_{K,M-L+1} - p_{K,M-L}}{\Delta t}\right]\right\} \tag{36}$$

where x_J is chosen at a 'midpoint', $(J-1/2)\Delta X$ and t_M at an 'endpoint', $(M-1)\Delta t$ of the space and time step respectively, with $\Delta Y = \Delta X = \sqrt{2}c\Delta t$, to allow the incident wave to sweep over one space step in one time step.

The solution of this set of linear algebraic equations is considerably simplified when they are recognised to be successive rather than simultaneous, i.e., the value p_{JM} does not appear on the right-hand side of eqn. (36) since β_{KLJ} and γ_{KLJ} are both zero for negative L and $p, \dot{p}$ are zero for zero and negative time, i.e., $M \leq L-1$. If t_M were chosen to be $M\Delta t$, i.e., the value of p at the end rather than the beginning of the mth time step, there would be a weak coupling in that β_{J1J} and γ_{J1J} would both multiply values of p_{JM} on the right-hand side of eqn. (36). These can be easily moved over to combine with the $(1/2)\, p_{JM}$ term; the remainder of the summation does not involve values of p at the present time step. Choosing $c\Delta t < \Delta X$ or ΔY should insure stability for an explicit scheme were this a finite difference scheme; the appropriate stability criteria here do not appear to have been studied. Clearly a large time step (covering several space steps) would lead to a strongly coupled (implicit) scheme while a small time step (covering a fraction of a space step) would lead to an uncoupled system. The time step and space step used in much of this work is matched to the incident wave; a step size ratio of $c\Delta t/\Delta X$ equal to or less than one should be used to remain within the classical Courant, Lewy, Fredricks stability criteria. Numerical results are given in Friedman & Shaw (1962) and agree very well with an analytical solution. Before leaving this problem however, it is of interest to note that the vertex (o, o) lies on 'both' surfaces and has $\partial R/\partial n_0$ equal to zero on both surfaces, thereby causing the surface integral in eqn. (29) to vanish *except* for the singular contribution at $R = 0$ which produces ε equal to 3/4 instead of 1/2 (the region excluded from the integration in Green's theorem is 3/4 of a circle rather than 1/2 of a circle for this boundary point). The analytical solution at this vertex is then seen to be $\phi(o, o, t) = (4/3)\phi_{w'}(o, o, t)$ for any form of incident wave. Furthermore, boundary points for $\sqrt{2}ct > X > ct$ have a solution $\phi(X, o, t) = 2\phi_{w'}(X, o, t)$ since the double integral is zero for these values even though these points have been reached by the incident wave; this is simply the reflection solution before corner diffraction effects reach the field point.

The simple numerical cases described above give the essence of the procedures used. Improvements clearly can be made in the form of the trial functions and boundary element shapes used, e.g., Cruse (1974) and, using finite element methods to solve boundary integral equations, Jeng & Wexler (1977).

Alternatively, Galerkin or other moment methods could be employed, e.g., Sharma (1967). These improvements, though an important part of the progress of this field, are not in themselves necessary for a basic understanding of the 'mechanics' of this numerical approach.

6.4 APPROXIMATE SOLUTIONS

Although the emphasis of this chapter, and indeed of this text is on numerical solution methods, this section appears appropriate in that it does represent a solution technique valid for certain types of problems. These problems generally fall in that class which lends itself to some form of expansion in terms of a physical parameter which is small. Typical of such solutions are the studies of time harmonic wave radiation by shells forms of slightly non-concentric cylindrical surfaces, submerged in an infinite acoustic fluid, Shaw & Tai (1974) and Shaw (1978*a*). Here an expansion of the problem may be made in powers of the non-concentricity distance; when this distance is zero, the outer and inner shell surfaces are concentric and the problem lends itself to separation of variables. This technique can be readily applied to other geometries, e.g., spherical, ellipsoidal, etc., in which the Helmholtz equation separates. The details of this procedure are left to the references; they basically involve an expansion of the fundamental solution into the modes of the separable solution and simultaneously a further expansion into powers of the separation distance. Consideration of terms of like order in this (small) distance leads to a successive system of equations which are of the same form for each order with only a modified 'forcing' term.

Another area of approximate solutions, which overlaps into the use of boundary integral equation methods for analytical purposes, is their use in deriving elastic plate or elastic shell theory from three-dimensional elasticity. If one considers that the ultimate goal of plate (or shell) theory is to provide a valid two-dimensional representation of an inherently three-dimensional problem, the role of boundary integral equation methods is apparent. They provide this service *without* approximation. Since there are in general two distinct surfaces (upper and lower or inner and outer) with possibly other edge surfaces over which the boundary integral is formed, it is clear that the plate (shell) thickness will enter into the kernels (Green's functions) through their dependence on the distance from field point to integration (source) point. An expansion of these kernels in terms of this thickness allows a hierarchy of plate (shell) theories to be developed from an essentially rigorous foundation in which this parameter occurs naturally. Since this application is somewhat removed from the main stream of this chapter, details are left to a reference, Shaw (1978*b*). It would be inappropriate however to leave this subject without indicating that St. Venant's principle follows very naturally from a boundary integral equation formulation, e.g., Shaw (1978*c*); indeed, a remarkable proof of this principle for the elastic half space was given by Sternberg (1954) using a form of Betti's theorem which in turn is a basis for a boundary integral equation formulation of elastostatics, Cruse & Rizzo (1968), Cruse (1968).

Equally far removed from this mainstream are the high frequency approximations of optics and acoustics, i.e., geometrical optics (acoustics). Where these techniques are employed in the treatment of wave propagation through inhomogeneous media, boundary integral equations as yet offer little competition in view of the lack of appropriate fundamental solutions.

However, boundary integral equation methods may play a significant role in the study of high frequency scattering by complex structures—especially in terms of an expansion of the integral equation kernels in terms of the (small) parameter, the wavenumber. This approach formed the basis of much classical analysis, e.g., Fresnel–Kirchhoff diffraction theory (which contained other assumptions as well).

6.5 CONCLUSIONS AND DIRECTIONS

Probably the most useful discussion of the application of boundary integral equation methods to wave propagation problems is that of the directions in which these seem to be heading. This is best stated in terms of where these methods have been used already and conclusions as to their present status. As with most of this chapter, these are subjective to some extent but should provide a reasonable overview.

In acoustics, two- and three-dimensional time harmonic exterior radiation and scattering problems seem well in hand e.g., reviews by Shaw (1970), Chertock (1971), Kleinman & Roach (1974). Primary advances here would seem to be in the areas of improved trial functions, since most examples to date use a simple constant form, and surface segmenting, and in the solution of the resulting system of algebraic equations—problems somewhat analogous to those currently found in the area of finite elements. It seems quite feasible that an all purpose program can be developed for arbitrarily shaped surfaces. Some care should be taken however in realising that the strength of numerical solution by boundary element discretisation lies in the area where classical high frequency and low frequency procedures are inadequate, i.e., for wavelengths comparable to the length scale of the obstacle. Finally more work is required on eliminating those spurious results which arise from the complementary interior problem eigenvalues, e.g., Ursell (1973), Schenck (1968). In fact the use of boundary integral equation methods to determine interior problem eigenvalues, e.g., Tai & Shaw (1974), DeMey (1976), seems another area worth further examination.

Transient acoustic problems also seem to be off to a good start with direct Kirchhoff retarded time integral equation solution methods suitable for early time, e.g., Friedman & Shaw (1962), Shaw & Friedman (1962), Mitzner (1967), and Fourier transform and corresponding Helmholtz integral equation solutions for longer time, e.g., Shippy (1975). More complex geometries and eventually a general numerical program for arbitrarily shaped obstacles seem to be the next steps.

The analogous elastic problems have a further dilemma. Should their formulation be in terms of displacement potentials, which have simple governing equations but complicated boundary relationships, e.g., Shaw (1968*a*), or directly in terms of tractions and displacements which have simpler boundary representations but more complicated governing equations (and thus kernel functions), e.g., Cruse & Rizzo, (1968), Cruse (1968)?

Coupled acoustic fluid–elastic structure problems seem a natural application of boundary integral equation methods, due to their common boundary, especially when the acoustic fluid is infinite in extent. This is true whether the elastic structure is represented by another boundary integral equation, e.g., Wilkinson & Liu (1970), Shaw (1973*a*, *b*) or by some other representation such as finite elements.

Water wave problems form still another class of problems only briefly touched on here. For long wave theory, the governing equation is analogous to a two-dimensional acoustic wave equation. However, the phase velocity is dependent on the water depth which clearly varies in such problems and thus the analogy is to acoustic propagation in an inhomogeneous medium. This is further complicated by the treatment of most of these as exterior problems, i.e., involving an infinite domain. This has led to the introduction of a coupled boundary integral equation—other numerical methods, e.g., Shaw (1974, 1975*a*), to finite differences, Shaw & Falby (1977), Shaw (1978*d*), to finite elements, etc., for problems where the infinite domain is represented by some constant depth (constant phase velocity) at some distance from the local inhomogeneity. Indeed, this coupling is an area of general interest, e.g., Zienkiewicz *et al.* (1977*a*, *b*). For waves over submerged obstacles, a boundary integral equation approach based on a somewhat complicated fundamental solution has been found useful and has been used by Garrison & Seetharama (1971), Garrison & Chow (1972), etc., with success. One of the primary difficulties here is that the governing equation is simple—indeed it is the Laplace equation—but the free surface boundary conditions are difficult. In fact, they may be taken as non-linear and still yield to a boundary integral equation approach since the governing differential equation is linear, e.g., Shaw (1975*b*).

ACKNOWLEDGEMENTS

The continuing support of the Office of Naval Research and in particular the Structural Mechanics division is gratefully acknowledged. Further acknowledgement is due to the École d'Été d'Analyse Numérique of E.D.F., I.R.I.A., and C.E.A. at which the author was an invited lecturer during the 1977 session on Boundary Integral Equation Methods.

REFERENCES

BAKER, B. B. & COPSON, E. T. (1939). *The Mathematical Theory of Huygen's Principle*, Oxford University Press, Oxford.

BANAUGH, R. P. & GOLDSMITH, W. (1963*a*). Diffraction of steady acoustic waves by surfaces of arbitrary shape, *J.A.S.A.*, **35,** 1590–601.

BANAUGH, R. P. & GOLDSMITH, W. (1963*b*). Diffraction of steady elastic waves by surfaces of arbitrary shape, *J. Appl. Mech.*, **30,** 589–95.

CHERTOCK, G. (1971). *Integral equation methods in sound radiation and scattering from arbitrary surfaces*, NSRDC Rep. 3538, Washington, DC.

CRUSE, T. A. (1968). A direct formulation and numerical solution of the general transient elastodynamic problem, II, *Jour. Math. Anal. and Appl.*, **22,** 341–55.

CRUSE, T. (1974). An improved boundary integral equation method for three-dimensional stress analysis, *Comp. and Structures*, **4,** 741–54.

CRUSE, T. A. & RIZZO, F. J. (1968). A direct formulation and numerical solution of the general transient elastodynamic problem, I *Jour. Math. Anal. and Appl.*, **22,** 244–59.

DELVES, L. M. & WALSH, J. (1974). *Numerical Solution of Integral Equations*, Clarendon Press, Oxford.

DEMEY, G. (1976). Calculation of eigenvalues of the Helmholtz equation by an integral equation, *Int. J. Num. Meth. Eng.*, **10,** 59–66.

DEMEY, G. (1977). A simplified integral equation method for the calculation of the eigenvalues of the Helmholtz equation, *Int. J. Num. Meth. Eng.*, **11,** 1340–2.

FOX, L. (1962). *Numerical Solution of Ordinary and Partial Differential Equations*, Pergamon Press, London.

FRIEDMAN, M. E. & SHAW, R. P. (1962). Diffraction of a plane shock wave by an arbitrary rigid cylindrical obstacle, *J. Appl. Mech.*, **29,** 40–46.

GARRISON, C. J. & CHOW, P. Y. (1972). Wave forces on submerged bodies, *ASCE*, **98**(WW3), 375–92.

GARRISON, C. J. & SEETHARAMA, R. V. (1971). Interaction of waves with submerged obstacles, *ASCE*, **97**(WW2), 259–77.

GREGOIRE, J. P., NEDELEC, J. C. & PLANCHARD, J. (1975). A method for computing the eigenfrequencies of an acoustic resonator, *IUTAM Proceedings—Marseille*, pp. 343–52.

HADAMARD, J. (1952). *Lectures on Cauchy's Problem in Linear Partial Differential Equations*, Dover Pub., New York (reprint of 1923 edition).

HESS, J. (1962*a*). *Calculation of potential flow about arbitrary three-dimensional bodies*, Douglas Aircraft Co. Rep. ES 40622.

HESS, J. (1962*b*). Calculation of potential flow about bodies of revolution having axes perpendicular to the free stream direction, *J. Aero. Sci.*, **29,** 726–42.

HUTCHINSON, J. R. (1978). Determination of membrane vibrational characteristics by the boundary integral equation methods. In: *Recent Advances in Boundary Element Methods*, Ed. C. Brebbia, Pentech Press, London, pp. 301–16.

JASWON, M. A. (1963). Integral equation methods in potential theory, I *Proc. Roy. Soc. London, Series A*, **275,** 23–32.

JENG, G. & WEXLER, A. (1977). Isoparametric finite element variational solution of integral equations for three dimensional fields, *Int. J. Num. Meth. Eng.*, **11,** 1455–74.

KIRCHHOFF. G. (1883). Zur Theorie der Lichstrahlen, *Ann. Physik*, **18,** 663–95.

KLEINMANN, R. E. & ROACH, G. F. (1974). Boundary integral equations for the three dimensional Helmholtz equation, *SIAM Review*, **16,** 214–36.

KUPRADZE, V. D. (1935). *Fundamental Problems in the Mathematical Theory of Diffraction*, RT 1260 (translation of Russian text).

KUPRADZE, V. D. (1963). Dynamical problems in elasticity. In: *Progress in Solid Mechanics*, Vol. III, eds. I. N. Sneddon and R. Hill, North-Holland, Amsterdam.

LAMB, H. (1932). *Hydrodynamics*, Cambridge University Press, Cambridge.

MEYER, R. E. (1971). Resonance of unbounded water bodies. In: *Mathematical Problems in the Geophysical Sciences*, 1*: Geophysical Fluid Dynamics*, Ed. W. H. Reid, Amer. Math. Soc., Providence, R.I.

MITTRA, R. (1973). *Computer Techniques in Electromagnetics*, Pergamon Press, New York.

MITZNER, K. M. (1967). Numerical solution for transient scattering from a hard surface of arbitrary shape—Retarded potential technique, *J.A.S.A.*, **42,** 391–7.

MORSE, P. & FESHBACH, H. (1953). *Methods of Theoretical Physics*, Vols. I and II, McGraw-Hill, New York.

NORTHOVER, F. H. (1971). *Applied Diffraction Theory*, American Elsevier Pub. Co., Inc., New York.

PAO, Y. H. & MOW, C. C. (1973). *Diffraction of Elastic Waves and Dynamic Stress Concentrations*, Crane Russak, New York.

SCHENCK, H. A. (1967). *Formulation and solution of acoustic radiation problems as integral equations—Selected references*, Naval Undersea Warfare Center, San Diego Cal.

SCHENCK, H. A. (1968). Improved integral formulation for acoustic radiation problems, *J.A.S.A.*, **44,** 41–58.

SHARMA, D. L. (1967). Scattering of steady elastic waves by surfaces of arbitrary shape, *Bull. Sis. Soc. Amer.*, **57,** 795–812.

SHAW, R. P. (1968*a*). Comments on 'Numerical solution for transient scattering from a hard surface of arbitrary shape—Retarded potential technique', *J.A.S.A.*, **43,** 638–9.

SHAW, R. P. (1968*b*). Diffraction of pulses by obstacles of arbitrary shape with an impedance boundary condition, *J.A.S.A.*, **44,** 1962–8.

SHAW, R. P. (1970). An integral equation approach to acoustic radiation and scattering. In: *Topics in Ocean Engineering II*, Ed. C. Bretschneider, Gulf Publ. Co., Houston, Texas, pp. 143–63.

SHAW, R. P. (1973*a*). Integral equation formulation of dynamic acoustic fluid–elastic solid interaction problems, *J.A.S.A.*, **53,** 514–20.

SHAW, R. P. (1973*b*). An integral equation formulation of acoustic fluid–elastic shell dynamic interaction problems, (presented at IASS Pacific Symposium on Hydromechanically Supported Shells, Honolulu, October 1971). In:

Hydromechanically Supported Shells, ed. R. Szilard, East West Center Press, Honolulu, Hi, pp. 615–32.

SHAW, R. P. (1974). Time harmonic scattering by obstacles surrounded by an inhomogeneous medium, (presented at the 87th Acoustical Society of America meeting, N.Y.C., April 1974), *J.A.S.A.*, **56,** 1354–60.

SHAW, R. P. (1975*a*). An outer boundary integral equation applied to transient wave scattering in an inhomogeneous medium, (presented at JSME/ASME Applied Mechanics Meeting, Honolulu, Hawaii, March 1975), *J. Appl. Mech.*, **42,** 147–52.

SHAW, R. P. (1975*b*). Boundary integral equation methods applied to water waves. In: *Boundary Integral Equation Method: Computational Applications in Applied Mechanics*, Eds. T. Cruse and F. Rizzo, AMD Vol. 11, ASME, New York, pp. 7–14.

SHAW, R. P. (1978*a*). Time harmonic acoustic radiation from a submerged elastic shell defined by nonconcentric circular cylinders, *J.A.S.A.*, **64,** 311–17.

SHAW, R. P. (1978*b*). *Elastic plate vibrations by boundary integral equations*, (presented at 15th Annual Meeting of the Society of Engineering Science, Gainesville, Fla., Dec. 3–6, 1978), Rep. No. 109, SUNY at Buffalo, New York.

SHAW, R. P. (1978*c*). *A boundary integral equation formulation of Saint Venant's principle*, Rep. No. 108, SUNY at Buffalo, New York.

SHAW, R. P. (1978*d*). Coupling boundary integral equation methods to 'other' numerical techniques. In: *Recent Advances in Boundary Element Methods*, Ed. C. Brebbia, Pentech Press, London, pp. 137–48.

SHAW, R. P. & ENGLISH, J. A. (1972). Transient acoustic scattering by a free sphere, *J. Sound Vibration*, **20,** 321–31.

SHAW, R. P. & FALBY, W. (1977). *FEBIE—A Combined Finite Element—Boundary Integral Equation Approach*, CETIM, Senlis, France, pp. 1.97–1.101. Also, *Computers and Fluids* (1978), **6,** 153–60.

SHAW, R. P. & FRIEDMAN, M. B. (1962). *Diffraction of a plane shock wave by a free cylindrical obstacle at a free surface*, 4th US National Congress of Applied Mechanics, pp. 371–9.

SHAW, R. P. & TAI, G. R. C. (1974). Time harmonic acoustic radiation from a nonconcentric circular cylinder, *J.A.S.A.*, **56**(5), 1437–43.

SHIPPY, D. J. (1975). Application of the boundary integral equation method to transient phenomena in solids. In: *Boundary Integral Equation Method: Computational Applications in Applied Mechanics*, Eds. T. Cruse and F. Rizzo, AMD Vol. 11, ASME, New York.

STERNBERG, E. (1954). On Saint Venant's principle, *Quart. Appl. Math.* **11,** 393–402.

SYMM, G. T. (1963). Integral equation methods in potential theory, II, *Proc. Roy. Soc. London, Series A*, **275,** 33–46.

TAI, G. R. C. & SHAW, R. P. (1974). Helmholtz equation eigenvalues and eigenmodes for arbitrary domains, *J.A.S.A.*, **56,** 796–804. Also, Rep. No. 90, SUNY at Buffalo, New York.

URSELL, F. (1973). On the exterior problems of acoustics, *Proc. Camb. Phil. Soc.*, **74,** 117–25.

VASTANO & REID, R. O. (1967). A numerical study of the Tsunami response at an island, *J. Mar. Res.*, **25,** 129–39.

WILKINSON, J. P. D. & LIU, H. K. (1970). *Integral-equation techniques for structural-acoustic interaction*, Rep. 70-C-233, R & D Center, Schenectady.

ZIENKIEWICZ, O. C., KELLY, D. W. & BETTESS, P. (1977*a*). The coupling of the finite element method and boundary solution procedures, *Int. J. Num. Meth. Eng.*, **11**, 355–75.

ZIENKIEWICZ, O. C., KELLY, D. W. & BETTESS, P. (1977*b*). Marriage à la Mode—The best of both worlds (finite elements and boundary integrals), *Proceedings of Int. Symp. on Innovative Numerical Analysis in Applied Engineering Science*, Ed. T. Cruse, J. C. Lachat, F. Rizzo and R. P. Shaw, CETIM, Senlis, France.

BIBLIOGRAPHY

General

ALAWNEH, A. D. & KANWAL, R. P. (1977). Singularity methods in mathematical physics, *SIAM Review*, **19**, 437–71.

BAKER, B. B. & COPSON, E. T. (1939). *The Mathematical Theory of Huygen's Principle*, Oxford University Press, Oxford.

CHERTOCK, G. (1968). Convergence of iterative solutions to integral equations for sound radiation, *Q.A.M.*, **26**, 268–71.

CHERTOCK, G. (1971). *Integral equation methods in sound radiation and scattering from arbitrary surfaces*, NSRDC Rep. 3538, Washington, DC.

CRUSE, T. (1974). An improved boundary integral equation method for three-dimensional stress analysis, *Comp. and Structures*, **4**, 741–54.

DELVES, L. M. & WALSH, J. (1974). *Numerical Solution of Integral Equations*, Clarendon Press, Oxford.

FOX, L. (1962). *Numerical Solution of Ordinary and Partial Differential Equations*, Pergamon Press, London.

HADAMARD, J. (1952). *Lectures on Cauchy's Problem in Linear Partial Differential Equations*, Dover Pub., New York (reprint of 1923 edition).

HESS, J. (1962). *Calculation of potential flow about arbitrary three-dimensional bodies*, Douglas Aircraft Co. Rep. ES 40622.

HESS, J. (1962). Calculation of potential flow about bodies of revolution having axes perpendicular to the free stream direction, *J. Aero. Sci.*, **29**, 726–42.

JASWON, M. A. (1963). Integral equation methods in potential theory, I *Proc. Roy. Soc. London, Series A*, **275**, 23–32.

JASWON, M. A. & SYMM, G. T. (1977). *Integral Equation Methods in Potential Theory and Elastostatics*, Academic Press, London.

KIRCHHOFF, G. (1883). Zur Theorie der Lichstrahlen, *Ann. Physik*, **18**, 663–95.

KUPRADZE, V. D. (1935). *Fundamental Problems in the Mathematical Theory of Diffraction*, RT 1260 (translation of Russian text).

KUPRADZE, V. D. (1963). Dynamic problems in elasticity. In: *Progress in Solid Mechanics*, Vol. III, ed. I. N. Sneddon and R. Hill, North Holland, Amsterdam.

LAMB, H. (1932). *Hydrodynamics*, Cambridge University Press, Cambridge.

MITTRA, R. (1973). *Computer Techniques in Electromagnetics*, Pergamon Press, New York.

MORSE, P. & FESHBACH, H. (1953). *Methods of Theoretical Physics*, Vols I and II McGraw-Hill, New York.

NORTHOVER, F. H. (1971). *Applied Diffraction Theory*, American Elsevier Pub. Co., Inc., New York.

PAO, Y. H. & MOW, C. C. (1973). *Diffraction of Elastic Waves and Dynamic Stress Concentrations*, Crane Russak, New York.

SHAW, R. P. (1970). An integral equation approach to acoustic radiation and scattering. In: *Topics in Ocean Engineering II*, Ed. C. Bretschneider, Gulf Publ. Co., Houston, Texas, pp. 143–63.

SHAW, R. P. (1971). Singularities in acoustic pulse scattering by free surface obstacles with sharp corners, *J. Appl. Mech.*, **38,** 526–8.

SHAW, R. P. (1975). Boundary integral equation methods applied to water waves. In: *Boundary Integral Equation Method: Computational Applications in Applied Mechanics*, Eds T. Cruse and F. Rizzo, AMD Vol. 11, ASME, N.Y.C., pp. 7–14.

STERNBERG, E. (1954). On Saint Venant's principle, *Quart. Appl. Math.*, **11,** 393–402.

SYMM, G. T. (1963). Integral equation methods in potential theory, II, *Proc. Roy. Soc. London, Series A*, **275,** 33–46.

Approximate

AR, E. & KLEINMAN, R. E. (1966). The exterior Neumann problem for the three dimensional Helmholtz equation, *Arch. Rat. Mech. Analy.*, **23,** 218–36.

CHERTOCK, G. (1964). Sound radiation from vibrating surfaces, *J.A.S.A.*, **36,** 1305–13.

ENGQUIST, B. & MAJDA, A. (1977). Absorbing boundary conditions for the numerical simulation of waves, *Math. of Computation*, **21**(139), 629–51.

JENG, G. & WEXLER, A. (1977). Isoparametric finite element variational solution of integral equations for three dimensional fields, *Int. J. Num. Meth. Eng.*, **11,** 1455–74.

NOBLE, B. (1962). Integral equation perturbation methods in low frequency diffraction. In: *Electromagnetic Waves*, Ed. R. E. Langer, Univ. of Wisc. Press, Madison, p. 323.

SHAW, R. P. (1978). Time harmonic acoustic radiation from a submerged elastic shell defined by nonconcentric circular cylinders, *J.A.S.A.*, **64,** 311–17.

SHAW, R. P. (1978). *Elastic plate vibrations by boundary integral equations*, (presented at 15th Annual Meeting of the Society of Engineering Science, Gainesville, Fla., Dec. 3–6, 1978), Rep. No. 109, SUNY at Buffalo, New York.

SHAW, R. P. (1978). *A boundary integral equation formulation of Saint Venant's principle*, Rep. No. 108, SUNY at Buffalo, New York.

SHAW, R. P. & TAI, G. R. C. (1974). Time harmonic acoustic radiation from a nonconcentric circular cylinder, *J.A.S.A.*, **56,** 1437–43.

Transient Wave Problems

Acoustic

FARN, C. L. S. & HUANG, H. (1968). Transient acoustic fields generated by a body of arbitrary shape, *J.A.S.A.*, **43,** 252–7.

FRIEDMAN, M. B. & SHAW, R. P. (1962). Diffraction of a plane shock wave by an arbitrary rigid cylindrical obstacle, *J. Appl. Mech.*, **29,** 40–46.

MITZNER, K. M. (1967). Numerical solution for transient scattering from a hard surface of arbitrary shape—Retarded potential technique, *J.A.S.A.*, **42,** 391–7.

SHAW, R. P. (1967). Diffraction of acoustic pulses by obstacles of arbitrary shape with a robin boundary condition—Part A, *J.A.S.A.*, **41,** 855–9.

SHAW, R. P. (1967). Scattering of plane acoustic pulses by an infinite plane with a general first order boundary condition, *J. Appl. Mech.*, **34,** 770–2.

SHAW, R. P. (1968). Comments on 'Numerical solution for transient scattering from a hard surface of arbitrary shape—Retarded potential technique', *J.A.S.A.*, **43,** 638–9.

SHAW, R. P. (1968). Diffraction of pulses by obstacles of arbitrary shape with an impedance boundary condition, *J.A.S.A.*, **44,** 1962–8.

SHAW, R. P. (1975). Transient scattering by a circular cylinder, *J. Sound Vibration*, **42,** 295–304.

SHAW, R. P. & COURTINE, D. F. (1969). Diffraction of a plane acoustic pulse by a free orthogonal trihedron, *J.A.S.A.*, **46,** 1382–4.

SHAW, R. P. & ENGLISH, J. A. (1972). Transient acoustic scattering by a free sphere, *J. Sound Vibration*, **20,** 321–31.

SHAW, R. P. & FRIEDMAN, M. B. (1962). *Diffraction of a plane shock wave by a free cylindrical obstacle at a free surface*, 4th US National Congress of Applied Mechanics, pp. 371–9.

SHIPPY, D. J. (1975). Application of the boundary integral equation method to transient phenomena in solids. In: *Boundary Integral Equation Method: Computational Applications in Applied Mechanics*, Eds T. Cruse and F. Rizzo, AMD Vol. 11, ASME, New York.

SOULES, G. W. & MITZNER, K. M. (1966). *Pulses in linear acoustics*, App. Res. Dep., Nortronics, Northrup Corp. Rep. ARD-66-60-R.

Elastic

SHAW, R. P. (1968). Retarded potential approach to the scattering of elastic waves by rigid obstacles of arbitrary shape, *J.A.S.A.*, **44,** 745–8.

Water

BOLOMEY, J. C., DURIX, C. & LESSELIER, D. (1978). Time domain integral equation approach for inhomogeneous and dispersive slab problems, *IEEE Trans., Antennas and Prop.*, **AP-26,** 658–67.

SHAW, R. P. (1975). An outer boundary integral equation applied to transient wave scattering in an inhomogeneous medium, *J. Appl. Mech.*, **42,** 147–52.

Electromagnetic

BENNETT, C. L. & WEEKS, W. L. (1970). Transient scattering from conducting cylinders, *IEEE Trans., Antennas and Prop.*, **AP-18,** 627–33.

Time Harmonic Wave Problems

Acoustic

BANAUGH, R. P. & GOLDSMITH, W. (1963). Diffraction of steady acoustic waves by surfaces of arbitrary shape, *J.A.S.A.*, **35**, 1590–601.

BRUNDIT, G. B. (1965). A solution to the problems of scalar scattering from a smooth bounded obstacle using integral equations, *Quar. Jour. Mech. App. Math*, **18**, 474.

BURTON, A. J. & MILLER, G. F. (1971). The application of integral equation methods to the numerical solution of some exterior boundary value problems, *Proc. Roy. Soc. London, Series A*, **323**, 201–10.

CHEN, L. H. & SCHWEIKERT, A. (1963). Sound radiation from arbitrary body, *J.A.S.A.*, **35**, 1626–32.

COLTON, D. (1978). The scattering of acoustic waves by a spherically stratified medium and an obstacle, *SIAM J. Math. Anal.*, **9**, 935–42.

COPLEY, L. G. (1967). Fundamental results concerning integral representations in acoustic radiation, *J.A.S.A.*, **44**, 28–32.

GREENSPAN, D. & WERNER, P. (1966). A numerical method for the exterior Dirichlet problem for the reduced wave equation, *Arch. Rat. Mech. Analy.*, **23**, 288–316.

JAIN, D. L. & KANWAL, R. P. (1970). Acoustic diffraction of a rigid annular disc, *Jour. Eng. Math.*, **4**, 219–28.

JAIN, D. L. & KANWAL, R. P. (1971). *Acoustic diffraction by a rigid annular spherical cap*, ASME 71-APM-CCC, New York.

JONES, D. S. (1974). Integral equations for the exterior acoustic problem, *Quar. Jour. Mech. App. Math.*, **27**, 129–42.

KLEINMANN, R. E. & ROACH, G. F. (1974). Boundary integral equations for the three dimensional Helmholtz equation, *SIAM Review*, **16**, 214–36.

MEYER, W. L., BELL, W. A. & ZINN, B. T. (1977). Integral solution of three dimensional acoustic radiation problems, *Proc. Int. Sym. Innovative Numerical Analysis in App. Eng. Sci., CETIM*, pp. 179–82.

SCHENCK, H. A., (1967). *Formulation and solution of acoustic radiation problems as integral equations—Selected references*, Naval Undersea Warfare Center, San Diego Cal.

SCHENCK, H. A. (1968). Improved integral formulation for acoustic radiation problems, *J.A.S.A.*, **44**, 41–58.

SHAW, R. P. (1975). Time harmonic scattering by circular cylinders with variable boundary conditions, *J. Sound Vibration*, **38**, 417–24.

URSELL, F. (1973). On the exterior problems of acoustics, *Proc. Camb. Phil. Soc.*, **74**, 117–25.

WATERMAN, P. C. (1969). New formulation of acoustic scattering, *J.A.S.A.*, **45**, 1417–29.

WELTON, P. J. (1970). *On the integral formulation of scalar wave scattering by arbitrary surfaces*, Rep. ARL-TM-70-39, App. Res. Lab., U. of Texas at Austin.

Elastic

BANAUGH, R. P. & GOLDSMITH, W. (1963). Diffraction of steady elastic waves by surfaces of arbitrary shape, *J. App. Mech.*, **30**, 589–95.

CRUSE, T. A. (1968). A direct formulation and numerical solution of the general transient elastodynamic problem, II, *Jour. Math. Anal. and Appl.*, **22,** 341–55.

CRUSE, T. A. & RIZZO, F. J. (1968). A direct formulation and numerical solution of the general transient elastodynamic problem, I *Jour. Math. Anal. and Appl.*, **22,** 244–59.

JAIN, D. L. & KANWAL, R. P. (1978). Scattering of P and S waves by spherical inclusions and cavities, *J. Sound Vibration*, **57,** 171–202.

KANWAL, R. P. (1969). Integral equations formulation of classical elasticity, *Q.A.M.*, **27,** 57–65.

SHARMA, D. L. (1967). Scattering of steady elastic waves by surfaces of arbitrary shape, *Bull. Sis. Soc. Amer.*, **57,** 795–812.

Water

DAVIS, A. M. J. & HOOD, M. J. (1976). Surface waves normally incident on a submerged horizontal cylinder, *SIAM J. Appl. Math*, **31,** 16–30.

GARRISON, C. J. & CHOW, P. Y. (1972). Wave forces on submerged bodies, *ASCE*, **98**(WW3), 375–92.

GARRISON, C. J. & SEETHARAMA, R. V. (1971). Interaction of waves with submerged obstacles, *ASCE*, **97**(WW2), 259–77.

HWANG, L. S. & TUCK, E. O. (1970). On the oscillations of harbors of arbitrary shape, *Jour. Fluid Mech.*, **42,** 447–64.

LEE, J. J. (1971). Wave induced oscillations in harbors of arbitrary shape, *Jour. Fluid Mech.*, **45,** 375–94.

LEE, J. J. & RAICHLEN, F. (1970). Resonance in harbors of arbitrary shape, *Proc. 12th Conf. on Coastal Eng., Washington, DC, Chap.* 131, *pp.* 2163–80.

LEE, J. J. & RAICHLEN, F. (1971). *Wave induced oscillations in harbors with connected basins*, Report KH-R-26, Cal. Inst. of Tech.

MEYER, R. E. (1971). Resonance of unbounded water bodies, In: *Mathematical Problems in the Geophysical Sciences*, 1*: Geophysical Fluid Dynamics*, Ed. W. H. Reid, Amer. Math. Soc., Providence, RI.

PROUDMAN, J. (1925). A theorem in tidal dynamics, *Philo. Mag.*, **49,** 570–3.

SHAW, R. P. (1975). Boundary integral equation methods applied to water waves. In: *Boundary Integral Equation Method:* Computational Applications in Applied Mechanics, Eds T. Cruse and F. Rizzo, AMD Vol. 11, ASME, New York, pp. 7–14.

VASTANO & REID, R. O. (1967). A numerical study of the Tsunami response at an island, *J. Mar. Res.*, **25,** 129–39.

Electromagnetic

JAIN, D. L. & KANWAL, R. P. (1971). Electromagnetic diffraction by a thin conducting annular disc, *Jour. Mth. Phy.*, **12,** 723–36.

MORITA, N. (1978). Surface integral representations for electromagnetic scattering from dielectric cylinders, *IEEE Trans., Antennas and Prop.*, **AP-26**(2), 261–6.

Eigenvalues

AGUIRRE-RAMIREZ, G. & WONG, J. P. (1977). Eigenvalues of the Helmholtz equation by the boundary integral equation—Finite element Method, *Proc. Int. Sym. Innovative Numerical Analysis in App. Eng. Sc., CETIM*, pp. 107–10.

DeMey, G. (1976). Calculation of eigenvalues of the Helmholtz equation by an integral equation, *Int. J. Num. Meth. Eng.*, **10,** 59–66.

DeMey, G. (1977). A simplified integral equation method for the calculation of the eigenvalues of the Helmholtz equation, *Int. J. Num. Meth. Eng.*, **11,** 1340–2.

Gregoire, J. P., Nedelec, J. C. & Planchard, J. (1975). A method for computing the eigenfrequencies of an acoustic resonator, *IUTAM Proceedings—Marseille*, pp. 343–52.

Hutchinson, J. R. (1978). Determination of membrane vibrational characteristics by the boundary integral equation methods. In: *Recent Advances in Boundary Element Methods*, Ed. C. A. Brebbia, Pentech Press, London, pp. 301–16.

Tai, G. R. C. & Shaw, R. P. (1974). Helmholtz equation eigenvalues and eigenmodes for arbitrary domains, *J.A.S.A.*, **56**(3), 796–804. Also Rep. No. 90, SUNY, Buffalo, N.Y.

Coupling

Baron, M. L., Bleich, H. H. & Matthews, A. T. (1965). Forced vibrations of an elastic circular cylindrical body of finite length submerged in an acoustic fluid, *Int. J. Solids Structures*, **1,** 3–22.

Shaw, R. P., (1973). Integral equation formulation of dynamic acoustic fluid–elastic solid interaction problems, *J.A.S.A.*, **53,** 514–20.

Shaw, R. P. (1973). An integral equation formulation of acoustic fluid–elastic shell dynamic interaction problems, (presented at IASS Pacific Symposium on Hydromechanically Supported Shells, Honolulu, October 1971). In: *Hydromechanically Supported Shells*, ed. R. Szilard, East West Center Press, Honolulu, Hi, pp. 615–32.

Shaw, R. P. (1974). Time harmonic scattering by obstacles surrounded by an inhomogeneous medium, (presented at the 87th Acoustical Society of America meeting, N.Y.C., April 1974), *J.A.S.A.*, **56,** 1354–60.

Shaw, R. P. (1975). An outer boundary integral equation applied to transient wave scattering in an inhomogeneous medium, (presented at JSME/ASME Applied Mechanics Meeting, Honolulu, Hawaii, March 1975), *J. Appl. Mech.*, **42,** 147–52.

Shaw, R. P. (1978). Coupling boundary integral equation methods to 'other' numerical techniques. In: *Recent Advances in Boundary Element Methods*, Ed. C. A. Brebbia, Pentech Press, London, pp. 137–48.

Shaw, R. P. & Falby, W. (1977). *FEBIE—A Combined Finite Element—Boundary Integral Equation Approach*, CETIM, Senlis, France, pp. 1.97–1.101. Also, *Computers and Fluids* (1978), **6,** 153–60.

Wilkinson, J. P. D. & Liu, H. K. (1970). *Integral-equation techniques for structural-acoustic interaction*, Rep. 70-C-233, R & D Center, Schenectady.

Zienkiewicz, O. C., Kelly, D. W. & Bettess, P. (1977). The coupling of the finite element method and boundary solution procedures, *Int. J. Num. Meth. Eng.*, **11,** 355–75.

Zienkiewicz, O. C., Kelly, D. W. & Bettess, P. (1977). Marriage à la Mode—The best of both worlds (finite elements and boundary integrals), *Proceedings of Int. Symp. on Innovative Numerical Analysis in Applied Engineering Science*, Ed. T. Cruse, J. C. Lachat, F. Rizzo and R. P. Shaw, CETIM, Senlis, France.

Chapter 7

THE BOUNDARY ELEMENT METHOD IN THERMOELASTICITY

F. J. Rizzo and D. J. Shippy

Department of Engineering Mechanics, University of Kentucky, USA

SUMMARY

A modern boundary element formulation and currently used numerical procedures are described for boundary value problems of linear, homogeneous, isotropic, steady-state thermoelasticity theory. Some numerical examples are presented for verification.

7.1 INTRODUCTION

Boundary element methods for the numerical solution of boundary value problems in a variety of areas of physics, applied mathematics, and engineering have achieved a considerable level of development in the last decade. The chapters in this volume and the various references cited are witness to that fact. The purpose of this chapter is to present a contemporary version of such a method for linear, homogeneous, isotropic, steady-state thermoelasticity theory.

The outstanding general feature of a boundary element method is that for many problems where it is used, the dimension of the problem is completely reduced by one. This is true for thermoelasticity also as is indicated here. The problem is formulated in two analytical steps, as shown in Section 7.2. Then currently used numerical procedures are outlined and several illustrative problems are given.

7.2 THE BOUNDARY ELEMENT FORMULATION

Consider an arbitrarily shaped solid body B in a thermomechanical environment which, as a consequence of conditions imposed on its surface S and possibly a body force distribution b_i per unit volume†, suffers a distribution of temperature θ, displacement u_i, and stress σ_{ij} from a uniform reference state. The stress, the displacement and temperature fields throughout the body are frequently of engineering interest.

The approach to such problems in classical thermoelasticity usually involves two steps. (1) Solve a heat transfer problem first to obtain the steady-state θ field based on well-posed thermal boundary conditions on S. (2) Then regard $\theta_{,i}$ as a type of body force which, together with well-posed mechanical boundary conditions, contributes to the u_i and σ_{ij} field in a familiar way (e.g., Boley & Weiner (1960), p. 77), and solve for u_i and/or σ_{ij}.

The boundary element method for such thermoelastic problems recognises the above two steps and begins with an integral formula pertinent to the heat transfer problem in the form (Cruse & Rizzo (1975), p. 1):

$$c(p)\theta(p) = \int_S [\theta'(Q)K(p,Q) - \theta(Q)K'(p,Q)]\,\mathrm{d}S(Q) \qquad (1)$$

In formula (1) the prime (′) indicates the (outward) normal derivative of the function, K is a two-point-dependent kernel function of differing form (see Appendix) according as B pertains to a three-dimensional (E^3) or two-dimensional (E^2) body, Q is a point $\in S$, and p is a point $\in B$ or S. If $p \in B$, $c(p) = 4\pi$ or $c(p) = 2\pi$, respectively, according as the body $B \in E^3$ or $B \in E^2$. If $p \in S$, $c(p)$ is a number dependent only upon the local geometry of S at p. In practice $c(p)$, $p \in S$, is calculated as the integral of K' over S. This is apparent from setting $\theta \equiv 1$ in formula (1). If S has a unique tangent at p, $c(p) = 2\pi$ for $B \in E^3$ and $c(p) = \pi$ for $B \in E^2$ as expected.

The significance of formula (1) is as follows. Let $p \in S$ and regard (1) as a relation among θ and θ' on S pertaining to one and the same temperature

† Standard cartesian tensor analysis is used. Latin subscripts have the range 1,2 or 1,2,3 according as B occupies, respectively, a region of two- or three-dimensional space.

field throughout B. Now since only part of the pair (θ, θ') (e.g θ or θ' but not both) may be specified at any $Q \in S$ according to a well-posed, steady-state, heat transfer problem, formula (1) with $p \in S$ provides, in principle, a means of obtaining that part of the pair (θ, θ') not initially prescribed. That is, we regard formula (1) with $p \in S$ as an integral equation to be solved for the unprescribed parts of the pair (θ, θ') on S. Once (θ, θ') are completely known everywhere on S, consider $p \in B$ in (1), and that formula yields θ at any point in the interior of B in terms of the indicated integral over S of known quantities. Thus, assuming the above-described process can be carried out, and the method for doing so *numerically* will be indicated subsequently, Step (1) in classical thermoelasticity has been completed.

Step (2) is similar to Step (1) and involves the following formula (Rizzo & Shippy, 1977) pertinent to the thermoelastic problem:

$$
\begin{aligned}
c_{ij}(p)u_i(p) = & \int_S [\{t_i(Q) - \psi(Q)n_i(Q)\}U_{ij}(p, Q) - u_i(Q)T_{ij}(p, Q)]\,\mathrm{d}S(Q) \\
& + \int_S [\{\psi(Q) + \gamma\theta(Q)\}W'_{,j}(p, Q) - \{\psi(Q) + \gamma\theta(Q)\}' \\
& \times W_{,j}(p, Q)]\,\mathrm{d}S(Q) + k_0 \int_S n_j(Q)W(p, Q)\,\mathrm{d}S(Q) \qquad (2)
\end{aligned}
$$

In formula (2) U_{ij}, T_{ij} and W are two-point-dependent kernel functions of differing forms (see Appendix) according as $B \in E^3$ or $B \in E^2$, $t_i(Q)$ is the surface traction, $r(p, Q) \equiv |Q - p|$, $n_j(Q)$ are unit (outward) normal components, $\psi(Q)$ is the surface value of a function ψ defined throughout B such that the mechanical body force, if any, is derivable from ψ according to

$$b_i = -\psi_{,i} \qquad \text{with } \psi_{,ii} = k_0 \qquad (3)$$

where k_0 is a constant, as is γ (see Appendix). In formula (2) p may be $\in B$ or S and the (′) signifies normal derivative as before. If $p \in S$, $c_{ij}(p)$ is a matrix of numbers determined in a manner similar to that for determining $c(p)$ in formula (1) (see e.g., Rizzo & Shippy (1977), eqns. (1), (14), (15)). If $p \in B$, $c_{ij}(p) \equiv \delta_{ij}$.

The significance of formula (2) is as follows. Let $p \in S$ and regard (2) as a relation among u_i and t_i on S and the functions $\theta, \theta', \psi, \psi'$ also on S, all of which pertain to one and the same thermoelastic stress state throughout B. Now θ and θ' are known on S following completion of step 1 as previously described. (Note that setting $p \in B$ and solving for interior θ with formula (1) is not required in using formula (2)). Variables ψ and ψ' are regarded as known boundary quantities obtainable from a potential for real mechanical body force. What is not known completely, of course, is the pair (u_i, t_i) on S, since only part of the pair can be prescribed corresponding to well-posed boundary conditions. For example, u_i may be prescribed on S (displacement type problem), or t_i may be prescribed on S (traction type problem), or parts of each may be prescribed over portions of S (various kinds of mixed problems), etc. Thus, the solution strategy via formula (2) with $p \in S$ is to solve the integral equation for that part of the pair (u_i, t_i) not initially prescribed. Once (u_i, t_i) are completely known everywhere on S, consider $p \in B$ in (2) and that formula yields u_i at any point in the interior of B in terms of the indicated integrals over S of known quantities. Finally, the stress components σ_{ij} at any $p \in B$ are given via formula (2) by first obtaining the displacement gradient $u_{i,k}$ at $p \in B$. This is done simply by differentiating the kernel function expressions with respect to x_k at p and evaluating the indicated integrals of the differentiated kernel functions. Based on Hooke's law, the stress field σ_{ij} is given in terms of $u_{i,j}$ according to

$$\sigma_{ij} = \frac{2\mu v}{1 - 2v} u_{m,m}\delta_{ij} + \mu[u_{i,j} + u_{j,i}] - \delta_{ij}\gamma\theta \tag{4}$$

wherein $p \in B$. If σ_{ij} on S is desired, it may be obtained directly in terms of (θ, u_i, t_i) on S (actually θ, t_i and 'in-surface' derivative(s) of u_i on S are required) in a familiar fashion (see e.g., Lachat, 1975; Cruse & Van Buren, 1971). Thus for *surface* σ_{ij} there is no need to refer to the interior p.

This completes the analytical formulation of the two-step thermoelastic solution process via the boundary element method.

7.3 NUMERICAL PROCEDURES

To implement the solution strategy outlined above, we first discretise the surface S of B, $B \in E^3$, into a total number m of surface elements of either quadrilateral or triangular shape as shown in Fig. 1(a) and locate eight or six nodes, respectively, on each element. The cartesian co-ordinates x_i^α of each node are specified, and the cartesian co-ordinates of a non-nodal point of an element are assumed to be given by (Rizzo & Shippy, 1977; Lachat & Watson, 1976)

$$x_i(\xi) = M^\alpha(\xi)x_i^\alpha \qquad \alpha = 1, 2 \ldots, 6 \text{ or } 8 \tag{5}$$

in which $M^\alpha(\xi)$ are second-order shape functions of intrinsic co-ordinates

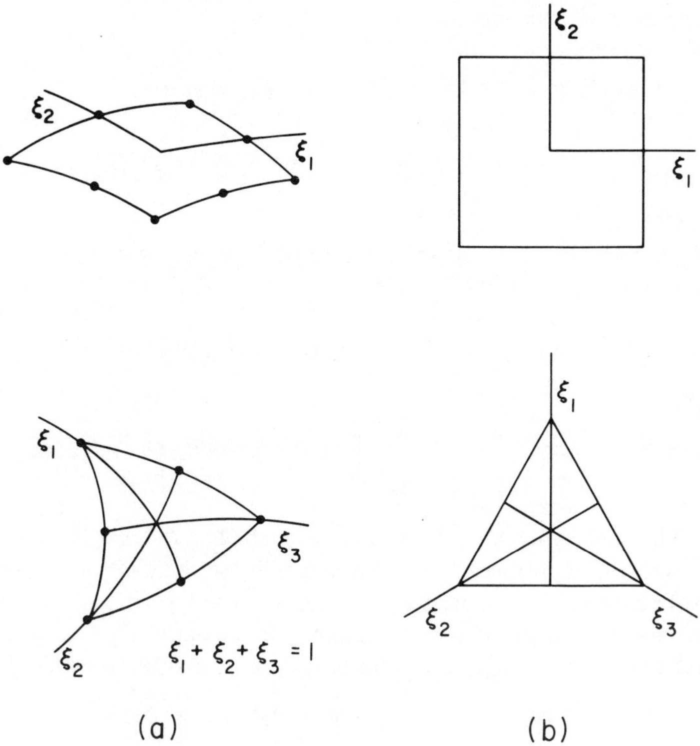

FIG. 1. Quadrilateral and triangular surface elements.

$(\xi) \equiv (\xi_1 \xi_2)$. We use one set M^{α}_{R} for quadrilateral shapes and another set M^{α}_{T} for triangular shapes (see Appendix). Next, we assume that any of the functions of Q alone in eqns. (1) and (2) varies over the elements according to

$$\phi(\xi) = M^{\alpha}(\xi)\phi^{\alpha} \tag{6}$$

where ϕ^{α} is the nodal value of the particular variable at issue and the shape functions are identical to those used in eqn. (5). Based on the discretising assumptions, formula (1) becomes

$$c(P_{\eta})\theta(P_{\eta}) = \theta'^{\alpha\sigma} \int_{S_{\sigma}} M(\xi) K(P_{\eta}, Q(\xi)) J(\xi)\,\mathrm{d}\xi - \theta^{\alpha\sigma} \int_{S_{\sigma}} M(\xi) K'(P_{\eta}, Q(\xi)) J(\xi)\,\mathrm{d}\xi \tag{7}$$

and formula (2) becomes

$$
\begin{aligned}
c_{ij}(P_\eta)u_i(P_\eta) = \Omega_i^{\alpha\sigma} & \int_{S_\sigma} M^\alpha(\xi) U_{ij}(P_\eta, Q(\xi)) J(\xi)\,\mathrm{d}\xi \\
& - u_i^{\alpha\sigma} \int_{S_\sigma} M^\alpha(\xi) T_{ij}(P_\eta, Q(\xi)) J(\xi)\,\mathrm{d}\xi \\
& + \mathrm{Y}^{\alpha\sigma} \int_{S_\sigma} M^\alpha(\xi) W'_{,j}(P_\eta, Q(\xi)) J(\xi)\,\mathrm{d}\xi \\
& - \mathrm{Y}'^{\alpha\sigma} \int_{S_\sigma} M^\alpha(\xi) W_{,j}(P_\eta, Q(\xi)) J(\xi)\,\mathrm{d}\xi \\
& + k_0 \sum_{\sigma=1}^{m} \int_{S_\sigma} n_j(Q(\xi)) W(P_\eta, Q(\xi)) J(\xi)\,\mathrm{d}\xi
\end{aligned} \tag{8}
$$

in which it is assumed $p \equiv P \in S$, and in which $\Omega_i = t_i - \psi n_i$; $\mathrm{Y} \equiv \psi + \gamma\theta$; and where the quantities $\phi^{\alpha\sigma}$ have the meaning 'the value of ϕ at local node α on surface element S_σ'. Note that $\sigma = 1, 2, \ldots, m$ and $\eta = 1, 2, \ldots, n$, where n is the total number of nodes; summation is implied on all repeated subscripts; $J(\xi)$ is simply the ratio $\mathrm{d}S(Q)/\mathrm{d}\xi$, which more explicitly is

$$J(\xi) = |\mathbf{a}_1 \times \mathbf{a}_2| \tag{9}$$

where the components of the vectors $\mathbf{a}_\mu$ in the cross product are given by

$$a_{\mu i} \equiv \frac{\partial M^\alpha}{\partial \xi_\mu} x_i^\alpha \tag{10}$$

Note that since a node always lies on at least two surface elements and frequently more, $\phi^{\alpha\sigma}$ is the same quantity for several sets of values of α and σ. This little book-keeping problem, when addressed, leads to viewing the system of eqns. (7) and (8) in the following matrix forms, respectively:

$$\mathbf{F}\boldsymbol{\theta} = \mathbf{G}\boldsymbol{\theta}' \tag{11}$$

$$\mathbf{Au} - \mathbf{B\Omega} = \mathbf{CY} - \mathbf{DY}' + \mathbf{E} \tag{12}$$

In eqn. (11) the column matrices $\boldsymbol{\theta}$ and $\boldsymbol{\theta}'$ are n elements long and contain the elements $\theta^\beta, \theta'^\beta, \beta = 1, 2, \ldots n$, which are, respectively, temperature and normal derivative of temperature at global node number β. The square matrices **F** and **G** are $n \times n$ in size and contain the integral coefficients (and the $c(P_\eta)$ term) as indicated in eqn. (7).

In eqn. (12), the column matrices **u** and **Ω** are $3n$ elements long and contain the elements $u_i^\beta, \Omega_i^\beta, \beta = 1, 2, \ldots, n$, which are, respectively, displacement and 'augmented' traction components at global node number β; **Y** and **Y′** are n elements long and contain Y^β, Y'^β, i.e., the surface contribution of temperature and body force potential and normal derivatives of these quantities, as indicated above, at the nodes. The square matrices **A** and **B** are $3n \times 3n$ elements in size and contain the integral coefficients of **u** and **Ω** as indicated in eqn. (8). Similar remarks hold for the $n \times 3n$ size matrices **C** and **D**. The column **E** contains the $3n$ values of the last integral in eqn. (8) as indicated.

The previously outlined solution strategy when referred to eqns. (11) and (12) above amounts to the following. Discretise the surface S of B into a number m of surface elements and form and store the matrices **F** and **G**. Specify an appropriate combination of elements $\boldsymbol{\theta}$ and $\boldsymbol{\theta}'$ at the nodes corresponding to a well-posed heat transfer problem. Perform the indicated matrix multiplications in (11) involving specified quantities and solve the system (11) for the elements of $\boldsymbol{\theta}, \boldsymbol{\theta}'$ not previously specified. At this point, the boundary set $\boldsymbol{\theta}, \boldsymbol{\theta}'$ are completely known. (If desired, although not necessary for Step (2) in the analysis, we could take $p \in B$ in formula (7) and generate the temperature θ at any desired interior p.)

To begin Step (2), form and store the matrices **A, B, C, D, E**. Specify the elements **Y** and **Y′** which involve the now-known nodal values of the set $\boldsymbol{\theta}, \boldsymbol{\theta}'$ and also (assumed) known nodal values of the body force potential ψ and its normal derivative ψ'. Next, perform the indicated matrix multiplications to yield a single known vector on the right side of eqn. (12). Then specify a sufficient number of the elements **u** and **t** corresponding to a well-posed elastostatics problem. Perform the remaining indicated matrix multiplications in (12) involving known or specified quantities at this stage, and solve the system (12) for the remaining unknowns. Now the set **u, t** are completely known such that the full complement of surface quantities $\mathbf{u}, \mathbf{t}, \boldsymbol{\theta}, \boldsymbol{\theta}', \boldsymbol{\psi}, \boldsymbol{\psi}'$ pertaining to the desired thermoelastic stress state in $\boldsymbol{B}$ are available at the nodes.

Putting $p \in B \neq P_\eta$ in formula (8) now would yield $u_i(p)$ at any desired p. $\sigma_{ij}(p)$ would be obtained similarly by first obtaining $u_{i,j}(p)$ from formula (8) as previously indicated and then using Hooke's law (4). Boundary σ_{ij} are obtainable in terms of $\boldsymbol{\theta}, \mathbf{u}, \mathbf{t}$ as indicated earlier.

For $B \in E^2$, i.e., a plane problem, S is a curve or the union of several curves, and most of the preceding remarks apply intact with the appropriate reduction in dimension. Note that S_σ is now an element of boundary curve with nodes at each end and one (usually midway) between. There are now three shape functions $M^{\alpha}(\xi)$, $\alpha = 1, 2, 3$, which have the form shown in the Appendix, and ξ is a single scalar variable. Matrices which were formerly $3n$ or $3n \times 3n$ in size are now $2n$ or $2n \times 2n$, respectively.

Having outlined the solution process for thermoelastic boundary value problems via the boundary element method above in some detail, it is apparent that success with the method depends, in large part, on being able to obtain sufficiently accurate values for the elements of the matrices **A**, **B**, **C**, **D**, **E**, **F**, **G**. Examination of eqns. (7) and (8) reveals that the mentioned matrix elements are sums of integrals of the form

$$I^{\alpha}_{\eta\sigma} \equiv \int_{S_\sigma} M^{\alpha}(\xi)\Gamma(P_\eta, Q(\xi))J(\xi)\,\mathrm{d}\xi \tag{13}$$

in which S_σ is a square or equilateral triangle as shown in Fig. 1(b) (or the interval -1 to 1 on the straight line for $B \in E^2$), since a mapping to those shapes is inherent in the change of variables implied by the introduction of the shape functions (5). Γ is any of the tensor, vector or scalar kernel functions indicated in eqns. (7) and (8). The quantities $c(P_\eta)$, $c_{ij}(P_\eta)$ in eqns. (7) and (8) respectively prove also to be the sum of certain integrals of the type (13) (see Rizzo & Shippy, (1977), eqns. (14) and (15) for details). Thus, it is necessary to obtain a sufficiently accurate value for all $I^{\alpha}_{\eta\sigma}$ by some numerical integration scheme. For a given choice of surface discretisation (number m and arrangement of surface elements and number of nodes n), the manner of obtaining values for $I^{\alpha}_{\eta\sigma}$ is perhaps the most important issue in the solution process. Total computer (CPU) time for a given problem is largely a function of how this is done.

To expand on this matter for $B \in E^3$, note first that two distinct cases in the evaluation of $I^{\alpha}_{\eta\sigma}$ arise: (a) $P_\eta \notin S_\sigma$, and (b) $P_\eta \in S_\sigma$. In case (b), note that since certain of the kernel functions are $0(1/r^2)$ and others at least $0(1/r)$, the integrands, while integrable, require special attention.

The first case, (a), is more straightforward. For integration over the square we use a Gaussian quadrature formula with weight function 1·0 to obtain

$$I^{\alpha}_{\eta\sigma} = \int_{S_\sigma} F^{\alpha}(P_\eta, \xi)\,\mathrm{d}\xi \simeq \sum_{\mu=1}^{\lambda_1}\sum_{\nu=1}^{\lambda_2} A^{(\lambda_1)}_{\mu} A^{(\lambda_2)}_{\nu} F^{\alpha}(P_\eta, \xi^{(\lambda_1)}\xi^{(\lambda_2)}) \tag{14}$$

where

$$F^{\alpha}(P_\eta, \xi) \equiv M^{\alpha}(\xi)\Gamma(P_\eta, Q(\xi))J(\xi) \tag{15}$$

and $A^{(\lambda_1)}_{1}$, $A^{(\lambda_2)}_{\nu}$ are weight factors corresponding to the order λ_1, λ_2 chosen for the quadrature formula (Stroud & Secrest, 1966). For integration over the triangle we use a special triangularly symmetric quadrature rule by Lyness & Jespersen (1975), which symbolically may be written in the same form as (14). In either case, accuracy is in principle a function of the order (size of λ_1, λ_2) of the quadrature rule but so also is the cost (CPU time), so that proper choice of the pair λ_1, λ_2 is an important practical issue for a given integration. After considerable study and experimentation (details will be available in Rizzo & Shippy (in preparation)) we have automated the choice of λ_1, λ_2 as a function primarily of the ratio of the distance of the point P_η to some characteristic size of the element S_σ. Secondarily, we increase the order λ_1, λ_2 for a relatively 'distorted' element over and above the order otherwise called for by the 'distance' criterion. Both criteria are an attempt to choose λ_1, λ_2 just large enough for a given integrand to give values of integrals, via formula (14), with consistent error.

For case (b), where $P_\eta \in S_\sigma$, we first take account of the fact that $c(P_\eta)$ and $c_{ij}(P_\eta)$ are calculated in a manner (Rizzo & Shippy, (1977), eqns. (14) and (15)) which analytically requires, for Γ of $0(1/r^2)$, that we evaluate

$$\hat{I}^{\alpha}_{\eta\sigma} \equiv \int_{S_\sigma} (M^\alpha(\xi) - \delta_{\alpha\eta})\Gamma(P_\eta, \xi)J(\xi)\,\mathrm{d}\xi \tag{16}$$

with $\delta_{\alpha\eta} = 1$ whenever node $\alpha \equiv P_\eta$; otherwise, $\delta_{\alpha\eta} = 0$. Thus, if we imagine a local polar co-ordinate system ρ, ϕ as shown in Fig. (2) for typical locations of P_η, we note that $M^\alpha(\xi)$ is $0(\rho)$ for $\alpha \neq P_\eta$ and that $(M^\alpha(\xi) - 1)$ is of the same order for $\alpha = P_\eta$. Thus since $\Gamma(P_\eta, \xi)$ is $0(1/\rho^2)$ and $\mathrm{d}\xi$ becomes $\rho\,\mathrm{d}\rho\,\mathrm{d}\phi$, the integrand in (16) is no longer improper and may be evaluated using the same types of quadrature rules previously indicated.

In case (b), since P_η occupies one of the nodes, the 'distance' criteria for choice of λ_1, λ_2 no longer has meaning and a threshold choice is made according to a pre-selected accuracy requirement. Finally, as with the $P_\eta \notin S_\sigma$ case, we can increase the order λ_1, λ_2 for a relatively 'distorted' element if necessary.

For $B \in E^2$, for $P_\eta \notin S_\sigma$, a one-dimensional version of formula (14) is used. Here, the choice of order of integration (size of λ) as a function of the proximity of P_η to S_σ is less critical than in three dimensions because of the comparatively fewer total integrations to be done. Hence, we presently pick a threshold size for λ (probably too high for most integrations) and compute all $P_\eta \notin S_\sigma$ integrals for constant λ. Efficiency measures, nevertheless, could be introduced as for three dimensions. Regarding the $P_\eta \in S_\sigma$ case, note that $M^\alpha(\xi) - \delta_{\alpha\eta}$ is $0(\xi)$ and since the kernel functions are at most $0(1/\xi)$ no transformation comparable to the polar co-ordinate

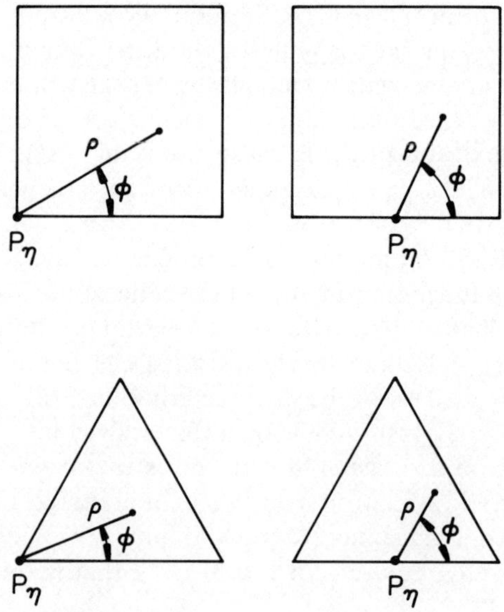

FIG. 2. Polar co-ordinate system for the case when $P_\eta \in S_\sigma$.

transformation is necessary in two dimensions. There is the problem, however of one integrand of 0(log ξ) which can be integrated easily enough using a 'log' weighted Gauss or by removing the singularity (Lachat, 1975; Wu, 1976; Fairweather *et al.*, 1979).

Based on the foregoing discussion, numerical values of all elements of the coefficient matrices **A**, **B**, etc., can be calculated automatically by a digital computer upon specifying the cartesian co-ordinates of all the nodes on the surface S of a body under investigation, the material parameters, and the quadrature parameters λ_1, λ_2.

7.4 SOME EXAMPLE PROBLEMS

Consider a hollow cylinder with inner and outer radii, a and b, respectively, with height h. A heat transfer problem is posed for the cylinder assuming the inner and outer surfaces are in contact with fluids of different temperatures so that on these surfaces we have thermal boundary conditions of the form

$$(k/h_c)\theta' = (\theta_0 - \theta) \tag{17}$$

in which k, h_c and θ_0 are known thermal conductivity and surface heat

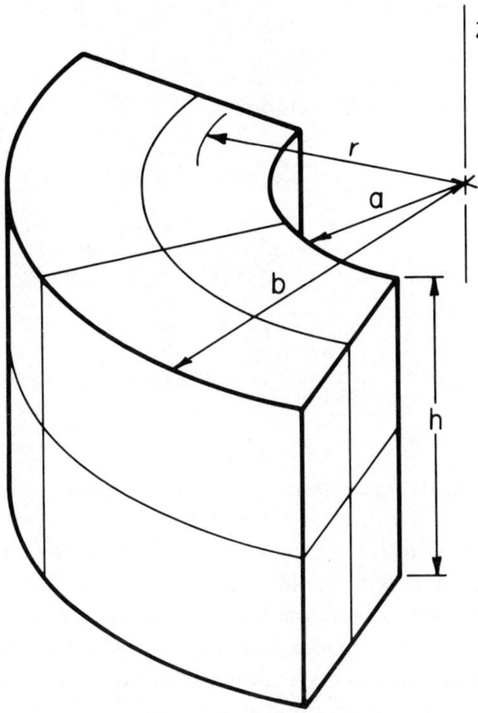

FIG. 3. Descretisation pattern for hollow cylinder, modified boundary conditions, zero normal displacement and zero shear traction on 'cut surfaces'.

transfer coefficients and fluid immersion temperatures, respectively. On the ends of the cylinder we assume $\theta' = 0$.

We model one quarter of the cylinder as shown in Fig. 3 with $m = 24$ boundary elements and $n = 74$ nodes. Using $a = 0{\cdot}4$, $b = 1{\cdot}0$, $h = 1{\cdot}0$ with $k/h_c = [ab/(a+b)]\ln(b/a) = 0{\cdot}2618$, and $\theta_0 = 0$ and $1{\cdot}0$ inside and outside, respectively, numerical results for thermal quantities compared with exact analytical values are as shown in Table 1.

For Step (2) of the analysis, the ends of the cylinder are held to zero normal displacement, while the curved lateral surfaces are traction free. (See Boley & Weiner, (1960), p. 290 for the specific analytical displacement field and associated thermoelastic stress field with which we compare results.) Using convenient values for all relevant parameters, numerical results for thermal stress and displacement compared with expected values are as shown in Table 2.

TABLE 1

THERMAL DATA FOR HOLLOW CYLINDER

Co-ordinate ρ	*Temperature* θ		*Flux* $\frac{\partial \theta}{\partial n}$	
	BEM	*Exact*	*BEM*	*Exact*
0·40	0·3572	0·3571	−1·3645	−1·3642
0·55	0·5312	0·5309	—	—
0·70	0·6626	0·6625	—	—
0·85	0·7689	0·7685	—	—
1·00	0·8571	0·8571	0·5456	0·5457

Our next example involves a portion of the cross-section of a film-cooled aircraft engine turbine blade as depicted in Fig. 4. The outer surface of the blade is in contact with hot combustion gases, while the inner (circular) chambers are in contact with cooling air pumped from a compressor. Of special interest is a sequence of small cooling holes bored from the outer blade surface at a sharp angle α (see Fig. 6) to the left-most circular chamber. This cooling hole is depicted with dashed lines in Fig. 4 since the cross-section of Fig. 4 is taken midway between cooling holes. Under the described circumstances the turbine blade is in a state of thermomechanical stress, and the level of stress in the vicinity of the 'feather edge' of the cooling hole is of particular interest. Analysis of this problem is the subject of a current research project at the University of Kentucky. The work is not yet finished but the following observations are perhaps of interest and provide

TABLE 2

STRESS AND DISPLACEMENT DATA FOR HOLLOW CYLINDER

Co-ordinate ρ	*Radial displacement*		*Transverse stress*		*Axial stress*	
	BEM	*Exact*	*BEM*	*Exact*	*BEM*	*Exact*
0·40	0·340	0·340	0·413	0·430	−0·263	−0·250
0·55	0·418	0·418	0·131	0·127	−0·476	−0·481
0·70	0·539	0·539	−0·044	−0·041	−0·662	−0·657
0·85	0·685	0·685	−0·152	−0·153	−0·797	−0·798
1·00	0·850	0·849	−0·236	−0·237	−0·918	−0·916

contemporary examples of how the thermoelastic boundary element solution process is being used.

Values of k, h_c, θ_0 were given by the engine manufacturer on a scale drawing from which we were able to state a thermal boundary condition of the type (17) for each part of each boundary shown in Fig. 4. Note that the cooling hole, being below or above the cross-section of Fig. 4, is not part of the boundary for this part of the analysis, which is planar. Using the

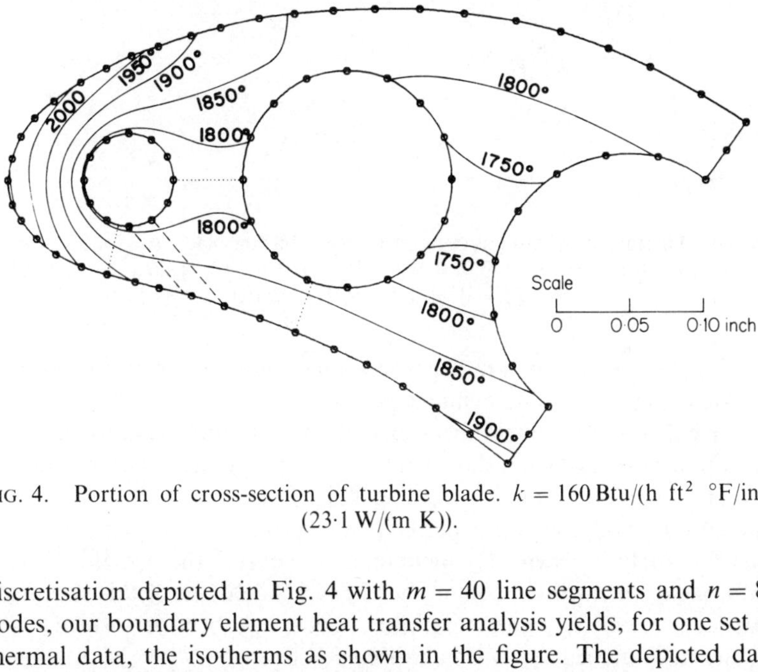

FIG. 4. Portion of cross-section of turbine blade. $k = 160$ Btu/(h ft^2 °F/in.), (23·1 W/(m K)).

discretisation depicted in Fig. 4 with $m = 40$ line segments and $n = 80$ nodes, our boundary element heat transfer analysis yields, for one set of thermal data, the isotherms as shown in the figure. The depicted data compares favourably with the manufacturer's temperatures obtained by other means.

Now assuming in-plane tractions of the body in Fig. 4 to be zero, we calculate, through Step (2) of the analysis, the stress distributions across the three dotted lines in Fig. 4. Results and material properties are shown in Fig. 5. Distribution of normal stress is shown in detail. Directions of shear stresses, which are an order of magnitude smaller than the normal stresses, are indicated by a single arrow on each 'cut' surface. Thus, the major effect of the thermal field in the blade cross-section is to produce bending, via the three 'thermal couples' shown in Fig. 5, on the piece of the blade indicated. While we have only sketchy stress data from the manufacturer with which

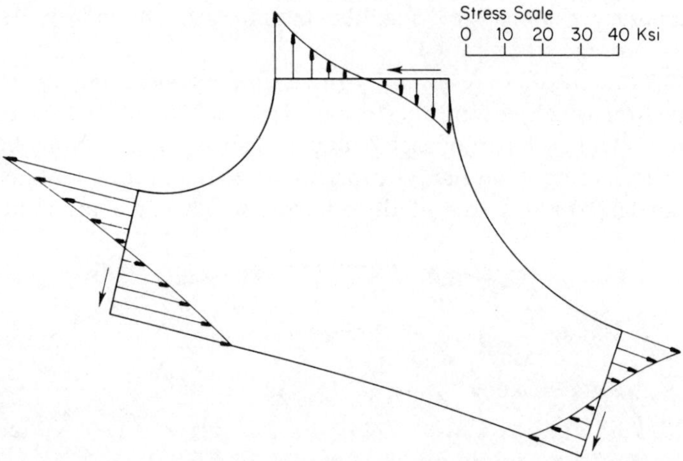

FIG. 5. Thermal-stress distributions on a piece of turbine blade. $E = 24{\cdot}3 \times 10^6$ psi, $(1{\cdot}68 \times 10^8\,\mathrm{kN/m^2})$; $\nu = 0{\cdot}333$; $\alpha = 9{\cdot}2 \times 10^{-6}$ in./(in. °F), $(1{\cdot}66 \times 10^{-5}\,\mathrm{m/m\,K})$); $k_0 = 0$. (1 Ksi $= 6895\,\mathrm{kN/m^2}$.)

to compare, our results are believable when compared with the essential features of the foregoing cylinder problem.

Consider now the three-dimensional piece of the blade as shown in Fig. 6, the top surface of which is shown in Fig. 5, and the bottom surface of which is a cross-section through a cooling hole. Regarding Fig. 6 as a three-dimensional body, we may discretise the surface(s), and perform a heat transfer analysis (Step (1)) including, of course, the specific thermal influence of the inclined cooling hole. Several analyses of this type using various discretisations and sets of thermal data have been performed and all results have agreed, to within expected tolerances, with manufacturers accepted thermal data.

The remaining step, Step (2), is, at the time of writing, being taken. This involves the approximation that the thermal couples of Fig. 5 are uniform through the thickness of the 'piece' of Fig. 6 such that these couples, together with the already determined θ field through the piece of Fig. 6, are responsible for the stress at the edges of the cooling hole. Thus we are in the process of experimenting with discretisations and various sets of thermal data and hole configurations (angles α in Fig. 6) to determine the thermal stresses along the edges, particularly at the feather edge, of the cooling hole.

We are confident in the outcome of our study of the body depicted in Fig.

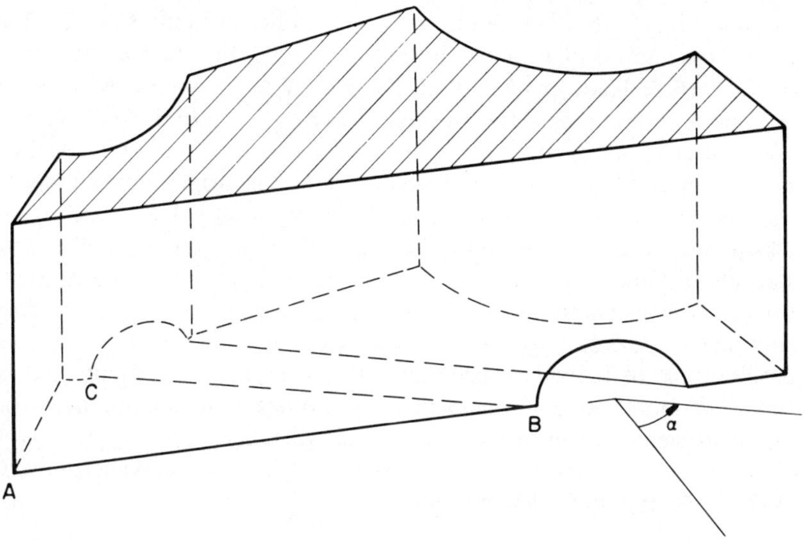

FIG. 6. Three-dimensional piece of turbine blade, showing cooling hole.

6, at least insofar as the integrity of our boundary element analysis computer programs is concerned, since an *isothermal* analysis simulating the effect of the centrifugal field has been successfully completed. This study yielded a mechanical stress concentration, at the feather edge of the cooling hole for the most extreme case of $\alpha = 60°$, of slightly less than 8·0. This result agrees with the manufacturer's isothermal expectations and our best expectations from the literature (Peterson, 1974; Ellyin, 1970; Ellyin & Izmiroglu, 1973).

7.5 SOME REMARKS

All computations were performed on an IBM 370-165 digital computer using single-precision arithmetic. The program and all data for a given problem (Step (1) or (2)) were retained in core. Note that only the matrix of coefficients of unknowns at the nodes need be formed and stored in a square array. The other elements of matrices of coefficients may be immediately multiplied by other quantities as they are generated and stored in column vectors. Individual coefficient arrays need not be formed and stored separately as implied earlier.

Regarding CPU times, our programs in their present form on the IBM machine required 13 s for the thermal solution and 81 s for the thermomechanical solution for the hollow cylinder. For the plane problem of the turbine blade cross-section, it required 5 s to obtain the thermal boundary solution plus the isotherms and 54 s to obtain the thermomechanical boundary solution plus the stress patterns of Fig. 5. However, we have not given full attention to efficiency measures in programming, order of approximate integration, type of algebraic equation solvers, etc., all of which can have a measurable effect on CPU times. Lachat (1975) and Lachat & Watson (1976) discuss some relevant optimisation and time-saving efficiency measures. Such matters are treated also in a recent volume (Brebbia, 1978). In any case, it appears that whatever the stage of development of individual programs, the boundary element method is becoming important and competitive with other numerical methods for many problems. It will be interesting to see what effect another decade of research will have on it alone and its marriage (e.g., Zienkiewicz *et al.* (1977)) with the finite element method.

ACKNOWLEDGEMENT

We gratefully acknowledge support of this investigation by the United States Air Force Office of Scientific Research under Grant No. AFOSR-75-2824.

REFERENCES

Boley, B. A. & Weiner, J. H. (1960). *Theory of Thermal Stresses*, Wiley, New York.

Brebbia, C. A. (Ed.) (1978). *Recent Advances in Boundary Element Methods*, Pentech Press, London.

Cruse, T. A. & Rizzo, F. J. (Eds), (1975). Boundary integral equation method: computational applications in applied mechanics, *ASME Proc. AMD*. Vol. 11, New York.

Cruse, T. A. & Van Buren, W., (1971). *Int. J. Frac. Mech.* **7,** 1.

Ellyin, F. (1970). *Exp. Mech.*, **10,** 195.

Ellyin, F. & Izmiroglu, U. M. (1973). *Trans. ASME, J. Eng. Ind.*, **95,** 151.

Fairweather, G., Rizzo, F. J., Shippy, D. J. & Wu, Y. S. (1979). On the numerical solution of two-dimensional potential problems by an improved boundary integral equation method, *J. Comp. Phys.* (in press).

Lachat, J. C. (1975). *A further development of the boundary integral technique for elastostatics*, Ph.D. Dissertation, Univ. of Southampton.

Lachat, J. C. & Watson, J. O. (1976). *Int. J. Num. Meth. Eng.*, **10,** 991.

LYNESS, J. N. & JESPERSEN, D. (1975). *J. Int. Maths. Applics.*, **15,** 19.
PETERSON, R. E. (1974). *Stress Concentration Factors*, Wiley, New York.
RIZZO, F. J. & SHIPPY, D. J. (1977). *Int. J. Num. Meth. Eng.*, **11,** 1753.
RIZZO, F. J. & SHIPPY, D. J. (in preparation). On the numerical integration of certain functions pertinent to the boundary integral method for three-dimensional elasticity.
STROUD, A. H. & SECREST, D. (1966). *Gaussian Quadrature Formulas*, Prentice-Hall, New York.
WU, Y. S. (1976). *The boundary integral equation method using various approximation techniques for problems governed by Laplace's equation*, Special Scientific Report AFOSR-TR-76-1313, Washington, DC.
ZIENKIEWICZ, O. C., KELLEY, D. W. & BETTESS, P. (1977). *Proceedings of the international symposium on innovative numerical analysis in applied engineering science*, Versailles, France, p. 19.

APPENDIX

Fundamental Kernel Functions

(i) $B \in E^3$

$$K = 1/r, \qquad K' = \frac{\partial}{\partial n}(1/r)$$

$$U_{ij} = \frac{(1+\nu)}{8\pi E(1-\nu)}\left(\frac{1}{r}\right)[(3-4\nu)\delta_{ij} + r_{,i}r_{,j}]$$

$$T_{ij} = \frac{-(1-2\nu)}{8\pi(1-\nu)}\left(\frac{1}{r^2}\right)\left[\frac{\partial r}{\partial n}\left\{\delta_{ij} + \frac{3}{(1-2\nu)}r_{,i}r_{,j}\right\} + r_{,i}n_j - r_{,j}n_i\right]$$

$$W = \frac{(1-2\nu)(1+\nu)}{8\pi E(1-\nu)}r$$

(ii) $B \in E^2$ (plane strain)

$$K = -\log r, \qquad K' = -\frac{\partial}{\partial n}\log r$$

$$U_{ij} = \frac{-(1+\nu)}{4\pi E(1-\nu)}[(3-4\nu)\log r\,\delta_{ij} - r_{,i}r_{,j}]$$

$$T_{ij} = \frac{-(1-2\nu)}{4\pi(1-\nu)}\left[\frac{\partial}{\partial n}\log r\left\{\delta_{ij} + \frac{2}{(1-2\nu)}r_{,i}r_{,j}\right\} + (\log r)_{,i}n_j - (\log r)_{,j}n_i\right]$$

$$W = \frac{-(1-2\nu)(1+\nu)}{8\pi E(1-\nu)}r^2(\log r - 1)$$

Material Parameters

E is Young's Modulus, ν is Poisson's ratio
k_0 is the Laplacian of the body force potential
$\Upsilon = E\alpha/(1 - 2\nu)$, α is the coefficient of thermal expansion

$$\mu = \frac{E}{2(1 + \nu)}$$

Shape Functions

(i) Quadrilateral

$$M_R^1 = \tfrac{1}{4}(\xi_1 + 1)(\xi_2 + 1)(\xi_1 + \xi_2 - 1), \qquad M_R^5 = \tfrac{1}{2}(\xi_1 + 1)(1 - \xi_2^2)$$
$$M_R^2 = \tfrac{1}{4}(\xi_1 - 1)(\xi_2 + 1)(\xi_1 - \xi_2 + 1), \qquad M_R^6 = \tfrac{1}{2}(\xi_2 + 1)(1 - \xi_1^2)$$
$$M_R^3 = \tfrac{1}{4}(1 - \xi_1)(\xi_2 - 1)(\xi_1 + \xi_2 + 1), \qquad M_R^7 = \tfrac{1}{2}(\xi_1 - 1)(\xi_2^2 - 1)$$
$$M_R^4 = \tfrac{1}{4}(\xi_1 + 1)(\xi_2 - 1)(\xi_2 - \xi_1 + 1), \qquad M_R^8 = \tfrac{1}{2}(1 - \xi_2)(1 - \xi_1^2)$$

(ii) Triangular

$$M_T^1 = \xi_1(2\xi_1 - 1), \qquad M_T^4 = 4\xi_1\xi_3$$
$$M_T^2 = \xi_2(2\xi_2 - 1), \qquad M_T^5 = 4\xi_1\xi_2 \qquad \xi_1 + \xi_2 + \xi_3 = 1$$
$$M_T^3 = \xi_3(2\xi_3 - 1), \qquad M_T^6 = 4\xi_2\xi_3$$

(iii) One-dimensional

$$M^1 = -\tfrac{1}{2}(\xi - \xi^2), \qquad M^2 = 1 - \xi^2, \qquad M^3 = \tfrac{1}{2}(\xi + \xi^2).$$

Chapter 8

THE BOUNDARY ELEMENT METHOD FOR PLATES AND SHELLS

H. TOTTENHAM

Hume, Tottenham and Bennett, Romsey, Hampshire, UK

SUMMARY

The various boundary integral methods for the analysis of plates in bending are discussed. A physical representation is used. It is seen that by using the reciprocal theorem, based on energy considerations, the higher order singularities present no problem. A similar approach is described for the bending of plates on elastic (Winkler) foundations and for very shallow shells.

A different treatment is outlined for the general shallow shell, here the stress function and normal displacement, together with their derivatives, are used as boundary variables.

Some results of numerical computations are given, which illustrate the various techniques described.

8.1 INTRODUCTION

Although the boundary integral equation method for the analysis of elastic plates has long been established, it is only in recent years that engineers have turned their attention to it. This delay in finding general practical application has been due, no doubt, to two main factors: first the classical integral methods were formulated in a mathematical rather than an engineering manner, and second because the development of the digital computer corresponded in time with the development of the finite element method.

After the pioneering work of Fredholm and Lauricella (see, e.g., Fredholm, 1900; Lauricella, 1909) at the beginning of this century no significant advance was made until the work of Muskhelishvili was published. The classical book of Muskhelishvili (1933) gave a thorough account of the solution of elasticity plane stress/strain and plate bending problems by means of complex potentials. Unfortunately the methods developed are not very readily implemented in numerical form. The fifth edition of this book (1966) contains a comprehensive review of the developments in this field. Kalandiya (1973) also gives a good account of the various complex potential methods as applied to plane stress and plate bending problems.

More recently Jaswon (see, e.g., Jaswon & Maiti, 1968; Jaswon *et al.*, 1967) has been developing similar methods but using a real (as distinct from complex) formulation. A presentation of his work, and of recent developments arising from it, is given by Jaswon & Symm (1977).

In this chapter we will be looking at a completely different formulation—one that is derived using essentially engineering principles and ideas. The quantities used in the expressions are the displacements, slopes, bending moments and shearing forces. The only concept introduced that is not normal to engineering courses on the bending of plates is that of 'higher order moments'. Although much of the work has not appeared in print before most of it has been included in lectures given by the author—the use of the reciprocal theorem and homogeneous solutions to the differential equation, for example, dating back to about 1960. Recently Bezine & Gamby (1978) have used a similar physical approach, comments on this work are given.

It is interesting to note that a similar physical approach is used by Bergman & Schiffer (1953) in their discussion of suitable 'Green's functions' for plates. The author (Tottenham, 1970) also used this approach in deriving internal integral equations for field problems, thus deriving computational patterns similar to the finite difference and finite element methods.

In the latter part of this chapter the analysis of shallow shells by means of the boundary integral formulation is described. Although this work is now ten years old it still appears to be the only successful approach to this field.

Finally, in two appendices, the formulae for the effects of point actions and for the coefficients for linear boundary elements are listed.

8.2 NOTATION

a	distance of point (x, y) from origin
e	a small distance
f_n	real part of Hankel function of first kind and order n
g_n	imaginary part of Hankel function of first kind and order n
k	constant, also coefficient of ground reaction
l	length of part of the boundary
m	applied moment distributed along curve
n	normal direction
p	applied distributed load
q	line load
r	distance between (x, y) and (ξ, η)
s	curvilinear co-ordinate
t	tangential direction
u_n	real part of Bessel function (argument $\sqrt{ix}$) order n
v_n	imaginary part of Bessel function (argument $\sqrt{ix}$) order n
x, y	co-ordinates
z	complex variable
A	region of plate
B_α	bi-couple in direction α
C	contour (closed)
D	bending stiffness
E	Young's modulus
M	edge bending moment
M_α	bending moment in direction α
P	concentrated load
Q	effective corner force
R	radius of circle
S	boundary of plate
V	boundary (Kirchoff) shear
α	arbitrary direction
β	arbitrary direction (also modified radial co-ordinate)
γ	angle between r and α direction
ε	error
θ	rotation of normal in plate
λ	ratio a/R
μ	Poisson's ratio (also used as a constant)
(ν, ζ)	arbitrary co-ordinates

(ξ, η) co-ordinates
ϕ angle (also Airy stress function)

In Section 8.7 a conventional shell notation is used, with additional symbols as defined in text.

8.3 INDIRECT METHODS FOR PLATES

Indirect methods, in boundary element terminology, are those in which the required stress–displacement state is built up from other known states. It is usual to assume that the plate is indefinitely extended and then to build up the required solution from a primal solution, which balances the applied loads but does not satisfy the boundary conditions, and other solutions corresponding to loads and moments distributed around the line of the boundary. These methods are often called 'fictitious load' methods. The boundary load method is a particular case of the more general 'contour load' method in which the known distributions correspond to loads distributed around a contour C which completely encloses, but does not coincide with, the boundary S of the real plate.

Let our basic problem be defined as: find the displacement $w(x, y)$ which is such that

$$\nabla^4 w = \frac{p(x, y)}{D} \text{ in } A \tag{1}$$

and has

$$\begin{aligned} w(s) &= \bar{w}(s) \text{ on } S_1 \qquad & \theta(s) &= \bar{\theta}(s) \text{ on } S_2 \\ M_n(s) &= \bar{M}(s) \text{ on } S_3 \qquad & V_n(s) &= \bar{V}(s) \text{ on } S_4 \end{aligned} \tag{2}$$

where s is a point on the boundary S, and $S_1 + S_4 = S_2 + S_3 = S$.

We consider the plate to be extended indefinitely, and that we know a solution, i.e. a particular integral, to the non-homogeneous equation (1). This particular integral which we denote by w_0 will generally not satisfy the boundary conditions. Let its boundary values be denoted by $\bar{w}_0$, $\bar{\theta}_0$, $\bar{M}_0$ and $\bar{V}_0$.

The displacement of a point (x, y) due to a concentrated load P acting at a point (ξ, η) can be expressed as

$$w(x, y) = \frac{P}{16\pi D}(r^2 \log r^2 - r^2) = P w_1(x, y, \xi, \eta) \tag{3}$$

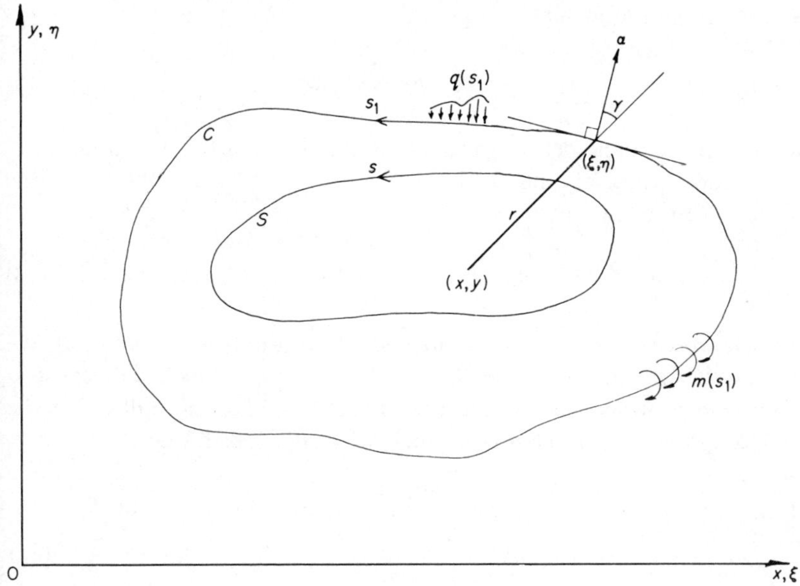

FIG. 1. External contour loads.

where

$$r^2 = (\xi - x)^2 + (\eta - y)^2$$

A load $q(s)$ distributed around some contour C will thus cause a displacement at a point (x, y), not on C, of

$$w(x,y) = \oint_C q(s) w_1(x, y, \xi, \eta)\, ds_1 \qquad [(\xi\eta) \text{ on } C] \tag{4}$$

Let the contour C completely enclose the region A and its boundary S (Fig. 1), then $w(x, y)$ as given by (4) will be a continuous function of x and y, with continuous derivatives of all orders, in the region A. We can thus differentiate with respect to x and y as required to find the slopes, bending moments and shears in the plate due to this contour load at all points in the plate and on its boundary.

By considering loads P and $-P$ at points (ξ, η) and $P(\xi + e\cos\alpha, \eta + e\sin\alpha)$ and letting e approach zero whilst the product Pe remains constant and equal to M_α, we will find the displacement due to a moment M_α at (ξ, η) we can write this as

$$w(x, y) = M_\alpha w_2(x, y, \xi, \eta)$$

and it will be found that

$$w_2(x, y, \xi, \eta) = \frac{\partial w_1}{\partial r} \cos \gamma \tag{5}$$

where γ is the angle between the radius vector r and the direction α.

If, then, we apply moments $m(s_1)$ normal to the contour C the displacement at (x, y) will be

$$w(x, y) = \oint_C m(s_1) w_2(x, y, \xi, \eta) \, \mathrm{d}s_1 \qquad [(\xi, \eta) \text{ on } C] \tag{6}$$

Again this expression is continuous in A (including the boundary S). We now seek to find the values of $q(s_1)$ and $m(s_1)$ which will cause a displacement which when added to our primal solution w_0 will satisfy the boundary conditions. That is we seek $m(s_1), q(s_1)$ such that

$$\oint_C \{q(s_1) w_1(s, s_1) + m(s_1) w_2(s, s_1)\} \, \mathrm{d}s_1 + \bar{w}_0(s) = \bar{w}(s), \qquad s \text{ on } S_1 \tag{7a}$$

$$\oint_C \left\{ q(s_1) \frac{\partial w_1}{\partial n}(s, s_1) + m(s_1) \frac{\partial w_2}{\partial n}(s, s_1) \right\} \mathrm{d}s_1 + \bar{\theta}_0(s) = \bar{\theta}(s), \qquad s \text{ on } S_2 \tag{7b}$$

$$\oint \left\{ q(s_1) \left(\frac{\partial^2}{\partial n^2} + \mu \frac{\partial^2}{\partial t^2} \right) w_1(s, s_1) + m(s_1) \right.$$
$$\left. \times \left(\frac{\partial^2}{\partial n^2} + \mu \frac{\partial^2}{\partial t^2} \right) w_2(s, s_1) \right\} \mathrm{d}s_1 + \bar{M}_0(s) = \bar{M}(s), \qquad s \text{ on } S_3 \tag{7c}$$

$$\oint \left\{ q(s_1) \left(\frac{\partial^3}{\partial n^3} + (2 - \mu) \frac{\partial^3}{\partial n \partial t^2} \right) w_1(s, s_1) + m(s_1) \right.$$
$$\left. \times \left(\frac{\partial^3}{\partial n^3} + (2 - \mu) \frac{\partial^3}{\partial n \partial t^2} \right) w_2(s, s_1) \right\} \mathrm{d}s_1 + \bar{V}_0(s) = \bar{V}(s), \qquad s \text{ on } S_4 \tag{7d}$$

Equations (7a)–(7d) form a set of 'contour integral' equations for the functions $q(s_1)$ and $m(s_1)$. At each point on the boundary two of these equations will apply. Various forms of 'contour' element equations can be set up. The simplest method is to subdivide the contour into n parts and introduce n nodal values q_i and m_i, having interpolation or shape functions α_i and β_i. We can then satisfy the boundary conditions at n points on S. A symmetric matrix of coefficients can be obtained by using the least square error on the boundary. Thus on S_1 the error $\varepsilon_1(s)$ will be

$$\varepsilon_1(s) = \int_C \left(\sum_{i=1}^{n} q_i \alpha_i(s_1) w_1(s, s_1) + \sum_{j=1}^{n} m_j \beta_j(s_1) w_2(s, s_1) \right) ds_1 + \bar{w}_0(s) - \bar{w}(s) \tag{8}$$

Similar terms obtain on S_2, S_3 and S_4. Summing these, squaring, differentiating with respect to a nodal value and setting the result to zero, leads to a symmetric matrix. Thus from $\partial\varepsilon/\partial q_m$ the coefficient of q_j is

$$\int_{S_1} \left\{ \oint_C \alpha_m(s_1) w_1(s, s_1)\, ds_1 \oint_C \alpha_j(s_1) w_1(s, s_1)\, ds_1 \right\} ds$$
$$+ \int_{S_2} \left\{ \oint_C \alpha_m(s_1) \frac{\partial w_1}{\partial n}(s, s_1)\, ds_1 \oint_C \alpha_j(s_1) \frac{\partial w_2}{\partial n}(s, s_1)\, ds_1 \right\} ds + \cdots \tag{9}$$

Although these expressions appear somewhat lengthy it must be remembered that the contour integral in fact only exist along the part of the boundary in which $q_i\alpha_i(s_1)$, for example, exists, i.e., only one or two element lengths. However such tests as have been carried out indicate that the computation time taken in evaluating these integrals is greater than the time saved due to a symmetric set of equations.

If the external contour C is a circle, radius R, and we use the centre of this circle as the origin for the co-ordinate systems, we can express $q(s_1)$ and $m(s_1)$ in the form of Fourier series,

$$q(s_1) = q_0 + \sum_{1}^{\infty} (q_i \cos i\theta + \bar{q}_i \sin i\theta)$$

$$m(s_1) = m_0 + \sum_{1}^{\infty} (m_i \cos i\theta + \bar{m}_i \sin i\theta)$$

Then using

$$\log(1 + \lambda^2 - 2\lambda \cos\psi) = 2 \sum_{1}^{\infty} \frac{\lambda^n}{n} \cos n\psi$$

where $\lambda = a/R, a^2 = x^2 + y^2$, we can integrate the above equations to find, after removing superfluous rigid body movements,

$$w(x, y) = \Bigg[(\lambda^2 \log R^2 + 2\lambda^2)q_0 + 2\left(1 - \frac{\lambda^2}{2}\right)(q_1 \cos\theta + \bar{q}_1 \sin\theta)$$

$$+ \sum_{j=2}^{\infty} \frac{\lambda^j}{j}\left(\frac{1}{j-1} - \frac{\lambda^2}{j+1}\right)(q_j \cos j\theta + \bar{q}_j \sin j\theta)$$

$$+ \frac{1}{R}\Bigg\{2(1 + \lambda^2)m_0 + \frac{\lambda^3}{2}(m_1 \cos\theta + \bar{m}_1 \sin\theta)$$

$$+ \sum_{j=2}^{\infty} \frac{\lambda^j}{j}\left(\frac{\lambda^2}{j+1} - \frac{j-2}{j(j-1)}\right)(m_j \cos j\theta + \bar{m}_j \sin j\theta)\Bigg\}\Bigg] + w_0(x, y) \tag{10}$$

and similar expressions for θ, M and V.

These equations bring out the relationship between the contour integral method and the boundary collocation method.

The conventional indirect method equations are derived from eqns. (7a)–(7d) by letting the contour C shrink onto the boundary S. This introduces some problems since in the limit there will be a point where s and s_1 coincide, i.e., we are examining the effects at the actual point of application of a load. Looking at the equations we can see that we wish to integrate $r^2 \log r^2$, $r \log r^2$, $\log r$, $1/r$ and $1/r^2$ along lines which include $r = 0$. No problems arise in the integration of $r^2 \log r^2$ and $r \log r^2$ along a line which includes $r = 0$. However terms in $\log r$, $1/r$ and $1/r^2$ occur in the expressions for the bending moments and shearing forces.

The simplest way to investigate these integrals is to use the equations of equilibrium. Let us consider a semi-circular region, centre of circle at $(0, 0)$, and apply a point load $16\pi P$ at $(0, 0)$, the displacement will then be $w_1 = r^2 \log r^2 - r^2$, r being measured from this point. Of the load $16\pi P$ only $8\pi P$ will be applied to the region we are considering, the other $8\pi P$ will be applied to a similar region in the negative range of x. For vertical equilibrium of the semi-circular disc we must have

$$-\int_{-R}^{R} V_x\big|_{x=0}\, \mathrm{d}y + \int_{-\pi/2}^{\pi/2} V_r R \cos\theta\, \mathrm{d}\theta - 2M_{xy}\big|_{\substack{x=0\\y=R}} + 2M_{xy}\big|_{\substack{x=0\\y=-R}} = 8\pi P$$

Putting in this

$$V_r = 8/r, \qquad M_{xy} = 0$$

we find that

$$\int_{-R}^{R} V_x\big|_{x=0} \, \mathrm{d}y = 0$$

or

$$\int_{S_1-\varepsilon}^{S_1+\varepsilon} \left(\frac{\partial^3 w_1}{\partial n^3}(s, s_1) + (2 - \mu) \frac{\partial^3 w}{\partial t^2 \partial n}(s, s_1) \right) \mathrm{d}s = 0$$

Considerations of moment equilibrium give

$$\int_{-R}^{R} M_x\big|_{x=0} \, \mathrm{d}y + \int_{-r_1/2}^{r_1/2} (V_r R \cos\theta + M_{r\theta} \sin\theta) R \, \mathrm{d}\theta = 0$$

or

$$\int_{-R}^{R} M_x\big|_{x=0} \, \mathrm{d}y = 4(1 + \mu) R \log R^2 - 8R$$

this is the same expression as we would obtain if we ignored the fact that the integration path passed through $r = 0$.

8.4 DIRECT METHODS

The direct methods of boundary element analysis are those that make use of the generalised Green's theorem In the case of the bending of elastic plates this theorem is the same as the Betti–Maxwell reciprocal theorem which states that for any two equilibrium states, A and B, of an elastic body the work that would be done by the forces A if given the displacements B is equal to the work that would be done by the forces B if given the displacements A. This means that for any elastic plate (or part thereof)

$$\iint_A p_{\mathrm{A}} w_{\mathrm{B}} \, \mathrm{d}A + \oint_S (V_{\mathrm{A}} w_{\mathrm{B}} - M_{\mathrm{A}} \theta_{\mathrm{B}}) \, \mathrm{d}s = \iint_A p_{\mathrm{B}} w_{\mathrm{A}} \, \mathrm{d}A + \oint_S (V_{\mathrm{B}} w_{\mathrm{A}} - M_{\mathrm{B}} \theta_{\mathrm{A}}) \, \mathrm{d}s \qquad (11)$$

If the boundary S has corners then it is necessary to add the terms $\sum Q_{\mathrm{A}} w_{\mathrm{B}}$ and $\sum Q_{\mathrm{B}} w_{\mathrm{A}}$ to the left- and right-hand sides of the equation. Here Q denotes the 'effective corner force' and the summation is made over all corners.

If we let p_{B} be a concentrated load P at (ξ, η), then w_{B} will be $P w_1(x, y, \xi, \eta)$, and this will be a continuous function of x and y except at the point of application of P. We can differentiate w_B as required to find the slopes, moments and shear forces, caused by P, at the boundary S. We can therefore write

$$\oint_S \left\{ Vw_1 - M\frac{\partial w_1}{\partial n} + \theta D\left(\frac{\partial^2 w_1}{\partial n^2} + \mu\frac{\partial^2 w_1}{\partial t^2}\right) - w\,.\,D\left(\frac{\partial^3 w_1}{\partial n^3} + (2-\mu)\frac{\partial^3 w_1}{\partial n \partial t^2}\right)\right\} \mathrm{d}s + \iint_A p w_1 \,\mathrm{d}A = kw(\xi, \eta) \tag{12}$$

where

$$\begin{aligned} k &= 1 \qquad (\xi,\eta) \text{ inside } A \\ &= \tfrac{1}{2} \qquad (\xi,\eta) \text{ on } S \\ &= 0 \qquad (\xi,\eta) \text{ outside } A \end{aligned} \tag{13}$$

The question of the meaning of the boundary integral terms when (ξ, η) is on S will be considered later.

We may also consider as our state B, the stressed state due to a concentrated moment M_α at (ξ, η). We have seen in the previous section that this gives a displacement $w_B = M_\alpha \partial w_1/\partial\alpha$. Again this is a continuous function of x and y except at the point (ξ, η), and hence we can differentiate eqn. (11) with respect to the direction α.

The state B may also represent that caused by higher order moments. Thus if we apply moments M_α and $-M_\alpha$ at points separated by e in the α-direction about (ξ, η) and let e approach zero, the product $M_\alpha e$ remaining constant at B_α, we will find that the displacement due to the bi-couple B_α is $B_\alpha \partial^2 w_1/\partial\alpha^2$. By taking bi-couples of magnitude D and μD in the α and β directions we will find that the resulting bi-couple $\bar{B}_\alpha$ will cause a displacement

$$w_3(x, y) = D\left(\frac{\partial^2 w_1}{\partial \alpha^2} + \mu\frac{\partial^2 w_1}{\partial \beta^2}\right) \cdot \bar{B}_2 = M_{1\alpha}\bar{B}_\alpha$$

In a like manner we can create a third order moment $\bar{C}_\alpha$ that gives a displacement

$$w_4(x, y) = D\left(\frac{\partial^3 w_1}{\partial \alpha^3} + \mu\frac{\partial^3 w_1}{\partial \alpha \partial \beta^2}\right) \cdot \bar{C}_\alpha = V_{1\alpha}\,.\,\bar{C}_\alpha$$

Using the Betti–Maxwell relationship with respect to the actual displacement–stress state and each of concentrated actions $P, M_\alpha, \bar{B}_\alpha$ and $\bar{C}_\alpha$ in turn leads to

$$k\mathbf{W} = \oint_S \mathbf{G}\mathbf{W}^U \mathrm{d}s + \iint_A p\mathbf{L}_\alpha^{\mathrm{T}} w_1 \,\mathrm{d}A \qquad (14)$$

where

$$\begin{aligned} \mathbf{W}^{\mathrm{T}} &= [w \quad \theta_\alpha \quad M_\alpha \quad V_\alpha] \\ \mathbf{W}^{U\mathrm{T}} &= [V \quad M \quad \theta \quad w] \\ \mathbf{G} &= \mathbf{L}_\alpha^{\mathrm{T}} \quad \mathbf{L}_n \quad w_1 \end{aligned}$$

$$\mathbf{L}\zeta = \left[1 \,\vdots\, \frac{\partial^2}{\partial \zeta} \,\vdots\, D\left(\frac{\partial^2}{\partial \zeta} + \mu \frac{\partial^2}{\partial v^2}\right) \,\vdots\, D\left(\frac{\partial^3}{\partial \zeta^3} + (2-\mu)\frac{\partial^3}{\partial \zeta \partial v^2}\right)\right]$$

when (ζ, v) form any orthogonal axis in the same sense as (x, y), (ξ, η), and (n, t).

The expression (14) represents four simultaneous boundary integral equations relating the boundary factors on S to the values of w, θ, M and V at any point inside S, on S, or simply to one another, according to whether (ξ, η) is inside A, on S, or outside A. At any point on the boundary only two of the four factors are unknown. Further if we use different values of (ξ, η) to solve these equations we must be sure that the equations are independent. Independence should be achieved if we take the points (ξ, η) on a closed contour C, external to S, since there is always a one-to-one mapping of one contour on another. This would give a four-fold set of equations whereas only a two-fold set can be independent. As yet no investigation has been reported as to how to select the equations to use. The author uses an intuitive idea that one can use only one from the pair of equations relating to w and V and one from those giving θ and M. So far no difficulties have been encountered in obtaining numerical solutions using such equations.

If we let the contour C, shrink onto S we will derive singular boundary integral equations for the boundary values w, θ, M and V. We must, however, consider whether the integrals are meaningful. We can, in fact, investigate these integrals in the same way as was carried out in the previous section, i.e., by considering the equilibrium of a semi-circular plate with the centre of the circle at (0, 0). If in these conditions we let the factors w, θ, M and V be unity in turn we will be evaluating the required term with a contour integral passing through the singularity. It will be found in all cases that no special term is needed to cater for this singularity, and that the integrals are, as would be expected, finite. In any calculation algorithm, however, it is necessary to remember that the singularity must not be at an end point of an integration. This can be accommodated by simply ignoring the nodal point when it is also the singular point.

A formulation using the first two equations of (14) was presented by Bezine & Gamby (1978). These authors use constant values over each boundary element and the element integrations were carried out numerically using the trapezium rule and 20 points. The present writer has found it more expeditious to use an analytical integration. The aforementioned authors also attempt a numerical integration by the singularities. This is unnecessary as the equilibrium conditions can be used. For example the highest order singularity occurs with the boundary shear V. A uniform displacement w shows that $\int V_B \mathrm{d}s = 0$. If then we put $\int V_B w \, \mathrm{d}s = \sum \alpha_i w_i$ we must have $\sum \alpha_i = 0$.

If the smooth contour S is approximated by a series of chords the angle at the junction between two chords will not be π. If then a point load is applied at such a junction the part of the load dispersed into the region of the problem will not be one half, but $\phi/2\pi$, where ϕ is the internal angle at the corner concerned. Thus when 'end' nodes are used on the elements the factor k in equation (14) should not be 1/2 but will vary from node to node as indicated above. Actual corners will be considered later.

The four possible equations (14) can be used to give two Fredholm equations of either the first kind or of the second kind according to whether the equations used relate to the specified boundary conditions or to the unknown boundary values. Although existence and uniqueness theorems do not exist generally for equations of the first kind there are computational advantages in using this formulation. For example we can construct a set of equations for the nodal values with symmetrical coefficients. In the simplest case, using constant value elements, we can obtain symmetry by integrating along the respective element. As a particular example consider a plate with fully fixed (clamped) edges. Since w and θ are then specified over the whole boundary we use only the first two of eqns. (14) throughout. These become

$$\tfrac{1}{2}w(\xi,\eta) = \oint\left[V(x,y)w_1(x,y,\xi,\eta) - M(x,y)\frac{\partial w_1}{\partial n}(x,y,\xi,\eta)\right]\mathrm{d}s + \iint_A p(x,y)w_1(x,y,\xi,\eta)\,\mathrm{d}x\,\mathrm{d}y = 0 \tag{15a}$$

$$\tfrac{1}{2}\theta(\xi,\eta) = \oint\left[V(x,y)\frac{\partial w_1}{\partial n_1}(x,y,\xi,\eta) - M(x,y)\frac{\partial^2 w_1}{\partial n \partial n_1}(x,y,\xi,\eta)\right]\mathrm{d}s + \iint_A p(x,y)\frac{\partial w_1}{\partial n_1}(x,y,\xi,\eta)\Bigg]\mathrm{d}x\,\mathrm{d}y = 0 \tag{15b}$$

In these n and n_1 are the directions of the normals at (x, y) and (ξ, η) respectively.

Conventionally we put $V = V_i$, $M = M_i$ on the element i and then satisfy the equations (15) at the node (ξ_i, η_i), usually the midpoint of the element. The first term in eqn. (15a) would then be

$$\sum_{j=1}^{n} V_j \int_{l_j} w_1(x, y, \xi_i, \eta_i)\, \mathrm{d}s$$

If now instead of putting $w(\xi_i, \eta_i) = 0$ we put

$$\int_{l_i} w(\xi, \eta)\, \mathrm{d}s = 0$$

we would obtain equations of the form

$$\sum_{j=1}^{N} a_{ji} V_j - \sum_{j=1}^{N} b_{ji} M_j + p_i = 0$$

$$\sum_{j=1}^{N} c_{ji} V_j - \sum_{j=1}^{N} d_{ji} M_j + q_i = 0$$

where, with (x, y) on s_j and (ξ, η) on s_i

$$a_{ji} = \int_{l_i}\int_{l_j} w_1(s_j, s_i)\, \mathrm{d}s_j\, \mathrm{d}s_i$$

$$b_{ji} = \int_{l_i}\int_{l_j} \frac{\partial w_1}{\partial n_j}(s_j, s_i)\, \mathrm{d}s_j\, \mathrm{d}s_i$$

$$c_{ji} = \int_{l_i}\int_{l_j} \frac{\partial w_1}{\partial n_i}(s_j, s_i)\, \mathrm{d}s_j\, \mathrm{d}s_i$$

$$d_{ji} = \int_{l_i}\int_{l_j} \frac{\partial^2 w_1}{\partial n_i \partial n_j}(s_j, s_i)\, \mathrm{d}s_j\, \mathrm{d}s_i$$

Since w, is symmetric with respect to the co-ordinates s_1 and s_2

$$a_{ij} = a_{ji} \qquad d_{ij} = d_{ji}$$

and also

$$c_{ij} = \int_{l_j}\int_{l_i} \frac{\partial w_1}{\partial n_j}(s_i, s_j)\, \mathrm{d}s_i\, \mathrm{d}s_j$$

$$= \int_{l_i}\int_{l_j} \frac{\partial w_1}{\partial n_i}(s_j, s_i)\, \mathrm{d}s_j\, \mathrm{d}s_i = b_{ji}$$

showing symmetry of the coefficients for the full set of unknowns V_i and $-M_i$.

A similar technique can be used for mixed boundary conditions and when higher order elements are used. At every point on the boundary one of each of the pairs (w, V) and (θ, M) will be specified and the complementary quantity will be unknown. At any point the two equations are selected from the four represented in (14) so that the left-hand sides are the specified quantities. These two equations are multiplied by the interpolation functions for the various nodal quantities of the complementary factor, and integrated around the boundary.

The equations obtained by the above procedure can be obtained directly from the Betti–Maxwell relationship. For this it is only necessary to replace the concentrated quantities P, M_α, B_α and C_α by distributed actions of the same type as the unknown quantities on the boundary. Thus on S_1, where w is specified, the normal shear V will be unknown, let here $V = \sum \alpha_i(s)V_i$. The 'state B' force will thus be a distributed line load, and for each nodal unknown we apply a line load $p(s) = \alpha_i(s)$. The work done by this force is then $k\int \alpha_i(s)w(s)$ with $k = \frac{1}{2}$. Similar on S_2 we apply distributed moments $\beta_i(s)$ where β_i are the interpolation funcations for the moment nodal values. On S_3 we apply bi-couples $\gamma_i(s)$, and on S_4 tri-couples $\delta_i(s)$, where γ_i and δ_i are the interpolation functions for the nodal values of θ_i and w_i respectively.

If the plate has corners there will be a further unknown quantity introduced; whatever the boundary conditions on either side of the corner there will be three unknown quantities at the corner itself. This means that we need to introduce one further equation for each corner. Such an equation can be obtained by using as state B of the Betti–Maxwell relationship, a bi-couple in the form of a warp of magnitude $+D(1-\mu)$ with respect to each of the sides meeting at the corner. Thus if the normals to the two sides are in directions n_2 and n_3 the bi-couple will have components $D(1-\mu)$ and $-D(1-\mu)$ so that the work done by this 'force' on a displacement w will be

$$D(1-\mu)\frac{\partial^2 w}{\partial n_2 \partial t_2} - D(1-\mu)\frac{\partial^2 w}{\partial n_3 \partial t_3} = M_{n2t2} - M_{n3t3} = Q^*$$

The displacement caused by this bi-couple will be

$$w_5 = D(1-\mu)\left(\frac{\partial^2 w_1}{\partial n_2 \partial t_2} - \frac{\partial^2 w_1}{\partial n_3 \partial t_3}\right) = L_\omega w_1 \tag{16}$$

and the boundary integral equation produced by this will be

$$Q^* = \int_S L_\omega \mathbf{L}_n \mathbf{W}^u \, \mathrm{d}s + \iint_A p L_\omega w_1 \, \mathrm{d}A \tag{17}$$

In the list of kernels in the appendices only one of the two parts of (16) is given, the second part is obtained by using the appropriate angle and a change sign for the expression.

It will be noted that the penalty for obtaining a symmetric matrix of coefficients is that we have to form a double integral for each coefficient. Even when numerical integration is used the saving in computation due to the fact that only $(n^2 + n)/2$ coefficients have to be determined in place of n^2, and that more efficient equation solvers can be used, makes the advantage outweigh the disadvantage.

Algebraic expressions are given for the kernels and also for the line integrals for constant and linear elements, in Appendix 1.

8.5 PLATES ON ELASTIC (WINKLER) FOUNDATIONS

The basic differential equation for a plate resting on an elastic foundation of the Winkler type is

$$\nabla^4 w + \frac{k}{D} w = \frac{p}{D} \tag{18}$$

where k is the reactive coefficient of the base, i.e., the reactive force at (x, y) is $kw(x, y)$. We may treat such plates by either the direct or indirect method. Putting $\mu^4 = k/D$ and changing the polar co-ordinates from (R, θ) to (β, θ) where $\beta = \mu R$ the differential equation becomes

$$\left(\frac{\partial^2}{\partial \beta^2} + \frac{1}{\beta}\frac{\partial}{\partial \beta} + \frac{1}{\beta^2}\frac{\partial^2}{\partial \theta^2}\right)^2 w + w = \frac{p}{\mu^4 D} \tag{19}$$

A solution representing a concentrated load P at $R = 0$ is

$$w_1 = \frac{P}{4D\mu^2} f_0(\beta) \tag{20}$$

where $f_0(\beta)$ is the real part of the Hankel function of the first kind with argument $(\sqrt{i}\beta)$. The displacement due to the concentrated moment M_α is

$$w_2 = \frac{M_\alpha}{4D\mu} f_0'(\beta) \cos \gamma \tag{21}$$

where primes denote differentiation with respect to β, and γ is the angle between the direction α and the radius vector R. It is possible to express f_0' in terms of the functions f_1 and g_1, i.e.,

$$f_0' = \frac{1}{\sqrt{2}}(f_1 - g_1)$$

Here we use the conventional notation

$$\begin{aligned} J_n(\sqrt{i}\beta) &= u_n(\beta) + iv_n(\beta) = ber_n(\beta) + i\,bei_n(\beta) \\ H_n(\sqrt{i}\beta) &= f_n(\beta) + ig_n(\beta) = ker_n(\beta) - i\,kei_n(\beta) \end{aligned} \tag{22}$$

However it is computationally advisable to keep to the functions f_0, g_0, u_0 and v_0 and their first derivatives. Higher order derivatives may be replaced by lower order ones using the fact that

$$\left(\frac{\partial^2}{\partial \beta^2} + \frac{1}{\beta}\frac{\partial}{\partial \beta} - n^2\right)\begin{bmatrix} u_n \\ v_n \\ f_n \\ g_n \end{bmatrix} = \begin{bmatrix} v_n \\ -u_n \\ g_n \\ -f_n \end{bmatrix} \tag{23}$$

Thus for example, the indirect boundary formulation leads to

$$\begin{aligned} w(a\theta) &= w(a_1\theta) + \frac{\mu^2}{4D}\oint\left[f_0(\beta)V(s) + \frac{1}{\mu}f_0'(p)\cos\gamma M(s)\right]\mathrm{d}s \\ \frac{\partial w}{\partial n}(a,\theta) &= \frac{\partial w(a\theta)}{\partial n} + \frac{\mu^2}{4D}\oint\left[\frac{\partial f_0(\beta)}{\partial n}V(s) + \frac{1}{\mu}\frac{\partial f_0'(\beta)}{\partial n}\cos\gamma M(s)\right]\mathrm{d}s \end{aligned} \tag{24}$$

Where (a, θ) is the polar co-ordinate of a point on the boundary, $V(s)$ and $M(s)$ are the fictitious boundary loads and

$$\beta^2 = \mu^2[a^2 + a_1^2 - 2aa_1\cos(\theta - \theta_1)]$$

(a_1, θ_1) being the co-ordinates of the running point s on S. The direct formulation follows an analogous pattern to that for simple plates with w_1 being replaced by f_0 and r being replaced by β/μ.

8.6 VERY SHALLOW SHELLS

It has been shown, for example by Vlasov (1949), that the equations for the normal displacement w and the membrane force function ϕ satisfy the equations

$$\nabla^4 w + \frac{12(1-\mu^2)}{h^2 R^2} w = \frac{p}{D} \tag{25}$$

$$\nabla^2 \phi + \frac{E_h}{R} w = 0 \tag{26}$$

The first of these equations has exactly the same form as that for a plate resting on elastic foundations (18). We can therefore treat very shallow shells in the same manner as plates on elastic foundations but now we have

$$\mu^4 = \frac{12(1-\mu^2)}{h^2 R^2} \tag{27}$$

The solution process is firstly to solve (25) using the boundary conditions associated with bending (i.e., those involving w, θ, M and V) and using the results to find ϕ from (26). It will be noted that with this formulation only one membrane type boundary condition can be applied. The most common type of boundary condition for civil engineering shallow shells is that there should be no outwards force—this in turn leads to the condition ϕ = constant on the boundary.

8.7 SHALLOW SHELLS

In the technical theory of thin shells the term 'shallow' has a special meaning. Although many authors have attempted definitions in terms of the actual physical description of the shell, each such description has its own shortcoming. On the whole the term is used to describe shells in which the rate of change of internal forces along any line is large compared with the rate of change of geometric coefficients. This is a rather vague prescription since the variability of the geometric coefficients will depend upon the co-ordinate system used and the rate of change of internal forces depends upon the loads applied. However shells on which one can set up a curvilinear co-ordinate system such that the coefficients of the first and second fundamental quadratic forms vary only slightly over the whole shell are shallow in the technical sense. A useful critique of the shallow shell assumptions and approximations has been presented by Korda (1964).

A summary of the basic relationships in shallow shell theory is given here for convenience.

A co-ordinate system (x, y) is used such that it is almost orthogonal and the curvatures R_x, R_y and warping R_{xy} are slowly varying and behave as constants in differentiation. The strain–displacement relationships are

$$\varepsilon_x = \frac{\partial u}{\partial x} - \frac{w}{R_x} \qquad \varepsilon_y = \frac{\partial v}{\partial x} - \frac{w}{R_y} \qquad \varepsilon_{xy} = \frac{\partial u}{\partial y} + \frac{\partial v}{\partial x} - \frac{2w}{R_{xy}}$$

$$\kappa_x = \frac{\partial^2 w}{\partial x^2} \qquad \kappa_y = \frac{\partial^2 w}{\partial y^2} \qquad \tau_{xy} = 2\frac{\partial^2 w}{\partial x \partial y} \tag{28}$$

The equilibrium equations are

$$\frac{\partial N_x}{\partial x} + \frac{\partial S}{\partial y} + q_x = 0 \qquad \frac{\partial S}{\partial x} + \frac{\partial N_y}{\partial y} + q_y = 0$$

$$\frac{\partial Q_x}{\partial x} + \frac{\partial Q_y}{\partial y} + \frac{N_x}{R_x} + \frac{N_y}{R_y} + \frac{2S}{R_{xy}} + q_z = 0$$

$$\frac{\partial H_2}{\partial x} + \frac{\partial M_{xy}}{\partial y} - Q_2 = 0 \qquad \frac{\partial M_{xy}}{\partial x} + \frac{\partial M_y}{\partial y} - Q_y = 0 \tag{29}$$

The elastic relationships are

$$N_x = B(\varepsilon_x + \mu\varepsilon_y) \qquad N_y = B(\varepsilon_y + \mu\varepsilon_x) \qquad S = \tfrac{1}{2}(1-\mu)B\varepsilon_{xy}$$

$$M_x = D(\kappa_x + \mu\kappa_y) \qquad M_y = D(\kappa_y + \mu\kappa_x) \qquad M_{xy} = (1-\mu)D\tau_{xy}$$

$$Q_x = D\frac{2}{\partial x}(\nabla^2 w) \qquad Q_y = D\frac{\partial}{\partial y}(\nabla^2 w) \tag{30}$$

Assuming that a particular integral is found for the first two equations of equilibrium the additional membrane stresses may be defined in terms of a stress function ϕ by

$$N_x = \frac{\partial^2 \phi}{\partial y^2} \qquad N_y = \frac{\partial^2 \phi}{\partial x^2} \qquad S = \frac{\partial^2 \phi}{\partial x \partial y} \tag{31}$$

We define the operator

$$\nabla_k^2 = R\left\{\frac{1}{R_x}\frac{\partial^2}{\partial x^2} - \frac{2}{R_{xy}}\frac{\partial^2}{\partial x \partial y} + \frac{1}{R_y}\frac{\partial^2}{\partial y^2}\right\} \tag{32}$$

where R is some measure of length. With the above we can readily derive the basic shallow shell equations

$$\nabla^4 \phi + \frac{Eh}{R}\nabla_k^2 w = 0 \tag{33}$$

$$\nabla^4 w - \frac{1}{RD}\nabla_k^2 \phi = \bar{p}$$

Here $\bar{p}$ is obtained from the applied load p_z and the particular (membrane) integral in accordance with the third equilibrium equation. Introducing new variables $\bar{w}$ and $\bar{\phi}$ given by

$$\bar{w} = \frac{Eh}{R^2} w \qquad \bar{\phi} = \frac{1}{R^2} \frac{\sqrt{Eh}}{D} \tag{34}$$

we obtain

$$\nabla^4 \begin{bmatrix} \bar{\phi} \\ \bar{w} \end{bmatrix} + \begin{bmatrix} 0 & -\alpha^2 \\ \alpha^2 & 0 \end{bmatrix} \nabla_k^2 \begin{bmatrix} \bar{\phi} \\ \bar{w} \end{bmatrix} = \begin{bmatrix} 0 \\ \alpha^4 \bar{p} \end{bmatrix} \tag{35}$$

or

$$\nabla^4 \mathbf{\Phi} + \mathbf{A} \nabla_k^2 \mathbf{\Phi} = \mathbf{p}$$

In many cases this equation can be factorised into

$$\left(\nabla^2 + \mathbf{C}_1 \frac{\partial}{\partial x} + \mathbf{C}_2 \frac{\partial}{\partial y}\right)\left(\nabla^2 + \mathbf{C}_3 \frac{\partial}{\partial x} + \mathbf{C}_1 \frac{\partial}{\partial y}\right) \mathbf{\Phi} = \mathbf{p} \tag{36}$$

This equation may then be solved using Vekua's theory of elliptic partial differential equation (see, e.g., Vekua, 1948). The solution will be built up from three parts, a particular integral $\mathbf{\Phi}_0$, and the solutions of the two homogeneous second order equations represented by the factors in the above, $\mathbf{\Phi}_1$ and $\mathbf{\Phi}_2$.

For simplicity we consider here a shallow spherical shell. The two second order equations are

$$\nabla^2 \mathbf{\Phi}_1 = 0 \tag{37a}$$

$$\nabla^2 \mathbf{\Phi}_2 + \mathbf{A} \mathbf{\Phi}_2 = 0 \tag{37b}$$

where

$$\mathbf{A} = \begin{bmatrix} 0 & -\alpha^2 \\ \alpha^2 & 0 \end{bmatrix}$$

Introducing complex variables $z, \bar{z} = x \pm iy$ we can transform (37b) into

$$\frac{\partial^2}{\partial z \partial \bar{z}} \mathbf{\Phi} + \tfrac{1}{4} \mathbf{A} \mathbf{\Phi} = 0$$

The Reimann function for this equation, defined by

$$\mathbf{R}(z, \bar{z}, t, \tau) + \tfrac{1}{4} A \int_t^z \int_\tau^{\bar{z}} \mathbf{R}(z_1, \bar{z}_1, t_1, \tau) \, dz_1 \, d\bar{z}_1 = \mathbf{I}$$

where $t = x_0 + iy_0$, $\tau = x_0 - iy_0$ are arbitrary points. This function is in fact the Bessel function of the first kind and of zero order

$$\mathbf{R}(z, \bar{z}, t, \tau) = J_0(\sqrt{\mathbf{A}(z - t)(\bar{z} - \tau)}) \tag{38}$$

Since eqn. (37a) can be obtained from (37b) by setting $\mathbf{A}$ as a nul matrix we find that the Reimann function for $\nabla^2\mathbf{\Phi} = \mathbf{0}$ is a unit matrix. We can thus write

$$\begin{aligned}\mathbf{\Phi}_1 &= R_e(\mathbf{\Phi}_3(z))\\ \mathbf{\Phi}_2 &= R_e\left(\mathbf{R}(z, 0, z, \bar{z})\mathbf{\Phi}_4(3) - \int_0^z \frac{\partial}{\mathrm{d}s}\mathbf{R}(s, 0, z, \bar{z})\mathbf{\Phi}_4(s)\,\mathrm{d}s\right)\end{aligned} \tag{39}$$

where $\mathbf{\Phi}_3$ and $\mathbf{\Phi}_4$ are arbitrary analytic complex vectors of order 2. We can combine these in the general solution

$$\mathbf{\Phi} = \mathbf{\Phi}_0 + R_e(|\mathbf{IR}_z|\bar{\psi}(z) - \int_0^z \frac{\partial}{\mathrm{d}s}|\mathbf{IR}_s|\psi(s)\,\mathrm{d}s \tag{40}$$

where

$$\mathbf{R}_\alpha = \mathbf{R}(\alpha, 0, z, \bar{z})$$

and ψ is a fourth order vector whose elements are arbitrary analytic functions. These functions are to be determined from the boundary conditions. These conditions will relate to w, ϕ and derivatives of these. Let m be the highest order derivative so that these conditions can be written as

$$\sum_{i,j=0}^{i+j=m} \mathbf{A}_{ij}\frac{\partial^{i+j}}{\partial x^i \partial y^j}\mathbf{\Phi} = \mathbf{f}(t) \tag{41}$$

where t is a point on the boundary. We then represent ψ as a Cauchy integral around the boundary

$$\psi(z) = \oint\left(1 - \frac{z}{t}\right)^m \log\left(1 - \frac{z}{t}\right)\mathbf{g}(t)\,\mathrm{d}s + \int \mathbf{g}(t)\,\mathrm{d}s \tag{42}$$

where $\mathbf{g}$ is a vector of arbitrary real functions. Substituting this expression into the right-hand side of (40) and applying the boundary conditions (41) leads to a singular boundary integral equation of the form

$$\mathbf{A}(t_0)\mathbf{g}(t_0) + \oint \mathbf{K}(t_0, t)\mathbf{g}(t)\,\mathrm{d}s = \mathbf{f}(t) \tag{43}$$

t_0 and t being points on the boundary S. Details of the application of this technique to singly connected shallow shell of arbitrary plan form have been given by Newton & Tottenham (1968). The more general case of shells with holes, as well as the numerical treatment of the singularities, will be found in Newton (1967).

8.8 NUMERICAL EXAMPLES

8.8.1 Contour Integral Method

As an example of the contour integral method we consider a uniform square plate, side 2, fixed on all four sides, subjected to a uniformly distributed load p per unit area (Fig. 2). We take as the contour a circle with radius R greater than $\sqrt{2}$ and having a common centre with the plate. The boundary conditions are

$$w = \frac{\partial w}{\partial x} = 0 \qquad x = \pm h$$

$$w = \frac{\partial w}{\partial y} = 0 \qquad y = \pm h$$

As a particular solution w_0 we use

$$w_0 = \frac{pr^4}{64D}$$

From the symmetry of the problem all the terms involving $\sin j\theta$ in eqn. (10) will be zero as will terms in $\cos j\theta$ except when j is a multiple of four. We thus have

$$w(x, y) = (\log R^2 + 1)\lambda^2 q_0 + 2(1 + \lambda^2)m_0/R$$

$$+ \sum_{i=4,8\ldots} \left\{\frac{1}{j}\left(\frac{1}{j-1} - \frac{\lambda^2}{j+1}\right)q_j + \left(\frac{\lambda^2}{j+1} - \frac{j-2}{j(j+1)}\right)\frac{m_j}{R}\right\}\lambda^j \cos j\theta$$

Along the edge $x = h$ the derivative $\partial w/\partial x$ will be

$$\frac{\partial w}{\partial x} = \cos\phi\frac{\partial w}{\partial a} + \frac{\sin\phi}{a}\frac{\partial w}{\partial \phi}$$

where ϕ is the angle between the radius vector a and the x axis. Limiting

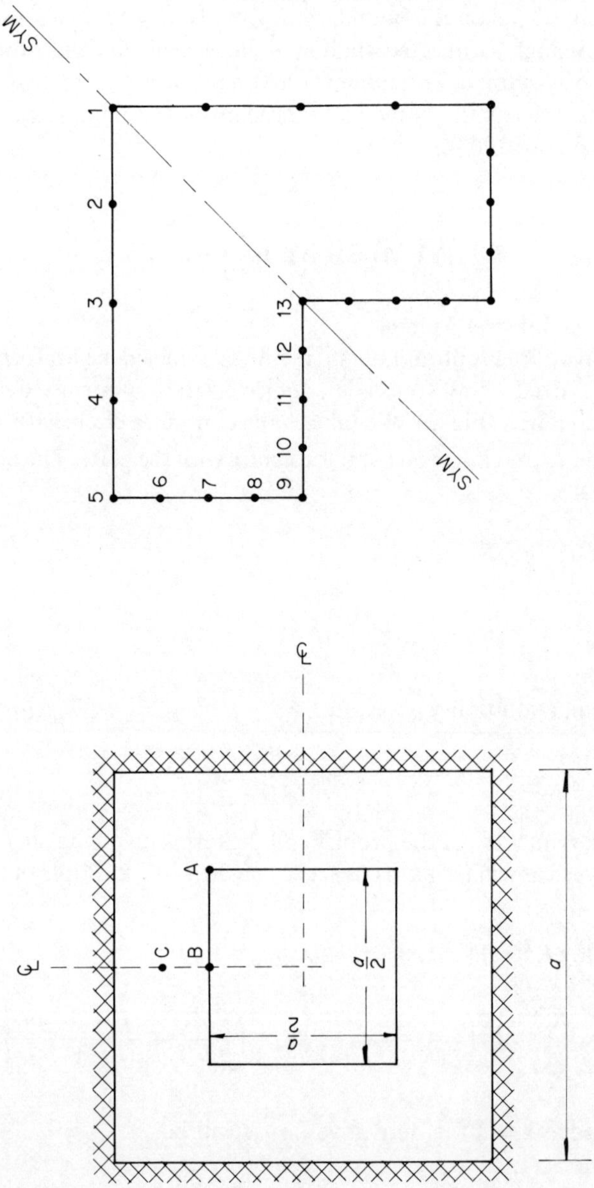

FIG. 2. Uniform square plate.

ourselves to the terms involving q_0, m_0, q_u and m_u, and satisfying the conditions $w = \partial w/\partial x = 0$ at the two points $(h, 0)$ and (h, h) gives

$$q_0 = -\frac{ph^4}{528D}\cdot\frac{2}{2+\log R^2}\cdot\left(5+\frac{9}{\gamma^2}\right) \qquad m_0 = \frac{ph^4}{578D}\cdot 5R$$

$$q_4 = \frac{ph^4}{528D}\cdot\frac{2}{\gamma^2}\left(\frac{20}{\gamma^2}-27\right) \qquad m_4 = \frac{ph^4}{528D}\cdot\frac{2R}{\gamma^2}\left(\frac{40}{\gamma^2}-27\right)$$

where $\gamma = h/R$. At the centre, where $\lambda = 0$ we find $w_0 = 2m_0/R = 0{\cdot}0189ph^4/D$, this compares with the value of $0{\cdot}0202\,ph^4/D$ given by Timoshenko (1940).

It will be noted that when those expressions are inserted into eqn. (10) the expression for $w(x, y)$ becomes independent of R.

8.8.2 Indirect Method

The indirect method was applied to the problem of the analysis of a uniform square plate with a square central opening. The external edges of the plate were simply supported and the internal edges free, the plate carried a uniformly distributed load. One quarter of the plate was considered, with nodes at quarter points along each side, giving a total of 24 nodes. The boundary conditions were satisfied at 13 points and symmetry used to

TABLE 1

Point	*Displacement* $(xqa^4/100D)$			
	Boundary integral		*Finite element*	*Finite difference*
	indirect	*direct*		
A	0·218 8	0·218 8	0·218 5	0·217 4
B	0·310 7	0·314 1	0·315 6	0·300 6
C	0·155 8	0·156 5	—	0·154 1

obtain twenty-six equations for the twenty-six fictitious loads. Corner forces were ignored and the fictitious moments at the corners of the plate applied in the diagonal plane. The results obtained are compared with those using finite element and variational finite difference methods in Table 1. In both of the comparative cases the mesh size was $a/16$. Poisson's ratio was 0·25.

The problem was solved using an Algol 60 computer program on the ICL Atlas computer.

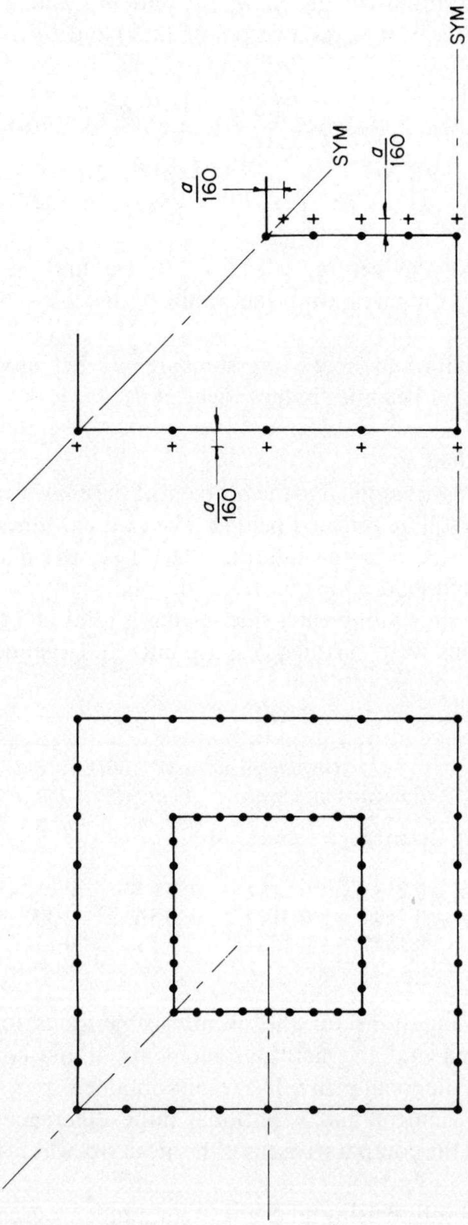

FIG. 3. Square plate with central opening.

8.8.3 Direct Methods

(a) Using 'Exterior' Points

The problem of the square plate with a central opening was also analysed using a direct method. For this the whole plate was considered, giving a total of 64 nodes, of these 56 had 2 nodal values (θ and V on the outside edge and M and V on the internal edge), whilst the 8 corner nodes had 3 nodal values (V on the two sides and the corner force at the external corners and the two slopes and displacement at the internal corners). After forming the necessary equations symmetry was introduced, reducing the number of unknowns to 20. The equations were formed by applying unit loads and moments at the positions marked in Fig. 3. A linear variation was assumed between nodes. The results obtained are shown in Table 1. An Algol 60 program was used and this was run on an ICL 1907.

(b) Direct Method Using Linear Elements

The direct boundary integral method was used to analyse uniform rectangular plates, simply supported along two opposite sides and fixed along the other two sides, subjected to a uniformly distributed load. Poisson's ratio was taken as zero in the initial stages of the computation, the displacements are independent of this quantity except for an overall factor of $(1 - \mu^2)$. The boundary was subdivided to give 12 elements along each edge, and after the formation of the necessary equations symmetry conditions were introduced to bring the number of nodal unknowns to 27. The problem was solved using an ICL 1907 computer, the run time being 13 s. The resulting central deflections for plates with different aspect ratios are compared with those given by Timoshenko (1940) in Table 2.

TABLE 2

a/b	*Deflection at centre* $= \alpha \dfrac{qa^4}{Eh^3}$ α	
	Timoshenko (1940)	*BEM*
1·0	0·020 9	0·021 3
1·2	0·034 0	0·034 6
1·4	0·050 2	0·050 3
1·6	0·065 8	0·065 8
1·8	0·079 9	0·080 0
2·0	0·098 7	0·099 0

TABLE 3

Point	θ $(x\ qR^3/D)$	V $(x\ qR)$	*Point*	M $(x\ qR^2)$	V $(x\ qR)$
5	0·0142	0·681	10	0·0366	0·281
6	0·0127	0·640	11	0·0553	0·358
7	0·0087	0·515	12	0·0650	0·401
8	0·0033	0·268	13	0·0702	0·416
			14	0·0742	0·420
			15	0·0759	0·422
				(0·0756)	

Point	w $(x\ qR^4/D)$	M_r $(x\ qR^2)$	M_θ $(x\ qR^2)$
21	0·002934	−0·04809	−0·02434
	(0·00293)	(−0·0473)	
22	0·003369	−0·04406	−0·02874
	(0·003370)	(−0·0446)	
23	0·001526	−0·0008	−0·01290
	(0·001529)	(−0·0016)	

(*c*) *Direct Method with Symmetric Equations, Linear Elements*
A semicircular uniform plate, simply supported along the straight edge and fixed along its curved edge, under uniformly distributed normal load, was analysed using the symmetric equation formulation. The coefficients of the element equations were computed using five Gauss point integration along the element and the explicit formulae (see Appendix 1) for integration along the 'equations' element.

The straight edge had eight equal elements and the curved edge was represented by twelve straight, chordal, elements, Symmetry was used to reduce the number of nodal unknowns to 23. Some results obtained are shown in Table 3, the figures in brackets are those given by Timoshenko (1940).

8.9 CONCLUSIONS

There are a variety of boundary methods which can be used for the static analysis of plates and shells. Those based upon the physical factors—displacement, slope, bending moment and shearing force—provide models suited to engineering computation. The use of the indirect method together with a fictitious external boundary leads to a very simple algorithm. The direct method can be formulated in such a manner as to give a symmetric array of coefficients—of particular importance when this method is to be used in conjunction with others such as finite elements.

These methods can all be applied to problems concerned with plates on elastic foundations and very shallow shells.

At present the general shallow shell cannot be treated quite so simply.

REFERENCES

BERGMAN, S. & SCHIFFER, M. (1953). *Kernel Functions and Elliptic Differential Equations in Mathematical Physics*, Academic Press, New York.

BEZINE, G. P. & GAMBY, D. A. (1978). A new integral equation formulation for plate bending problems. In: *Recent Advances in Boundary Element Methods*, Ed. C. A. Brebbia, Pentech Press, London.

FREDHOLM, I. (1900). Sur les equations de l'equilibre d'un corps solide elastic, *Acta Mathematica*, **23,** 1–42.

JASWON, M. A. & MAITI, M. (1968). An integral equation formulation of plate bending problems, *J. Eng. Math.*, **2,** 83–93.

JASWON, M. A. & SYMM, G. T. (1977). *Integral Methods in Potential Theory and Elastostatics*, Academic Press, London.

JASWON, M. A., MAITI, M. and SYMM, G. T. (1967). Numerical biharmonic analysis and some applications, *Int. J. Sol. Struct.*, **3,** 309–32.

KALANDIYA, A. I. (1973). *Mathematical Methods of Two-dimensional Elasticity*, Nauka, Moscow (English translation, Moscow, 1975).

KORDA, P. E. (1964). *A critical evaluation of the theory of shallow shells, and application of digital computers to mixed boundary value problems*, Ohio State Univ. Res. Lab. Report RF-1699-1.

LAURICELLA, G. (1909). Sur l'integration de l'equation relative a l'equilibre des plagues elastique encastree, *Acta Mathematica*, **32,** 201–56.

MUSKHELISHVILI, N. I. (1933). *Some Basic Problems of the Mathematical Theory of Elasticity*, 1st edition Acad Sciences, Moscow (English translation of 3rd edition, Translation Noordhoff, Groningen, 1953. 5th Russian edition, Nauka, Moscow, 1966).

NEWTON, D. A. (1967). *Contribution to the analysis of boundary value problems in thin shallow shells*, Ph.D. Thesis, Southampton University.

NEWTON, D. A. & TOTTENHAM, H. (1968). Boundary value problems in thin shallow shells of arbitrary plan form, *J. Eng. Maths.*, **2,** 211–24.

TIMOSHENKO, S. P. (1940). *Theory of Plates and Shells*, McGraw-Hill, New York.

TOTTENHAM, H. (1970). A direct numerical method for the solution of field problems, *Int. J. Num. Meth. Eng.*, **2,** 117–31.

VEKUA, I. N. (1948). *New Methods of Solving Elliptic Equations*, Gostekhizdat, Moscow.

VLASOV, V. Z. (1949). *The General Theory of Shells and its Application in Engineering*, G.I.T.T.L., Moscow (English translation, N.A.S.A., Washington, 1964).

APPENDIX 1

Influence Functions (Fig. 4)

(a) Load P at (ξ, η)

$$w_1(x, y, \xi, \eta) = \frac{P}{16\pi D} |r^2 \log r^2 - r^2|$$

$$\theta_1(x, y) = \frac{P}{16\pi D} \cdot 2r \log r^2 \cos \phi$$

$$M_1(x, y) = \frac{P}{8\pi} |(1 + \mu)(1 + \log r^2) + (1 - \mu) \cos 2\phi|$$

$$V_1(x, y) = \frac{P}{8\pi} \cdot \frac{1}{r} |(5 - \mu) \cos \phi + (1 - \mu) \cos 3\phi|$$

$$M_{nt1}(x, y) = \frac{P}{8\pi} \cdot (1 - \mu) \sin 2\phi$$

$$Q_{n1}(xiy) = \frac{P}{2\pi} \cdot \frac{\cos \phi}{r}$$

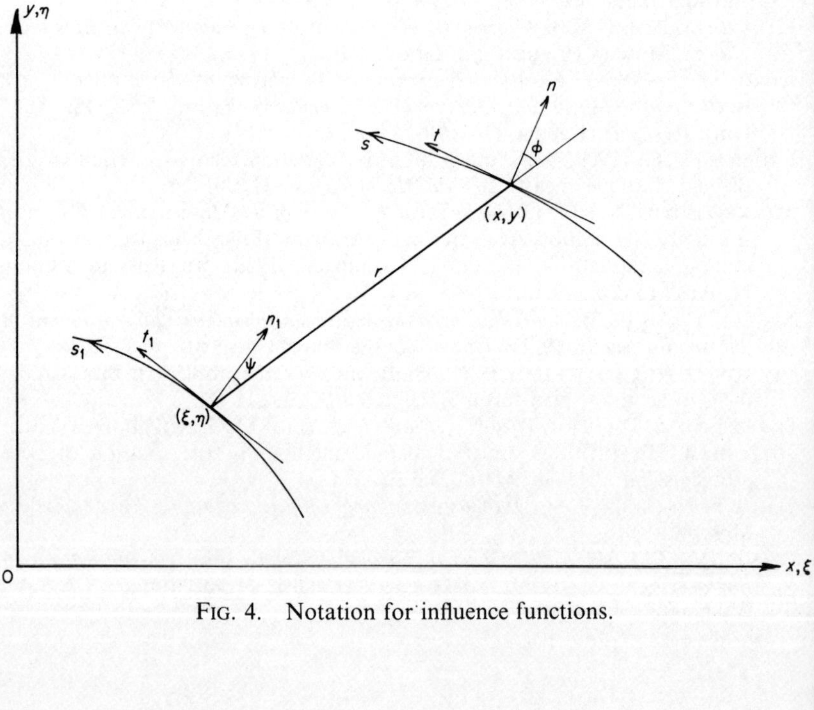

FIG. 4. Notation for influence functions.

(b) Moment M_{n1} at (ξ, η)

$$w_2(x, y) = \frac{-M_{n1}}{8\pi D} \cdot r \log r^2 . \cos\psi$$

$$\theta_2(x, y) = \frac{-M_{n1}}{8\pi D} \cdot |(1 + \log r^2)\cos(\psi - \phi) + \cos(\phi + \psi)|$$

$$M_2(x, y) = \frac{-M_{n1}}{8\pi}\frac{1}{r}|2(1 + \mu)\cos\psi + (1 - \mu)\{\cos(2\phi - \psi) - \cos(2\phi + \psi)\}|$$

$$V_2(x, y) = \frac{M_{ni}}{8\pi r^2}|(5 - \mu)\cos(\phi + \psi) + 2(1 - \mu)\cos(3\phi + \psi) - (1 - \mu)\cos(3\phi - \psi)|$$

$$M_{nt2}(x, y) = \frac{M_{n1}}{8\pi}\frac{(1 - \mu)}{r}|2\sin\psi - \sin(2\phi - \psi) - \sin(2\phi + \psi)|$$

$$Q_{n2}(x, y) = \frac{-M_{n1}}{2\pi}\frac{\cos(\phi + \psi)}{r^2}$$

(c) Bi-couple B_{n1} at (ξ, η)

$$w_3(x, y) = \frac{B_{n1}}{8\pi D}|(1 + \mu)(1 + \log r^2) + (1 - \mu)\cos 2\psi|$$

$$\theta_3(x, y) = \frac{B_{n1}}{8\pi}\frac{1}{r}|2(1 + \mu)\cos\phi + (1 - \mu)\{\cos(2\psi - \phi) - \cos(2\psi + \phi)\}|$$

$$M_3(x, y) = \frac{-B_{n1}}{4\pi}\frac{1}{r^2}|(1 - \mu^2)(\cos 2\phi + \cos 2\psi) - (1 - \mu)^2 \times \cos 2(\psi + \phi)|$$

$$V_3(x, y) = \frac{B_{n1}}{4\pi}\frac{(1 - \mu)}{r^3}|-2(1 + \mu)\cos 3\phi + (5 - \mu)\cos(2\psi + \phi) + 3(1 - \mu)\cos(2\psi + 3\phi)|$$

$$M_{nt3}(x, y) = \frac{B_{n1}}{4\pi}\frac{(1 - \mu)}{r^2}|(1 + \mu)\sin(\phi + \psi) - (1 - \mu)\sin 2(\phi + \psi)|$$

$$Q_{n3}(x, y) = \frac{B_n}{\pi}\frac{(1 - \mu)}{r^3}\cos(2\psi + \phi)$$

(d) Tri-couple C_{n1} at (ξ, η)

$$w_4(x, y) = \frac{-C_{n1}}{8\pi D} \cdot \frac{1}{r}|(5 - \mu)\cos\psi + (1 - \mu)\cos 3\psi|$$

$$\theta_4(x, y) = \frac{C_{n1}}{8\pi D}\frac{1}{r^2}|(5 - \mu)\cos(\psi + \phi) + 2(1 - \mu)\cos(3\psi + \phi) - (1 - \mu)\cos(3\psi - \phi)|$$

$$M_4(x,y) = \frac{-C_{n1}}{4\pi}\frac{(1-\mu)}{r^3}|(5-\mu)\cos(\psi+2\phi) + 3(1-\mu) \times \cos(3\psi+2\phi) - 2(1+\mu)\cos 3\psi|$$

$$V_4(x,y) = \frac{3C_{n1}}{4\pi}\cdot\frac{(1-\mu)}{r^4}|(5-\mu)\{\cos(\psi+3\phi) + \cos(3\psi+\phi)\} + 4(1-\mu)\cos 3(\psi+\phi)|$$

$$M_{nt4}(x,y) = \frac{C_{n1}}{4\pi}\frac{(1-\mu)}{r^2}|(5-\mu)\sin(\psi+2\phi) + 3(1-\mu)\sin(3\psi+2\phi)|$$

$$Q_{n4}(x,y) = 3C_{n1}\frac{(1-\mu)}{r^4}\cos(3\psi+\phi)$$

(e) Warping bi-couple B_{nt} at (ξ,η)

$$w_5(x,y) = \frac{B_{nt}}{8\pi D}(1-\mu)\sin 2\psi$$

$$\theta_5(x,y) = -\frac{B_{nt}}{8\pi D}\frac{(1-\mu)}{r}[\sin(2\psi+\phi) - \sin(2\psi-\phi)]$$

$$M_5(x,y) = \frac{B_{nt}}{4\pi}\frac{(1-\mu)}{r^2}[(1+\mu)\sin 2\psi - (1-\mu)\sin 2(\psi+\phi)]$$

$$V_5(x,y) = \frac{B_{nt}}{4\pi}\frac{(1-\mu)}{r^3}[(5-\mu)\sin(2\psi+\phi) + 3(1-\mu)\sin(2\psi+3\phi)]$$

$$M_{nt5}(x,y) = \frac{B_{nt}}{4\pi}\frac{(1-\mu)^2}{r^2}\cos 2(\phi+\psi)$$

$$Q_{nt}(x,y) = \frac{B_{nt}}{2\pi}\frac{(1-\mu)}{r^2}\sin(2\psi+\phi)$$

APPENDIX 2

Expressions for Boundary Element with Linearly Varying Fields (Fig. 5)

(a) With unit load at P displacements on AB

$$w_1 = \frac{1}{16\pi D}|(x^2+h^2)\log(x^2+h^2) - (x^2+h^2)|$$

$$\theta_1 = \frac{1}{8\pi D}h\log(x^2+h^2)$$

$$M_1 = \frac{1}{8\pi}\left|(1+\mu)\log(x^2+h^2) + \frac{2(h^2+\mu x^2)}{x^2+h^2}\right|$$

$$V_1 = \frac{1}{8\pi}\left|\frac{2h(3-\mu)}{(x^2+h^2)} - \frac{4(1-\mu)x^2h}{(x^2+h^2)^2}\right|$$

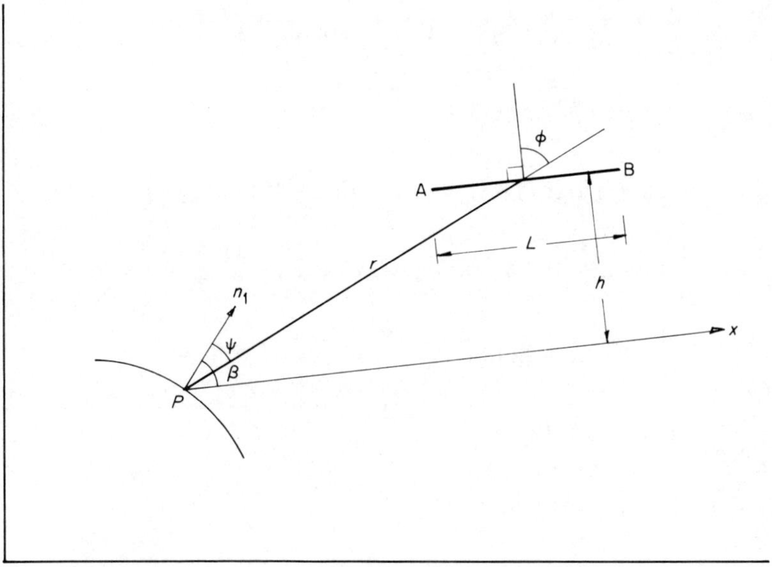

FIG. 5. Notation for linear element coefficients.

Field variables are

$$w = \frac{1}{L}|(x_B - x)w_A + (x - x_A)w_B|$$

$$\theta = \frac{1}{L}|(x_B - x)\theta_A + (x - x_A)\theta_B|$$

$$M = \frac{1}{L}|(x_B - x)M_A + (x - x_B)M_B|$$

$$V = \frac{1}{L}|(x_B - x)V_A + (x - x_B)V_B|$$

$$\int_A^B (Vw_1 - M\theta_1 + \theta M_1 - wV_1)\,dx$$

$$= (x_B I_1 - I_2)V_A + (I_2 - x_A I_1)V_B - (x_B I_3 - I_4)M_A - (I_4 - x_A I_1)M_B$$
$$+ (x_B I_5 - I_6)\theta_A + (I_6 - x_A I_5)\theta_B - (x_B I_7 - I_8)w_A - (I_8 - x_A I_7)w_B$$

$$I_1 = \frac{1}{16\pi D}\left|\frac{x}{3}(x^2 + 3h^2)\log(x^2 + h^2) - \frac{5}{9}x^3 - \frac{7xh^2}{3} + \frac{4}{5}h^3\tan^{-1}\left(\frac{x}{h}\right)\right|_A^B$$

$$I_2 = \frac{1}{16\pi D}\left|\frac{(x^2 + h^2)}{4}\log(x^2 + h^2) - \frac{(x^2 + h^2)^2}{8} - \frac{x^4}{4} - \frac{x^2h^2}{2}\right|_A^B$$

$$I_3 = \frac{1}{16\pi D}\left|\frac{xh}{4}(\log(x^2+h^2)-1) + \frac{h^2}{2}\tan^{-1}\left(\frac{x}{h}\right)\right|_A^B$$

$$I_4 = \frac{1}{16\pi D}\left|h(x^2+h^2)\log(x^2+h^2)\right|_A^B$$

$$I_5 = \frac{1}{8\pi}\left|(1+\mu)x\log(x^2+h^2) - (1-\mu)x + 4h\tan^{-1}\left(\frac{x}{h}\right)\right|_A^B$$

$$I_6 = \frac{1}{8\pi}\left|\frac{1}{2}[(1+\mu)x^2 + (3-\mu)h^2]\log(x^2+h^2) - \frac{(1-\mu)}{2}x^2\right|_A^B$$

$$I_7 = \frac{1}{8\pi}\left|\frac{2(1-\mu)xh}{x^2+h^2} + 2\tan^{-1}\left(\frac{x}{h}\right)\right|_A^B$$

$$I_8 = \frac{1}{8\pi}\left|(1+\mu)h\log(x^2+h^2) - \frac{2(1-\mu)h^3}{(x^2+h^2)}\right|_A^B$$

(b) With unit load M_{n1} at P

We use

$$\begin{bmatrix} w_2 \\ \theta_2 \\ M_2 \\ V_2 \end{bmatrix} = \frac{\partial}{\partial n_1}\begin{bmatrix} w_1 \\ \theta_1 \\ M_1 \\ V_1 \end{bmatrix} = -\left(\cos\beta\frac{\partial}{\partial x} + \sin\beta\frac{\partial}{\partial y}\right)\begin{bmatrix} w_1 \\ \theta_1 \\ M_1 \\ V_1 \end{bmatrix}$$

so that

$$\int_A^B\left(\frac{(x_B - x)}{L}V_A + \frac{(x-x_B)}{L}V_B\right)w_2\,dx$$

$$= -\Big||x_B\cos\beta w_1|_A^B - |x\cos\beta w_1|_A^B + \cos\beta\int_A^B w_1\,dx$$
$$+ \sin\beta x_B\int_A^B\frac{\partial w_1}{\partial y}dx - \sin\beta\int_A^B x\frac{\partial w_1}{\partial y}dx\Big|V_A$$

$$-\Big||x\cos\beta w_1|_A^B - \int_A^B\cos\beta w_1\,dx - |x_A\cos\beta w_1|_A^B$$
$$+ \sin\beta\int_A^B x\frac{\partial w_1}{\partial y}dx - x_A\sin\beta\int_A^B\frac{\partial w_1}{\partial y}dx\Big|V_B$$

$$= -|L\cos\beta w_{1A} + \cos\beta I_1 + x_B\sin\beta I_9 - \sin\beta I_{10}|V_A$$
$$-|L\cos\beta w_{1B} - \cos\beta I_1 + \sin\beta I_{10} - x_A\sin\beta I_9|V_B$$

Similar expressions apply for the coefficients of M_A, M_B, θ_A, θ_B, w_A and w_B, involving integrals I_{11} to I_{16} in place of I_9 and I_{10}, and I_3, I_5 and I_7 in place of I_1.

$$I_9 = \frac{1}{8\pi D}\left| xh(\log(x^2 + h^2) - 2) + 2h^2 \tan^{-1}\left(\frac{x}{h}\right)\right|_A^B$$

$$I_{10} = \frac{1}{16\pi D}\left| h(x^2 + h^2)\log(x^2 + h^2) - x^2 \right|_A^B$$

$$I_{11} = \frac{1}{16\pi D}\left| x(\log(x^2 + h^2) - 2) + 4h \tan^{-1}\left(\frac{x}{h}\right)\right|_A^B$$

$$I_{12} = \frac{1}{16\pi D}\left| (x^2 + 3h^2)\log(x^2 + h^2) - 2x^2 \right|_A^B$$

$$I_{13} = \frac{1}{4\pi}\left| 2\tan^{-1}\left(\frac{x}{h}\right) - (1 - \mu)\frac{xh}{(x^2 + h^2)}\right|_A^B$$

$$I_{14} = \frac{1}{4\pi}\left| 2\mu h \log(x^2 + h^2) + (1 - \mu)\frac{h^3}{(x^2 + h^2)}\right|_A^B$$

$$I_{15} = -\frac{1}{4\pi}\left| (1 + \mu)\frac{x}{(x^2 + h^2)} - 2(1 - \mu)\frac{xh^2}{(x^2 + h^2)^2}\right|_A^B$$

$$I_{16} = -\frac{1}{4\pi}\left| \frac{(3 - \mu)}{2}\log(x^2 + h^2) + \frac{(1 - 3\mu)x^2h^2 - 2(1 + \mu)h^4}{(x^2 + h^2)^2}\right|_A^B$$

Chapter 9

APPLICATION OF THE BOUNDARY ELEMENT METHOD IN ELECTRICAL ENGINEERING PROBLEMS

M. H. LEAN

Department of Electrical Engineering, University of Manitoba, Canada

M. FRIEDMAN†

Nuclear Research Center-Negev and Department of Electrical Engineering, Ben Gurion University, Beer-Sheva, Israel

and

A. WEXLER

Department of Electrical Engineering, University of Manitoba, Canada

SUMMARY

This chapter attempts to bridge the gap between electrical and non-electrical engineers working on boundary element methods. It introduces the formulation of integral equations in electrical terms, describes various projective methods of solution (including moment, Galerkin, Rayleigh–Ritz, and point-matching methods), and the isoparametric boundary element algorithm. Electrostatic, magnetostatic and electromagnetic wave-scattering problems are considered. An approach to the mutual constraint of partial differential and integral equation formulations is outlined and reference is made to its application in discretisation of integral equations in block-sparse form. A considerable number of pertinent references are included.

† On sabbatical at the University of Manitoba, Winnipeg, R3T 2N2, Canada.

9.1 INTRODUCTION

This chapter is intended to fulfil a two-fold purpose: to introduce the notation and techniques of integral equation methods to electrical engineers, while drawing parallels with their application in other fields in order to assist electrical engineers in their understanding of other chapters of this book; and to aid non-electrical engineers to read the extensive literature available on integral equation applications to electrostatics, magnetostatics and electromagnetic scattering in open regions. The mathematical study of electrical technology has a long history of boundary integral equation solution by numerical methods, mainly by pulse-expansion and point matching techniques. These, of course, were the precursors to those highly accurate developments that are the subject of this volume. In assisting this cross fertilisation, the authors hope to further encourage development of the boundary element method.

9.2 INTEGRAL EQUATION FORMULATIONS

The application of integral equations in electrical engineering is best shown via potential theory and surface layers. A detailed discussion that consists of both mathematical rigour and engineering aspects was given by Stakgold (1968). A variety of methods can be applied for solving differential or integral operator equations. Those of variational and moment types were described by Mikhlin (1964) and Harrington (1968), and applied by many authors.

9.2.1 Surface Layers

If a unit positive charge is concentrated at a fixed point $\mathbf{r}'$, in free space, its corresponding potential $G(\mathbf{r}, \mathbf{r}')$ satisfies the differential equation

$$-\nabla^2 G = \delta(\mathbf{r} - \mathbf{r}') \tag{1}$$

If G is required to possess spherical symmetry and to vanish for $|\mathbf{r}| = \infty$, eqn. (1) is uniquely solved by the free space Green's function

$$G(\mathbf{r}, \mathbf{r}') = \frac{1}{4\pi|\mathbf{r} - \mathbf{r}'|} \tag{2}$$

This expression is valid for an n-dimensional space, $n \geq 3$ (except for the coefficient $1/4\pi$). For $n = 2$ the solution is $-(\log|\mathbf{r} - \mathbf{r}'|)/2\pi$.

Assume now that charge is distributed in free space with volume density $\rho(\mathbf{r})$. Then by the superposition principle, the potential at $\mathbf{r}$ is given by

$$\phi(\mathbf{r}) = \int_V G(\mathbf{r}, \mathbf{r}')\rho(\mathbf{r}')\,\mathrm{d}v' \tag{3}$$

We shall assume $n = 3$, and then by differentiating under the integral sign one gets

$$-\nabla^2 G = -\rho(\mathbf{r})*\nabla^2 \frac{1}{4\pi|\mathbf{r}|} = \rho*\delta = \rho(\mathbf{r}) \tag{4}$$

where * is the notation for convolution.

The next basic element to the point charge is the dipole. Let h be a small positive number, and let charges $-1/h$ and $1/h$ be located at $\mathbf{r}'$ and $\mathbf{r}' + h\hat{\mathbf{l}}$ respectively, where $\hat{\mathbf{l}}$ is a given unit vector. The limiting charge configuration when $h \to 0$ is called a unit dipole with axis $\mathbf{l}$. By using the previous mathematical notations one can show that the dipole potential is given by

$$d_l(\mathbf{r}, \mathbf{r}') = \frac{\cos\theta}{4\pi|\mathbf{r} - \mathbf{r}'|^2} \tag{5}$$

where θ is the angle between $\hat{\mathbf{l}}$ and $\mathbf{r} - \mathbf{r}'$.

Assume now that charge is distributed on a surface S with a surface charge density $\sigma(\mathbf{s})$. This configuration is defined as a simple layer resulting with the potential

$$\phi(\mathbf{r}) = \int_S \frac{1}{4\pi|\mathbf{r} - \mathbf{s}'|}\sigma(\mathbf{s}')\,\mathrm{d}s' \tag{6}$$

This integral is well defined for all $\mathbf{r} \notin S$ and satisfies $\nabla^2\phi = 0$. It can be shown however that it converges even for $\mathbf{r} = \mathbf{s} \in S$. In doing this one is simply using the fact that a circular surface element $\mathrm{d}s'$, whose centre is at s, equals $\pi|\mathbf{s} - \mathbf{s}'|^2$ and thus cancels out the singularity of the integral.

If, instead of a simple layer, one considers a surface distribution of dipoles normal at each point to the surface S, one gets a configuration known as a double layer. If the surface density is $\eta(\mathbf{s})$, the total potential due to the double layer is given by

$$d(\mathbf{r}) = \int_S \eta(\mathbf{s}')\frac{\cos(\mathbf{r} - \mathbf{s}', \mathbf{n}')}{4\pi|\mathbf{r} - \mathbf{s}'|^2}\,\mathrm{d}s' \tag{7}$$

Again, $d(\mathbf{r})$ is well defined out of S, and satisfies there $\nabla^2 d = 0$. If $\mathbf{r} = \mathbf{s} \in S$, the right-hand side of (7) still converges. This is due to the fact that $\cos(\mathbf{s} - \mathbf{s}', \mathbf{n}') \to 0$ when $\mathbf{s}' \to \mathbf{s}$.

The obvious immediate questions regarding $\phi(\mathbf{r})$ and $d(\mathbf{r})$ relate to their behaviour and the behaviour of their derivatives as the observation point $\mathbf{r}$ approaches a point $\mathbf{s}$ on the surface S. We already know that the right-hand sides of (6) and (7) converge to some values for $\mathbf{r} = \mathbf{s}$. This definitely does not mean that those values are the limits of both integrals as $\mathbf{r} \to \mathbf{s}$. However, by using some delicate techniques one can rigorously prove that the potential generated by a simple layer is continuous everywhere, i.e.,

$$\lim_{\mathbf{r} \to \mathbf{s}} \phi(\mathbf{r}) = \int_S \sigma(\mathbf{s}') \frac{1}{4\pi|\mathbf{s} - \mathbf{s}'|} \mathrm{d}s' \tag{8}$$

Let us now define the positive and negative sides of S (if S is closed, its exterior is defined as the positive side), and let $\boldsymbol{\nu}$ denote the normal to S at a fixed point s. By agreement $\boldsymbol{\nu}$ will always point to the positive side of S. Then one can show that $\partial\phi/\partial\nu$ is not continuous at the surface, but rather produces a jump given by

$$\lim_{\mathbf{r} \to \mathbf{s}\pm} \frac{\partial \phi}{\partial \nu}(\mathbf{r}) = \mp\tfrac{1}{2}\sigma(\mathbf{s}) + \int_S \sigma(\mathbf{s}') \frac{\cos(\mathbf{s}' - \mathbf{s}, \boldsymbol{\nu})}{4\pi|\mathbf{s} - \mathbf{s}'|^2} \mathrm{d}s' \tag{9}$$

An obvious conclusion from (9) is that the normal component of the electric field has a jump discontinuity at the surface S equal to the surface charge density there.

The discussion of continuity at the surface regarding the double layer potential and its normal derivative provides opposite answers. The potential $d(\mathbf{r})$ satisfies

$$\lim_{\mathbf{r} \to \mathbf{s}\pm} d(\mathbf{r}) = \pm\tfrac{1}{2}\eta(\mathbf{s}) + \int_S \eta(\mathbf{s}') \frac{\cos(\mathbf{s} - \mathbf{s}', \mathbf{n})}{4\pi|\mathbf{s} - \mathbf{s}'|^2} \mathrm{d}s' \tag{10}$$

i.e., it possesses a jump discontinuity upon crossing a double layer. Its normal derivative $\partial d/\partial\nu$ is however continuous everywhere.

Define now the 'kernel'

$$k(\mathbf{s}, \mathbf{r}') = \frac{\cos(\mathbf{s} - \mathbf{r}', \mathbf{n})}{4\pi|\mathbf{s} - \mathbf{r}'|^2} \tag{11}$$

Then (10) becomes

$$\lim_{\mathbf{r} \to \mathbf{s}\pm} d(\mathbf{r}) = \pm\tfrac{1}{2}\eta(\mathbf{s}) + \int_S k(\mathbf{s}, \mathbf{s}')\eta(\mathbf{s}') \, \mathrm{d}s' \tag{12}$$

while the equivalent of (9) is

$$\lim_{\mathbf{r}\to\mathbf{s}\pm} \frac{\partial \phi}{\partial v}(\mathbf{r}) \mp \tfrac{1}{2}\sigma(\mathbf{s}) + \int_S k(\mathbf{s}', \mathbf{s})\sigma(\mathbf{s}')\,\mathrm{d}s' \tag{13}$$

Once the concepts of simple and double layers are fully introduced, one is ready to establish the integral equations of potential theory. This is done via what is known as the 'Dirichlet and Neumann problems'.

9.2.2 Integral Equations in Potential Theory

We will start with what is known as the classical formulation of an interior Dirichlet problem.

Let v be an open bounded region with a smooth boundary S, and consider the problem

$$-\nabla^2 w = 0, \qquad \mathbf{r} \in V; \qquad w = f, \qquad \mathbf{r} \in S \tag{14}$$

where f is a given continuous function over S. The classical approach is to find a solution in terms of a double-layer spread on S with unknown density $\eta(\mathbf{s})$. The solution $w(\mathbf{r})$ must then satisfy

$$w(\mathbf{r}) = \int_S \frac{\cos(\mathbf{r} - \mathbf{s}', \mathbf{n}')}{4\pi|\mathbf{r} - \mathbf{s}'|^2}\eta(\mathbf{s}')\,\mathrm{d}s' \tag{15}$$

where $\mathbf{n}'$ is the outward normal to S at $\mathbf{s}'$.

If we look for a solution within the interior of V, the boundary condition given by (14) is, in view of (10), equivalent to

$$f(\mathbf{s}) = -\tfrac{1}{2}\eta(\mathbf{s}) + \int_S k(\mathbf{s}, \mathbf{s}')\eta(\mathbf{s}')\,\mathrm{d}s', \qquad \mathbf{s} \in S \tag{16}$$

which is a Fredholm equation of the second kind for $\eta(\mathbf{s}')$. Here $k(\mathbf{s}, \mathbf{s}')$ is given by (11).

It is well known that $\lambda = \frac{1}{2}$ is not an eigenvalue of

$$\lambda\phi(\mathbf{s}) = \int_S k(\mathbf{s}, \mathbf{s}')\phi(\mathbf{s}')\,\mathrm{d}s' \tag{17}$$

Therefore, eqn. (16) has a unique solution $\eta(\mathbf{s}')$. Consequently, (14) has a unique solution of the form (15) where $\eta(\mathbf{s}')$ is the unique solution of (16). Using the maximum principle for harmonic functions one can easily establish the fact that (14) has no more than one solution. Thus (14) has a unique solution given by (15).

We next move to formulate both interior and exterior Dirichlet problems as integral equations of the first kind.

One has to keep in mind that upon solving an exterior Dirichlet problem, one should always impose some boundary condition at infinity to assure uniqueness of the solution. This condition is, in general, a requirement for a vanishing potential at infinity.

Let V_e be the exterior of a bounded open region V_i whose boundary is S. The exterior Dirichlet problem is defined as the following boundary value problem

$$\nabla^2 v = 0, \qquad \mathbf{r} \in V_e; \qquad v|_S = f, \qquad v|_\infty = 0 \tag{18}$$

Then one can show the following:

(1) A solution of (18) must satisfy

$$v|_\infty = 0\left(\frac{1}{|\mathbf{r}|}\right), \qquad \left.\frac{\partial v}{\partial r}\right|_\infty = 0\left(\frac{1}{|\mathbf{r}|^2}\right) \tag{19}$$

(2) Let $G(\mathbf{r}, \mathbf{r}') = 1/(4\pi|\mathbf{r} - \mathbf{r}'|)$, v a solution to (18), $\mathbf{r}'$ a fixed point, and S_R the surface of a sphere of radius R centred at any arbitrary origin. Then

$$\lim_{R\to\infty} \int_{S_R} \left(G\frac{\partial v}{\partial n} - v\frac{\partial G}{\partial n}\right) \mathrm{d}s = 0 \tag{20}$$

The interior Dirichlet problem associated with (18) is

$$\nabla^2 w = 0, \qquad \mathbf{r} \in V_i; \qquad w|_S = f \tag{21}$$

where f in (18) and (21) are identical.

As previously mentioned, the free space fundamental solution $G(\mathbf{r}, \mathbf{r}')$ satisfies

$$-\nabla^2 G = \delta(\mathbf{r} - \mathbf{r}'), \qquad \text{for all } \mathbf{r}, \mathbf{r}' \tag{22}$$

By multiplying (22) by w and (21) by G, adding and integrating over V_i, one gets

$$\int_{V_i} (G\nabla^2 w - w\nabla^2 G)\, \mathrm{d}v_i = \begin{cases} w(\mathbf{r}') & \mathbf{r}' \in V_i \\ 0 & \mathbf{r}' \in V_e \end{cases} \tag{23}$$

Then by applying Green's theorem one easily gets

$$\int_S \left[G(\mathbf{s}, \mathbf{r}')\frac{\partial w}{\partial n} - w\frac{\partial G(\mathbf{s}, \mathbf{r}')}{\partial n}\right] \mathrm{d}s = \begin{cases} w(\mathbf{r}') & \mathbf{r}' \in V_i \\ 0 & \mathbf{r}' \in V_e \end{cases} \tag{24}$$

where $\mathbf{n}$ is pointing outward over S.

Similarly one may repeat this procedure, regarding v and G, except that the surface integral is now composed of one integral over S, and a second integral over S_R. This second integral as mentioned above vanishes as $R \to \infty$. Thus

$$\int_S \left(-G\frac{\partial v}{\partial n} + v\frac{\partial G}{\partial n} \right) \mathrm{d}s = \begin{cases} 0 & \mathbf{r}' \in V_\mathrm{i} \\ v(\mathbf{r}') & \mathbf{r}' \in V_\mathrm{e} \end{cases} \tag{25}$$

We now add (25) and (24). Both w and v equal to f on S, and this leads to

$$\int_S G(\mathbf{s}, \mathbf{r}')z(\mathbf{s})\,\mathrm{d}s = \begin{cases} w(\mathbf{r}') & \mathbf{r}' \in V_\mathrm{i} \\ v(\mathbf{r}') & \mathbf{r}' \in V_\mathrm{e} \end{cases}$$

where

$$z(\mathbf{s}) = \frac{\partial w}{\partial n} - \frac{\partial v}{\partial n}, \qquad \mathbf{s} \in S \tag{26}$$

Due to the symmetry of G in $\mathbf{r}, \mathbf{r}'$ one finally gets

$$\int_S G(\mathbf{r}, \mathbf{s}')z(\mathbf{s}')\,\mathrm{d}s' = \begin{cases} w(\mathbf{r}) & \mathbf{r} \in V_\mathrm{i} \\ V(\mathbf{r}) & \mathbf{r} \in V_\mathrm{e} \end{cases} \tag{27}$$

In other words, both w and v can be expressed as a simple layer potential. Since this kind of potential is continuous at the surface S, the equation for the unknown function $z(\mathbf{s}')$ is a Fredholm integral equation of the first kind, namely

$$\int_S \frac{1}{4\pi|\mathbf{s} - \mathbf{s}'|} z(\mathbf{s}')\,\mathrm{d}s' = f(\mathbf{s}), \qquad \mathbf{s} \in S \tag{28}$$

Once z is found, w and v are calculated by (27).

We next move to the interior Neumann problem formulated by

$$-\nabla^2 w = 0, \qquad \mathbf{r} \in V_\mathrm{i}; \qquad \frac{\partial w}{\partial n} = f, \qquad \mathbf{r} \in S \tag{29}$$

where $\mathbf{n}$ is the outward normal and f is continuous over S. First, it is easily seen that (29) will not always possess a solution. A necessary condition for (29) to be well posed is

$$\int_S f(\mathbf{s}')\,\mathrm{d}s' = 0 \tag{30}$$

Secondly, if (30) is satisfied it can be shown that (29) has a solution, but not a unique one, since one can add any arbitrary constant to it and get another solution. The exterior Neumann problem has no restriction like (30) and is defined by

$$\nabla^2 v = 0, \qquad \mathbf{r} \in V_e; \qquad \frac{\partial v}{\partial n} = f, \qquad \mathbf{r} \in S; \qquad v|_\infty = 0 \tag{31}$$

It can be shown that (31) has a unique solution. Finally, the Neumann problem can be solved in terms of an integral equation in a similar manner to the Dirichlet problem.

9.2.3 Green's Function for a Given Region

Green's function is an important feature related to solutions of integral equations. Previously, the most commonly used free-space Green's function was defined by (1) and (2) as the potential distribution in space due to a positive point charge. Next we shall define the Green's function of a given region V as the potential distribution due to that same charge provided that the boundary S of V is a grounded metallic shell.

It is therefore the solution $G(\mathbf{r}, \mathbf{r}')$ of the boundary value problem

$$-\nabla^2 G = \delta(\mathbf{r} - \mathbf{r}'), \qquad \mathbf{r}, \mathbf{r}' \in V; \qquad G = 0, \qquad \mathbf{r} \in S \tag{32}$$

Upon discussing together both Green's functions, the free-space one will be denoted as $G_0(\mathbf{r}, \mathbf{r}')$.

One can always separate and write

$$G(\mathbf{r}, \mathbf{r}') = G_0(\mathbf{r}, \mathbf{r}') + w(\mathbf{r}, \mathbf{r}') \tag{33}$$

where G_0 is the free-space fundamental solution satisfying (1) and w solves

$$\nabla^2 w = 0, \qquad \mathbf{r} \in V; \qquad w = -G_0, \qquad \mathbf{r} \in S \tag{34}$$

To find G is thus equivalent to finding a harmonic function w in V given its boundary values on S.

The main properties of G are the following:

(1) G is unique;
(2) G is symmetric, i.e., $G(\mathbf{r}, \mathbf{r}') = G(\mathbf{r}', \mathbf{r})$;
(3) G is positive in V;
(4) for three or more dimensions $0 < G(\mathbf{r}, \mathbf{r}') < G_0(\mathbf{r}, \mathbf{r}')$; and
(5) the integral operator generated by G is completely continuous.

Green's function plays a major role in solving integral equations. This is demonstrated for example in solving a Dirichlet problem for Poisson equation

$$-\nabla^2 w = \rho(\mathbf{r}), \qquad \mathbf{r} \in V; \qquad w = f, \qquad \mathbf{r} \in S \tag{35}$$

Here $\rho(\mathbf{r})$ is the source function specified over V while f is given on the boundary S. By multiplying (32) by w and (35) by G, then subtracting and integrating over V we get

$$\int_V (-w\nabla^2 G + G\nabla^2 w)\,\mathrm{d}v = \int_V (w\delta(\mathbf{r}-\mathbf{r}') - G\rho(\mathbf{r}))\,\mathrm{d}v \tag{36}$$

Using Green's theorem regarding the left side of (36) one deduces

$$w(\mathbf{r}') = \int_V G(\mathbf{r},\mathbf{r}')\rho(\mathbf{r})\,\mathrm{d}v - \int_S f(\mathbf{s})\frac{\partial G(\mathbf{s},\mathbf{r}')}{\partial n}\,\mathrm{d}s$$

since G vanishes on S. By changing the roles of $\mathbf{r},\mathbf{r}'$ we finally have

$$w(\mathbf{r}) = \int_V G(\mathbf{r},\mathbf{r}')\rho(\mathbf{r}')\,\mathrm{d}v' - \int_S f(\mathbf{s}')\frac{\partial G(\mathbf{r},\mathbf{s}')}{\partial n}\,\mathrm{d}s' \tag{37}$$

One can interpret (37) as follows: the solution of (35) is related to two influence functions, $G(\mathbf{r},\mathbf{r}')$ that works for the interior sources $\rho(\mathbf{r})$ and

$$Z(\mathbf{r},\mathbf{r}') \triangleq -\frac{\partial G(\mathbf{r},\mathbf{r}')}{\partial n} \tag{38}$$

that works for the 'boundary sources' $f(\mathbf{s})$.

Now consider the case $\rho \equiv 0, f \equiv 1$. Here $w(\mathbf{r})$ must be identically 1. Thus one gets

$$\int_S Z(\mathbf{r},\mathbf{r}')\,\mathrm{d}s' = 1, \qquad \mathbf{r} \in V \tag{39}$$

This has an obvious physical interpretation: $Z(\mathbf{r},\mathbf{r}')$ is the negative of the charge density induced over S, due to a unit source at $\mathbf{r}$ when S is grounded. Clearly, the total induced charge must be -1. Further implementations of the Green's function will be discussed later, regarding the solution of various problems pertaining to static and time-dependent fields.

We now turn to a brief description of the most common methods for solving linear operator equations in general and integral operator equations in particular.

9.3 METHODS OF SOLUTION

We shall briefly discuss the method of moments, the Rayleigh–Ritz method, the Galerkin method and the point matching (collocation) method. One should notice that the Rayleigh–Ritz and Galerkin procedures are, in some cases, identical and that the method of moments—in spite of its 'general' formulation—is basically a variational method as well.

9.3.1 Method of Moments

Consider the inhomogeneous equation

$$Lf = g \tag{40}$$

where L is a linear operator and g is a known function. Assuming that f be written in terms of expansion functions $\{f_n\}$ as

$$f = \sum a_n f_n \tag{41}$$

where $\{a_n\}$ are constants, one gets

$$\sum a_n L f_n = g \tag{42}$$

If now a suitable inner product $\langle\cdot,\cdot\rangle$ has been determined, one can define a set of test functions $\{w_m\}$ and have

$$\sum a_n \langle w_m, Lf_n\rangle = \langle w_m, g\rangle \tag{43}$$

for all m.

In matrix notation (43) will be expressed as

$$\mathbf{l}_{mn}\underline{a}_a = \underline{g}_m \tag{44}$$

where

$$\mathbf{l}_{mn} = \begin{bmatrix} \langle w_1, Lf_1\rangle\langle w_1, Lf_2\rangle \cdots \\ \langle w_2, Lf_1\rangle\langle w_2, Lf_2\rangle \cdots \\ \cdots\cdots\cdots\cdots\cdots\cdots \end{bmatrix} \tag{45}$$

$$a_n = \begin{bmatrix} \alpha_1 \\ \alpha_2 \\ \vdots \end{bmatrix} \qquad g_m = \begin{bmatrix} \langle w_1, g\rangle \\ \langle w_2, g\rangle \\ \vdots \end{bmatrix} \tag{46}$$

The solution of (44) is simply

$$\underline{a}_n = \mathbf{l}_{nm}^{-1}\underline{g}_m \tag{47}$$

and upon defining

$$\underline{f}_n^{\mathrm{T}} = [f_1, f_2, \ldots] \tag{48}$$

one finally gets

$$f = \underline{f}_n^{\mathrm{T}}\, \underline{a}_n = \underline{f}_n^{\mathrm{T}}\mathbf{l}_{nm}^{-1}\underline{g}_m \tag{49}$$

as either an exact or an approximate solution, depending upon the sets $\{\mathrm{f}_\mathrm{n}\}$, $\{\mathrm{w}_\mathrm{n}\}$.

9.3.2 The Galerkin Method

In applying the method of moments, two sets of functions namely, the expansion and the test functions are involved. The particular choice of identical sets, i.e.,

$$\underline{w}_n = \underline{f}_n \tag{50}$$

for all n, leads to a scheme known as the Galerkin method. Naturally it is the simplest one but not always the most convenient. However, the condition in which the Galerkin method leads to convergence are quite clear and will be specified later.

9.3.3 The Rayleigh–Ritz Method

This approach concerns a specific class of operators, namely 'positive definite bounded below'. A linear operator L will be called positive definite, if it is symmetric, i.e.,

$$\langle Lu, v \rangle = \langle v, Lu \rangle \tag{51}$$

for all u, v, and if

$$\langle Lu, u \rangle > 0, \qquad u \neq 0. \tag{52}$$

(if L is not symmetric but has an adjoint L^a, a variational formulation for (40) can still be worked out (Jeng & Wexler, 1978). If, further, there exists a positive number γ^2 that satisfies

$$\langle Lu, u \rangle \geq \gamma^2 \langle u, u \rangle \tag{53}$$

for all u, the operator L will be considered as 'bounded below'. Now, if L of (40) satisfies (51)–(53) then (40) has a unique solution f, and this solution is the minimum solution for the functional

$$F(u) \triangleq \langle Lu, u \rangle - 2\langle u, g \rangle \tag{54}$$

The Rayleigh–Ritz procedure provides a minimising sequence $\{u_n\}$ to $F(u)$, that converges 'in energy' to the unique solution f:

$$\|u_n - f\| \triangleq \langle L(u_n - f), u_n - f \rangle \underset{n \to \infty}{\to} 0 \tag{55}$$

This is done by first choosing an expansion set $\{f_n\}$ that is 'complete in energy' and whose finite subsets are all independent. Then, the 'nth order approximation' is expressed by

$$u_n = \sum_{i=1}^{n} a_i f_i \tag{56}$$

where the a_i are chosen to provide a minimum to $F(u_n)$. Finally the resulting sequence is shown to converge in energy to f.

At the nth step, the system to solve is a set of n linear equations in n unknowns given by

$$\sum_{k=1}^{n} \langle Lf_i, f_k \rangle a_k = \langle g, f_i \rangle, \qquad i = 1, \ldots, n \tag{57}$$

The system (57) is identical to (43), if $\underline{w}_n = \underline{f}_n$. This means that the Galerkin and the Rayleigh–Ritz methods are the same provided that the operator L is positive definite and bonded below. Other methods of constructing the minimising sequence obviously differ from the Galerkin scheme. If L does not fully satisfy (51)–(53), Galerkin's method can still be applicable. A fairly general sufficiency criterion for the convergence of Galerkin's approach can be formulated as follows: assume that L can be written as

$$L = L_0 + K \tag{58}$$

where L_0 is a positive definite bounded below operator. If (40) has no more than one solution, and if the operator $L^{-1}K$ is completely continuous in the energy space of L_0, then Galerkin's approximate solutions converge in the energy of L_0 to the exact solution of (40). The Rayleigh–Ritz scheme, for a positive definite operator bounded below, is quite powerful. Its performance for integral equations was demonstrated by McDonald *et al.* (1973, 1974).

9.3.4 Point-matching Method

Since the integrals $\mathbf{l}_{mn} = \langle \underline{w}_m, L\underline{f}_n \rangle$ are often heavily time-consuming when computed, an alternative way—known as a point-matching method—for obtaining approximate solutions is suggested. Equation (3) must be satisfied only at discrete points whose total number N must be identical to the number of expansion functions. One then solves a system similar to (44) to get the unknown $a_1, \ldots, a_n$. In terms of the method of moments, collocation is identical to using Dirac delta functions as the test functions. Point-matching solutions, as expected, are in general less accurate than the corresponding Galerkin approximations. For low orders of solution they are quite sensitive to the particular choice of the matching points. This was effectively demonstrated by Jeng & Wexler (1977).

So far we concluded that the method of moments provides the Galerkin and the point-matching schemes as special cases, while the variational approach through the Rayleigh–Ritz minimising sequence is a particular case of the Galerkin method. However, the general method of moments has by itself a variational interpretation (e.g., Jeng & Wexler, 1978).

9.4 ISOPARAMETRIC BEM ALGORITHM

The solution to integral equations via the Boundary Element Method (BEM) involves: (1) discretisation of the original continuous problem through proper modelling of boundary surfaces and sources; and (2) approximation of the integral operator. The outcome is a set of linear equations that may then be solved by standard procedures; viz., elimination, inversion or iterative techniques.

9.4.1 Modelling Algorithm

Boundary surfaces are represented by triangular elements which are referred to a two-dimensional simplex defined by the vertices (0, 1), (0, 0), (1, 0) in the 'local' ξ–η plane. A mapping from the 'local' to the original 'global' frame is given by appropriate shape functions (expansion functions) which are polynomials $\{\alpha_i\}$ of order N.

The degree N of these polynomials determines the 'curvature' of the element and the number of nodes required to specify it. The number of nodes is given by

$$M = \tfrac{1}{2}(N + 1)(N + 2) \tag{59}$$

The global co-ordinates may then be expressed as

$$\begin{aligned} x &= \sum_{i=1}^{M} \alpha_i(\xi, \eta)x_i \\ y &= \sum_{i=1}^{M} \alpha_i(\xi, \eta)y_i \\ z &= \sum_{i=1}^{M} \alpha_i(\xi, \eta)z_i \end{aligned} \tag{60}$$

where α_i are defined on the ξ–η plane and satisfy

$$\alpha_i = \begin{cases} 1 & \text{at node } i \\ 0 & \text{at all other nodes} \end{cases} \tag{61}$$

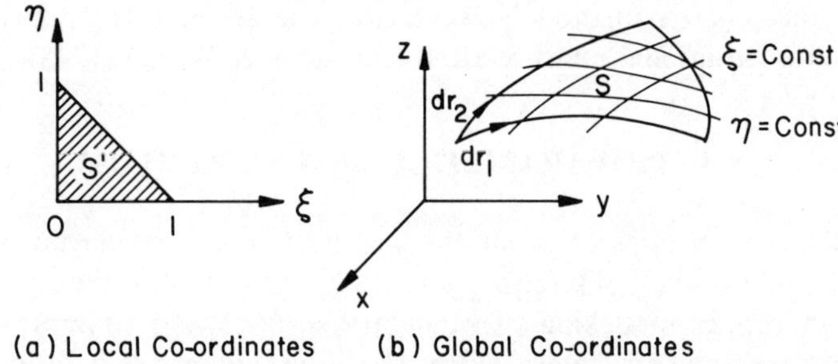

FIG. 1. (a) The simplex element and (b) its transformation to three-space.

Obviously α_i needs to be generated only once. The same set may be applied to each element thus saving on both storage and computing costs. Equation (60) essentially describes a family of surfaces in space (Fig. 1). Clearly $N = 1$ would correspond to a linear approximation of the surface element, $N = 2$ to a quadratic one, etc. In theory, N can be increased indefinitely for better accuracy, but in practice $N \leq 3$ has usually been found to be satisfactory. Expansion functions up to $N = 3$ are given in Table 1. In Fig. 2 we show the spatial disposition of nodes in the simplex and their local node numbering schemes for various values of N.

Boundary sources may also be approximated in terms of the same set of expansion functions. In so doing, an isoparametric approach is implicitly assumed. Most researchers have used this concept in the sense that both sources and surface geometry are modelled to the same degree. Recently, Lachat & Watson (1975) used quadratic variation in geometry together with linear, quadratic or cubic variation of the unknowns with respect to intrinsic co-ordinates. Respectively, these correspond to superparametric, isoparametric and subparametric implementation schemes. Further discussion of this approach was given by Nedelec (1976). Obviously a

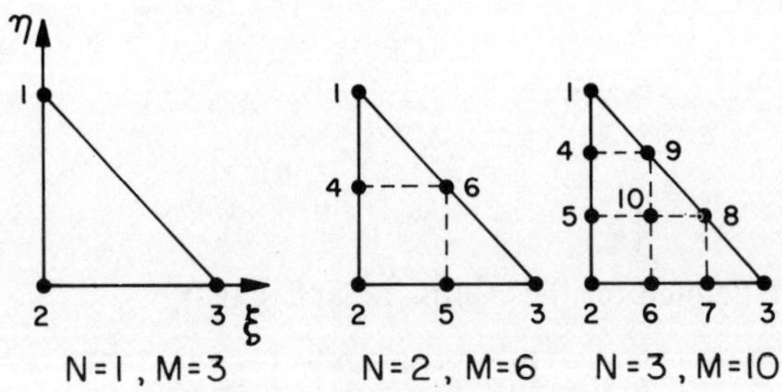

FIG. 2. The 'local' nodes.

TABLE 1
SHAPE FUNCTIONS $\alpha_i(\xi, \eta)$. $L_1 = \eta$, $L_2 = 1 - \eta - \xi$, $L_3 = \xi$

i	$N = 1$	$N = 2$	$N = 3$
1	L_1	$L_1(2L_1 - 1)$	$\frac{1}{2}L_1(3L_1 - 1)(3L_1 - 2)$
2	L_2	$L_2(2L_2 - 1)$	$\frac{1}{2}L_2(3L_2 - 1)(3L_2 - 2)$
3	L_3	$L_3(2L_3 - 1)$	$\frac{1}{2}L_3(3L_3 - 1)(3L_3 - 2)$
4		$4L_1L_2$	$\frac{9}{2}L_1L_2(3L_1 - 1)$
5		$4L_2L_3$	$\frac{9}{2}L_1L_2(3L_2 - 1)$
6		$4L_3L_1$	$\frac{9}{2}L_2L_3(3L_2 - 1)$
7			$\frac{9}{2}L_2L_3(3L_3 - 1)$
8			$\frac{9}{2}L_3L_1(3L_3 - 1)$
9			$\frac{9}{2}L_3L_1(3L_1 - 1)$
10			$27\,L_1L_2L_3$

reduction in the number of variables can be achieved by using the superparametric scheme.

Usually, by a shrewd choice of element partitioning to fit the problem, it may be possible to optimise the number of variables required. This measure would mean grading and clustering of elements around areas with intense activity (e.g., geometrically singular points), and less emphasis paid elsewhere. Also, where some regularity in geometry occurs, it is possible to take advantage of symmetry, thus reducing the number of unknowns required. Recent advances in modelling techniques have given rise to the use of bicubic surface patches defined by B-splines (Coons, 1974; Clark, 1976). These B-spline curves used as boundaries of an element allow discontinuities in geometry without causing 'folds' or 'creases' within the element. This will prove advantageous in the modelling of curved surfaces with large elements. The present algorithm used by the authors has to rely on at least a third degree representation (i.e., $N = 3$) in order to model, for example, one-eighth the surface of a sphere by one element. The fidelity of the representation achieved (to within 0·5% deviation) is at the expense of more nodes per element and, at the same time, the need for a higher-order quadrature scheme.

9.4.2 The Functional

Consider the general form of an integral problem

$$K\sigma = g \tag{62}$$

where K is the integral operator, σ is the unknown and g is the known forcing function or boundary condition. While any method of moments is 'projective' in nature, the variational Rayleigh–Ritz approach is justified through a 'stationarity' of the energy functional F at discrete points. For K self-adjoint, the functional is given as

$$F = \langle K\sigma, \sigma \rangle - 2\langle \sigma, g \rangle \tag{63}$$

for the real case, or as

$$F = \langle K\sigma, \sigma \rangle - \langle \sigma, g \rangle - \langle g, \sigma \rangle \tag{64}$$

for the complex one. The pair of brackets denote an inner product defined by

$$\langle u, v \rangle = \int_R uv^* \, \mathrm{d}\Omega \tag{65}$$

This definition could include a weighting function. Inherent in both (63) and (64) is the assumption that the Galerkin scheme is implied, i.e., expansion and testing functions are chosen to be identical. Hence, by substituting

$$\sigma = \underline{\sigma}^{\mathrm{T}}\underline{\alpha} = \underline{\alpha}^{\mathrm{T}}\underline{\sigma} \tag{66}$$

Where the underlining bars and the superscript 'T' denote column matrix and matrix transpose respectively, eqn. (63), for example, may be re-written as

$$F = \underline{\sigma}^{\mathrm{T}}\langle K\underline{\alpha}, \underline{\alpha}^{\mathrm{T}} \rangle \underline{\sigma} - 2\underline{\sigma}^{\mathrm{T}}\langle \underline{\alpha}, g \rangle \tag{67}$$

Considering the variational of the functional with respect to $\underline{\sigma}$, one gets

$$\begin{aligned} \frac{\partial F}{\partial \underline{\sigma}} &= \frac{\partial \underline{\sigma}^{\mathrm{T}}}{\partial \underline{\sigma}} \{\langle K\underline{\alpha}, \underline{\alpha}^{\mathrm{T}} \rangle \underline{\sigma}\} + \frac{\partial}{\partial \underline{\sigma}} \{\langle K\underline{\alpha}, \underline{\alpha}^{\mathrm{T}} \rangle \underline{\sigma}\}^{\mathrm{T}} \underline{\sigma} - 2\frac{\partial \underline{\sigma}^{\mathrm{T}}}{\partial \underline{\sigma}} \langle \underline{\alpha}, g \rangle \\ &= \langle \underline{\alpha}, \underline{\alpha}^{\mathrm{T}} \rangle \underline{\sigma} + \frac{\partial}{\partial \underline{\sigma}} \{\underline{\sigma}^{\mathrm{T}} \langle K\underline{\alpha}, \underline{\alpha}^{\mathrm{T}} \rangle^{\mathrm{T}}\} \underline{\sigma} - 2\langle \underline{\alpha}, g \rangle \\ &= \langle K\underline{\alpha}, \underline{\alpha}^{\mathrm{T}} \rangle \underline{\sigma} + \langle \underline{\alpha}, \underline{\alpha}^{\mathrm{T}} \rangle \underline{\sigma} - 2\langle \underline{\alpha}, g \rangle \end{aligned} \tag{68}$$

Since K is self-adjoint, one finally gets

$$\frac{\partial F}{\partial \underline{\sigma}} = 2\langle \underline{\alpha}, \underline{\alpha}^{\mathrm{T}} \rangle \underline{\sigma} - 2\langle \underline{\alpha}, g \rangle \tag{69}$$

Enforcing the functional to be stationary at the solution point, we get

$$\langle K\underline{\alpha}, \underline{\alpha}^{\mathrm{T}} \rangle \underline{\sigma} = \langle \underline{\alpha}, g \rangle \tag{70}$$

Recognising that the inner product on the left-hand side of (70) is a square matrix, and that on the right-hand side there is a column matrix, one may re-write it as

$$\mathbf{S}\underline{\sigma} = \underline{b} \tag{71}$$

The integral form of (71) is

$$\int_S \underline{\alpha}(\mathbf{s}) \int_S K(\mathbf{s}, \mathbf{s}')\underline{\alpha}^{\mathrm{T}}(\mathbf{s}')\,\mathrm{d}s'\,\mathrm{d}s\underline{\sigma} = \int_S \underline{\alpha}(\mathbf{s})g\,\mathrm{d}s \tag{72}$$

where $K(\mathbf{s}, \mathbf{s}')$ may be the pertinent Green's function of the problem. It is obvious that (72) could have been obtained by a direct application of the Galerkin scheme to (62), subsequent to substitution of eqn. (66) into it.

The equivalence of the Galerkin and the variational algorithms is not confined to K being self-adjoint. As previously mentioned, Jeng & Wexler (1978) have demonstrated this equivalence for non-self-adjoint operators as well.

9.4.3 Numerical Integration

Evaluation of the matrix elements posed by (72) involve both single and double surface integrals over the boundary. Standard Gaussian quadrature is used so that the exact integral is approximated by a finite Riemann sum. Since computation time increases drastically as one moves to higher-order quadrature, one has to compromise especially in the evaluation of the double integrals. Nevertheless, the trade-off involved in using a higher-order scheme with non-planar elements, rather than the low-order rectangular rule with planar patches seems advantageous.

The actual integration is performed over the two-dimensional simplex, thus contributing to the algorithmic simplicity as the set of Gaussian points associated with it needs to be generated just once. To obtain an expression for the differential area in the global frame, consider the differential changes in position in it due to changes in the simplex. With reference to Fig. 1, we get

$$\mathrm{d}\mathbf{r}_1(\xi) = \frac{\partial x}{\partial \xi}\mathrm{d}\xi\hat{\mathbf{i}} + \frac{\partial y}{\partial \xi}\mathrm{d}\xi\hat{\mathbf{j}} + \frac{\partial z}{\partial \xi}\mathrm{d}\xi\hat{\mathbf{k}} \tag{73}$$

and

$$\mathrm{d}\mathbf{r}_2(\eta) = \frac{\partial x}{\partial \eta}\mathrm{d}\eta\hat{\mathbf{i}} + \frac{\partial y}{\partial \eta}\mathrm{d}\eta\hat{\mathbf{j}} + \frac{\partial z}{\partial \eta}\mathrm{d}\eta\hat{\mathbf{k}} \tag{74}$$

The differential area may then be expressed as

$$\mathrm{d}\mathbf{s} = \mathrm{d}\mathbf{r}_1 \times \mathrm{d}\mathbf{r}_2 \tag{75}$$

The Jacobian of the transformation is calculated from (73) and (74) and given by

$$J = \sqrt{M_{31}^2 + M_{32}^2 + M_{33}^2} \tag{76}$$

Where M_{ij} are the minors of

$$A_{ij} = \begin{bmatrix} \dfrac{\partial x}{\partial \xi} & \dfrac{\partial y}{\partial \xi} & \dfrac{\partial z}{\partial \xi} \\ \dfrac{\partial x}{\partial \eta} & \dfrac{\partial y}{\partial \eta} & \dfrac{\partial z}{\partial \eta} \\ 1 & 1 & 1 \end{bmatrix} \tag{77}$$

taken along the bottom row.

Regarding the double surface integral, one should note that the inner integral is singular at the Gaussian points of the outer integral, since the same points are being used. To cater for this functional singularity, the 'addition–subtraction' method (McDonald *et al.*, 1974) is used. To further illustrate this approach, consider the inner integral on the left-hand side of (72), namely

$$H(\mathbf{s}) = \int_S \alpha(\mathbf{s}')K(\mathbf{s}, \mathbf{s}')\,\mathrm{d}s' \tag{78}$$

where $H(s)$ is the kernel of the singularity. Re-write this expression as

$$H(\mathbf{s}) = \int_S \{\alpha(\mathbf{s}') - \alpha(\mathbf{s})\}K(\mathbf{s}, \mathbf{s}')\,\mathrm{d}s' + \alpha(\mathbf{s})\int_S K(\mathbf{s}, \mathbf{s}')\,\mathrm{d}s' \tag{79}$$

The first term on the right-hand side is now an integral with a non-singular integrand at $\mathbf{s}' = \mathbf{s}$, and may be computed accurately with any low-order integration scheme. The second integral has to be calculated analytically over the surface element. However, even for a quadratic surface element, this is impossible. Jeng & Wexler (1977) approximated this second integral by dividing the quadratic element into four planar subsections altogether described by the usual six nodes that are attached to each quadratic element. In spite of this manipulation being *ad hoc*, the numerical results have justified its implementation. The main shortcoming of this approach is that the second integral may not always be calculated by a closed form, and resort to another approximation in the form of a series truncation is always possible.

Lachat & Watson (1976) employed an alternative approach whereby recourse is made to the evaluation of Cauchy principal values and the selection of quadratic schemes according to upper bounds for error in terms of derivatives of the integrands at singular points. Watson, in an unpublished report, has found that accuracy is maintained even when integrating over regions adjacent to functional singularities. It appears that this alternative method is quite applicable. Other treatments for functional singularities are given elsewhere (Burton & Miller, 1971).

Geometrical singularities arise at corners and sharp edges on surface boundaries, leading to singular charge distributions. Care must be exercised when both types of singularities appear simultaneously. Under certain circumstances, the geometrical discontinuity may be treated by representing the charge distribution by a functional quantity dependent on spatial co-ordinates. For example, McDonald *et al.* (1974) represented the edge discontinuity for a thin plate in two dimensions by

$$\sigma(x) \cong \frac{1}{\sqrt{1-x}} \tag{80}$$

But, in most cases, treatment to alleviate the problem is left to a judicious choice of the spatial disposition of elements—using finely graded elements near the singularities.

9.5 ELECTROSTATICS

In scalar potential theory, one is concerned with the solution of the Laplace or Poisson equation in a three-dimensional space subject to Dirichlet and/or Neumann boundary conditions. Consider the exterior Dirichlet problem posed by a surface S surrounded by an infinite region V_e in free space. Let the potential ϕ be harmonic in V_e subject to the boundary condition $\phi(s) = g(s)$, and with asymptotic behaviour

$$\phi \sim \frac{1}{|\mathbf{r}|} \qquad \text{as} \qquad |\mathbf{r}| \to \infty$$

Invoking the free-space Green's function

$$G(\mathbf{r}, \mathbf{r}') = \frac{1}{4\pi|\mathbf{r} - \mathbf{r}'|} \tag{81}$$

the above Laplacian problem may be posed as

$$\phi(\mathbf{r}) = \frac{1}{\varepsilon_0} \int_S G(\mathbf{r}, \mathbf{s}')\sigma(\mathbf{s}')\,\mathrm{d}s', \qquad \mathbf{r} \in (S' + V_e) \tag{82}$$

where $\sigma(\mathbf{s}')$ is an assumed simple-layer distribution of charges, $\varepsilon_0 \simeq 8{\cdot}854 \times 10^{-12}$ F/m the permittivity of the free space, and $\mathbf{r}, \mathbf{r}'$ the observer and source locations respectively. Formally, the Green's function satisfies Poisson's equation

$$-\nabla^2 G(\mathbf{r}, \mathbf{r}') = \delta(\mathbf{r} - \mathbf{r}') \tag{83}$$

as previously stated in (1). In applying the specified boundary conditions, eqn. (82) may be re-written as

$$\frac{1}{\varepsilon_0} \int_S G(\mathbf{r}, \mathbf{s}') \sigma(\mathbf{s}') \, \mathrm{d}s' = g(\mathbf{s}), \qquad \mathbf{r} \in S \tag{84}$$

or as an operator equation

$$K\sigma = g \tag{85}$$

Equation (84) is a Fredholm integral equation of the first kind and may be solved for $\sigma(\mathbf{s})$, thence for ϕ everywhere. Alternatively, the same problem could have been posed as

$$\phi(\mathbf{r}) = \frac{1}{\varepsilon_0} \int_S \frac{\partial G(\mathbf{r}, \mathbf{s}')}{\partial n'} \mu(\mathbf{s}') \, \mathrm{d}s', \qquad \mathbf{r} \in V_{\mathrm{e}} \tag{86}$$

where $\mu(s')$ would be an assumed double-layer (dipole) source distribution over S. Normal derivatives on S would point outwards into V_{e}. Enforcing the given boundary conditions on S, (86) takes the form

$$\frac{1}{\varepsilon_0} \int_S \frac{\partial G(\mathbf{s}, \mathbf{s}')}{\partial n'} \mu(\mathbf{s}') \, \mathrm{d}s' + \frac{\mu(\mathbf{s})}{2\varepsilon_0} = g(s), \qquad \mathbf{s} \in S \tag{87}$$

which is a Fredholm integral equation of the second kind in $\mu(\mathbf{s})$. Both formulations given above are particular cases of the so-called 'indirect method' often alluded to in other fields of engineering. The main conceptual disadvantage in their implementation lies in the formalism attached to the fictitious sources. All other quantities of interest will have to be found indirectly subsequent to the knowledge of the source distribution.

A better formulation in terms of a complete set of boundary data (both essential and natural) is possible, and results in the 'direct method'. Equation (83) is coupled to the Laplacian in ϕ by an algebraic manipulation and integrated over the region V_{e} to obtain

$$\phi(\mathbf{r}) = \int_{V_{\mathrm{e}}} [\varepsilon_0 G(\mathbf{r}, \mathbf{r}') \nabla^2 \phi(\mathbf{r}') - \phi(\mathbf{r}') \nabla^2 G(\mathbf{r}, \mathbf{r}')] \, \mathrm{d}v'; \qquad \mathbf{r}, \mathbf{r}' \in V_{\mathrm{e}} \tag{88}$$

Using Green's theorem the volume integral is transferred into

$$\phi(\mathbf{r}) = \gamma \int_S \left(\varepsilon_0 G(\mathbf{r}, \mathbf{s}') \frac{\partial \phi(\mathbf{s}')}{\partial n'} - \phi(\mathbf{s}') \frac{\partial G(\mathbf{r}, \mathbf{s}')}{\partial n'} \right) \mathrm{d}s', \qquad \mathbf{r} \in V_e \quad (89)$$

where the contribution from the other boundary surface at infinity is negligible due to asymptotic considerations. As previously mentioned (Stakgold, 1968)

$$\gamma = \begin{cases} 0 & \mathbf{r} \notin V_e \\ 1 & \mathbf{r} \in V_e \\ 2 & \mathbf{r} \in S \end{cases} \quad (90)$$

Applying the Dirichlet condition to equation (89), and keeping in mind that $\gamma = 2$ on S, we then get

$$\varepsilon_0 \int_S G(\mathbf{r}, \mathbf{s}') \frac{\partial \phi(\mathbf{s}')}{\partial n'} \mathrm{d}s' = \tfrac{1}{2} g(\mathbf{s}) + \int_S g(\mathbf{s}') \frac{\partial G(\mathbf{s}, \mathbf{s}')}{\partial n'} \mathrm{d}s', \qquad \mathbf{s} \in S \quad (91)$$

which yields directly the true charge, $\varepsilon_0(\partial \phi(\mathbf{s}')/\partial n')$, on the surface. For the single-layer formulation, the actual charge distribution can be obtained from

$$\varepsilon_0 \frac{\partial \phi(\mathbf{s})}{\partial n} = \tfrac{1}{2}\sigma(\mathbf{s}) - \int_S \sigma(\mathbf{s}') \frac{\partial G(\mathbf{s}, \mathbf{s}')}{\partial n'} \mathrm{d}s', \qquad \mathbf{s} \in S \quad (92)$$

subsequent to the determination of $\sigma(\mathbf{s})$ in the initial stage.

The preceding discussion has centred on problems concerning impermeable boundaries. Extension to permeable media problems is straightforward—needing only the consideration of interface conditions. This fact will be further illustrated in Section (9.6).

To get an insight into the solution process one may consider the formulation given by (84). Applying the Galerkin scheme, one gets

$$\int_S \underline{\alpha}(\mathbf{s}) \frac{1}{\varepsilon_0} \int_S G(\mathbf{s}, \mathbf{s}') \underline{\alpha}^{\mathrm{T}}(\mathbf{s}') \, \mathrm{d}s' \, \mathrm{d}s \underline{\sigma} = \int_S \underline{\alpha}(\mathbf{s}) g(s) \, \mathrm{d}s \quad (93)$$

where the polynomic expansion of $\sigma(\mathbf{s})$ is given by (66). Equation (93) is easily recognised to be a special case of (72). The singularity of the inner integral at Gaussian points of the outer integral is catered for by the 'addition–subtraction' technique previously mentioned, with the analytical expression for the modified kernel of singularity given by Jeng & Wexler (1977).

Both parts of the functional defined by (63) have physical significance at the solution point. The first term within the first pair of brackets, namely $K\sigma$, is the potential $\phi(\mathbf{s})$. The product of potential and charge integrated over the surface yields the energy of the system. The second inner product yields twice the energy. From an energy point of view, the parallel with the partial differential equation approach is exact. Based on this energy concept, the electrostatic capacitance of the system may be obtained. Restating the functional in eqn. (63) in integral form, we get

$$F = \underline{\sigma}^{\mathrm{T}} \left\{ \frac{1}{\varepsilon_0} \int_S \underline{\alpha}(\mathbf{s}) \int_S G(\mathbf{s}, \mathbf{s}') \underline{\alpha}^{\mathrm{T}}(\mathbf{s}') \, \mathrm{d}s' \, \mathrm{d}s \right\} \underline{\sigma} - 2\underline{\sigma}^{\mathrm{T}} \int_S \underline{\alpha}(\mathbf{s}) g \, \mathrm{d}s \quad (94)$$

Combining eqns. (93) and (94), one gets for the approximated solution

$$\begin{aligned} F &= \underline{\sigma}^{\mathrm{T}} S \underline{\sigma} - 2\underline{\sigma}^{\mathrm{T}} \underline{b} \\ &= \underline{\sigma}^{\mathrm{T}} \underline{b} - 2\underline{\sigma}^{\mathrm{T}} \underline{b} \\ &= -\underline{\sigma}^{\mathrm{T}} \underline{b} \end{aligned} \quad (95)$$

which is simply the multiplication of a row and a column matrix. This may be rewritten as

$$F = -\int_S g(s) \sigma(\mathbf{s}) \, \mathrm{d}s \quad (96)$$

in which $\sigma(\mathbf{s})$ is understood to be the approximation described by eqn. (66).

In the particular case of $g(s) =$ constant, one gets $F = -QV_0$ where Q denotes the total surface charge and V_0 denotes the constant potential over the surface. Realising that the electrostatic energy E equals $\frac{1}{2}QV_0$, one deduces that $F = -2E$. Hence the capacitance of the system, C, is given by

$$C = -F/V_0^2 \quad (97)$$

The previous equation relates the value of the functional at stationary points, to the capacitance of the system under charge equilibrium. Let us denote the exact and approximate solutions to $K\sigma = g$ by σ_0, σ_1 respectively, and let $K\sigma_1 = g_1$. Obviously $g_1 \neq g$ or σ_1 would be the exact solution. By (95) the value of F at σ_1 is

$$F_1 = (K\sigma_1, \sigma_1) - 2(\sigma_1, g) = -(\sigma_1, g) \quad (98)$$

that leads to

$$(g_1, \sigma_1) = (g, \sigma_1) \quad (99)$$

The approximated energy at the boundary g_1 therefore 'averages' the exact energy g, the boundary condition over S.

9.5.1 Example Problem

The aforementioned problem is formally solved using one-eighth symmetry, i.e., the boundary surface is represented only in the positive sector of a three-dimensional frame. This results with the Green's function being represented by the expression

$$G(x, y, z; x', y', z') = \frac{1}{4\pi}\left\{\frac{1}{[(x \pm x')^2 + (y \pm y')^2 + (z \pm z')^2]^{1/2}}\right\} \tag{100}$$

so that the the right-hand side is expanded to incorporate eight terms. Values of capacitance for three different geometries—the cube, the sphere and the prolate spheroid—are computed via the previously outlined functional concept. With the exception of the cube, exact values are available for the remaining two conic surfaces. As an attempt to correlate data, all three shapes are made to share a common dimension d, where d represents: the side length for the cube; the diameter for the sphere; the semi-major axis of the prolate spheroid. Results for $d = 2m$ are presented in Table 2. All the results are valid for a second degree approximation, isoparametric curved elements are used where applicable.

TABLE 2
CALCULATED CAPACITANCE

	No. of elements	*No. of nodes*	*Capacitance (pF)*	
			Numerical	*Exact*
Cube	12	31	146·06	—
Sphere	4	15	112·07	111·26
Prolate spheroid	14	39	146·39	146·33

9.5.2 Comparison and Interpretation of the Numerical Results

The capacitance of a unit cube (in CGS units) has previously been computed by Reitan & Higgins (1951) using their 'method of subareas'. Enforcing symmetry within cube faces, they used 6 unknowns to obtain a value of 0·6555 e.s.u.—well within their cited upper and lower bounds for a cube of side a (in centimeters): $0{\cdot}622\,11a < c < 0{\cdot}710\,55a$. Jaswon & Symm (1977) reported that Laskar (1974), and Noble (1971) obtained 0·6603 e.s.u. and 0·6607 e.s.u. respectively, the former using 66 variables. From Table 2, our calculated value is 146·06 pF, for a cube of side 2 m, using 12 quadratic elements and a total number of 31 variables. Since the capacitances of geometrically similar solids are in proportion to the ratio of corresponding linear dimensions, it follows that our equivalent value would be 0·6573 e.s.u. It should be noted that upon using linear elements resulting in 10 variables, we obtained values of 0·6539 e.s.u. and 0·6566 e.s.u.; the latter with element outlay designed to cater for the corner singularity. In going from 10 to 31 variables, we have improved our accuracy by 0·1 %. A finer mesh of elements around the corner might have produced a better result.

The exact value of capacitance for a sphere of radius R is given by

$$4\pi\varepsilon_0\left(\frac{1}{r}\right)F$$

The modelling of the surface curvature may be unnecessarily crude, using only four elements. Also, the fact that the estimated value is higher than the exact one does not quite agree with the expected trend of convergence for the variational formulation. This discrepancy may in part or in whole be explained through the nine-point quadrature scheme used for integration.

Numerical values for the ellipsoid have previously been investigated by Jeng & Wexler (1977) employing one sixteenth symmetry. With 7 elements and 22 modes (second degree approximation), they obtained a value of 145·72 pF compared with the exact value of 146·33 pF. The scheme used here employed a slightly different arrangement of elements—14 in all—and 39 modes to describe one-eighth the surface. The capacitance obtained is 146·39 pF which is closer to the exact value. The same reasoning as for the sphere may be used to argue why the approximated value is slightly higher than the exact number. As a point of interest, the potential at $(x, y, z) = (0, 2, 0)$ is found to be 0·594 76V as compared to the exact value of 0·594 84V. This presents a deviation of only 0·01 %.

9.6 MAGNETOSTATICS

The law of Biot-Savart

$$\mathbf{B}(\mathbf{r}) = \mu \int_V \frac{\mathbf{J}(\mathbf{r}') \times (\mathbf{r} - \mathbf{r}')}{4\pi|\mathbf{r} - \mathbf{r}'|^3} \mathrm{d}v' \tag{101}$$

permits one to compute the magnetic flux density field $\mathbf{B}(\mathbf{r})$, in a homogeneous medium of permeability μ, due to a known electrical current $\mathbf{J}(\mathbf{r})$. The permeability of the homogeneous medium is

$$\mu = \mu_0\mu_r \tag{102}$$

where $\mu_0 = 4\pi \times 10^{-7}$ H/m is the permeability of vacuum (and is fixed by the choice of the system of units) and μ_r is the relative permeability of the medium. The relationship between **B** and the magnetic field intensity **H** is expressed by

$$\mathbf{B} = \mu\mathbf{H} \tag{102a}$$

9.6.1 Standard Formulation

As is known by physical observation, an elemental current produces a magnetic field in a plane at right angles to the current. This action is fulfilled, mathematically, by the cross product in (101). The relationship between two vector functions is very conveniently expressed by a dyadic function in a manner analogous to the relationship between column matrices that may be expressed by a matrix.

In particular, the dyad

$$\begin{aligned}\bar{\bar{\mathbf{G}}}(\mathbf{r},\mathbf{r}') = \frac{1}{4\pi|\mathbf{r}-\mathbf{r}'|^3}&((r_x - r'_x)(\hat{u}_y\hat{u}_z - \hat{u}_z\hat{u}_y)\\ &+ (r_y - r'_y)(\hat{u}_z\hat{u}_x - \hat{u}_x\hat{u}_z)\\ &+ (r_z - r'_z)(\hat{u}_x\hat{u}_y - \hat{u}_y\hat{u}_x))\end{aligned} \tag{103}$$

postmultiplied by the vector

$$\mathbf{J}(\mathbf{r}') = \hat{u}_x J_x(\mathbf{r}') + \hat{u}_y J_y(\mathbf{r}') + \hat{u}_z J_z(\mathbf{r}') \tag{104}$$

results in

$$\begin{aligned}\bar{\bar{\mathbf{G}}}(\mathbf{r},\mathbf{r}') \,.\, J(r) = \frac{1}{4\pi|\mathbf{r}-\mathbf{r}'|^3}&((r_x - r'_x)(\hat{u}_y J_z - \hat{u}_z J_y)\\ &+ (r_y - r'_y)(\hat{u}_z J_x - \hat{u}_x J_z)\\ &+ (r_z - r'_z)(\hat{u}_x J_y - \hat{u}_y J_x))\end{aligned} \tag{105}$$

Formally, this multiplication is accomplished by performing the dot products between adjacent unit vectors. Expansion of the integrand of (101) is easily seen to yield (105). And so, (101) can be expressed in vector form of (3), i.e.

$$\mathbf{B}(\mathbf{r}) = \mu \int_V \bar{\bar{\mathbf{G}}}(\mathbf{r},\mathbf{r}') \,.\, \mathbf{J}(\mathbf{r}')\,\mathrm{d}v' \tag{106}$$

Thus, in an analogous fashion, given a specification of **B(r)** over a surface S, the equivalent surface currents required to sustain that field may be obtained from the solution of the simple source integral equation

$$\mathbf{B}(\mathbf{s}) = \mu \int_S \bar{\bar{\mathbf{G}}}(\mathbf{s}, \mathbf{s}') \,.\, \mathbf{J}(\mathbf{s}') \, \mathrm{d}s' \tag{107}$$

Application of the BEM method, with three variables at each node point, is fairly direct.

Thorough discussions of dyadic Green's functions, including various integral equation formulations involving vector generalisations of the scalar equation (89) are available in Tai (1971) and Harrington (1961).

An alternative formulation, involving the magnetic vector potential (Paris & Hurd, 1969) is given by

$$\mathbf{A}(\mathbf{s}) = \mu \int_V G(\mathbf{s}, \mathbf{s}')\mathbf{J}(\mathbf{s}') \, \mathrm{d}v' \tag{108}$$

where $G(\mathbf{r}, \mathbf{r}')$ is the scalar Green's function given by (2). The magnetic flux density field may be obtained by employing

$$\mathbf{B} = \nabla \times \mathbf{A} \tag{109}$$

whenever required. However, computationally it is more convenient and accurate to employ (106)—having obtained the equivalent surface currents—rather than to numerically differentiate the computed magnetic vector potential.

Within regions that contain no current, **H** may be derived from a scalar potential

$$\mathbf{H} = -\nabla\phi$$

where ϕ is known as the magnetomotive force or mmf. When one is free to use a scalar formulation, the mode of solution is identical to that employed for the solution of electrostatic problems.

9.6.2 Permeable Media†

Thus far, problems involving totally homogeneous regions or homogeneous and impermeable regions have been considered. In the latter case, impermeable regions were replaced by homogeneous regions enclosed in equivalent source distributions: simple layer $\sigma(\mathbf{s})$ or double layer (represented by a simple source distribution $G(\mathbf{r}, \mathbf{s}')$ and a dipole layer $\partial G(\mathbf{r}, \mathbf{s}')/\partial n$.

In the solution of permeable-media problems it is necessary to formulate

† This subsection has been substantially published in Wexler (1978).

each region as though each were a homogeneous-region problem and then to impose interface conditions between the regions. Most frequently, interface conditions specify the continuity of potential and of normal flux density. And so

$$\phi_1(s) = \phi_2(s) \tag{111}$$

and

$$\mu_1 \frac{\partial \phi_1(s)}{\partial n} = \mu_2 \frac{\partial \phi_2(s)}{\partial n} \tag{112}$$

need to be enforced.

For convenience define

$$\mu_r = \mu_1/\mu_2 \tag{113}$$

which is the relative permittivity of region 1 with respect to that of region 2. In this case eqn. (112) becomes

$$\mu_r \frac{\partial \phi_1(s)}{\partial n} = \frac{\partial \phi_2(s)}{\partial n} \tag{114}$$

Consider an example in which a dielectric material is denoted as region 1. Taking the unit normal to point out of the object, eqn. (89), multiplied by μ_r, is

$$\frac{\mu_r}{2}\phi(s) = \oint_S \left(G(\mathbf{s}, \mathbf{s}')\mu_r \frac{\partial \phi_1(\mathbf{s}')}{\partial n'} - \mu_r \phi(\mathbf{s}') \frac{\partial G(\mathbf{s}, \mathbf{s}')}{\partial n'} \right) \mathrm{d}s' \tag{115}$$

Consider the outer region to be region 2. Assume that a source exists in region 2, perhaps in the form of free charge or perhaps as a surface having a specified voltage (i.e., a Dirichlet condition). In any case, as indicated by eqn. (28), the Dirichlet condition can be represented by an equivalent source distribution which, when integrated, results in a forcing function $f(\mathbf{r})$. Therefore

$$\phi_2(\mathbf{r}) = \gamma \oint_S \left(-G(\mathbf{r}, \mathbf{s}') \frac{\partial \phi_2(\mathbf{s}')}{\partial n'} + \phi_2(\mathbf{s}') \frac{\partial G(\mathbf{r}, \mathbf{s}')}{\partial n'} \right) \mathrm{d}s' + f(\mathbf{r}) \tag{116}$$

in which the unit normal is taken to be that used in eqn. (114). Letting $\mathbf{r} = \mathbf{s}$,

$$\tfrac{1}{2}\phi(\mathbf{s}) = \oint_S \left(-G(\mathbf{s}, \mathbf{s}') \frac{\partial \phi_2(\mathbf{s}')}{\partial n'} + \phi(\mathbf{s}') \frac{\partial G(\mathbf{s}, \mathbf{s}')}{\partial n'} \right) \mathrm{d}s' + f(\mathbf{s}) \tag{117}$$

in which, by deletion of subscripts of the potential, we are implying continuity of potential. Note that (117) tells us that the total potential is

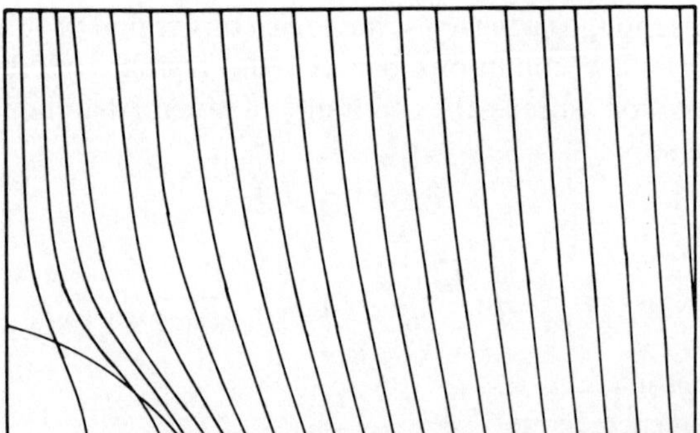

FIG. 3. Equipotential contours for a prolate spheroid introduced into a uniform field. $\mu_r = 10$. (From Jeng & Wexler, to be published.)

given by the sum of a contribution due to the Dirichlet boundary condition in region 2 plus a contribution due to the introduction of the dielectric material into an otherwise homogeneous region.

Adding eqns. (115) and (117), making use of eqn. (114) and re-arranging, we get

$$\frac{\mu_r + 1}{2}\phi(\mathbf{s}) + (\mu_r - 1)\oint_S \phi(\mathbf{s}')\frac{\partial G(\mathbf{s}, \mathbf{s}')}{\partial n'}\mathrm{d}s' = f(\mathbf{s}) \tag{118}$$

which is a Fredholm integral equation of the second kind.

Using eqn. (118), the solution for a prolate spheroid introduced into a previously uniform field, is illustrated in Fig. 3. Employing symmetry, the problem was solved using fourteen quadratic boundary elements in one

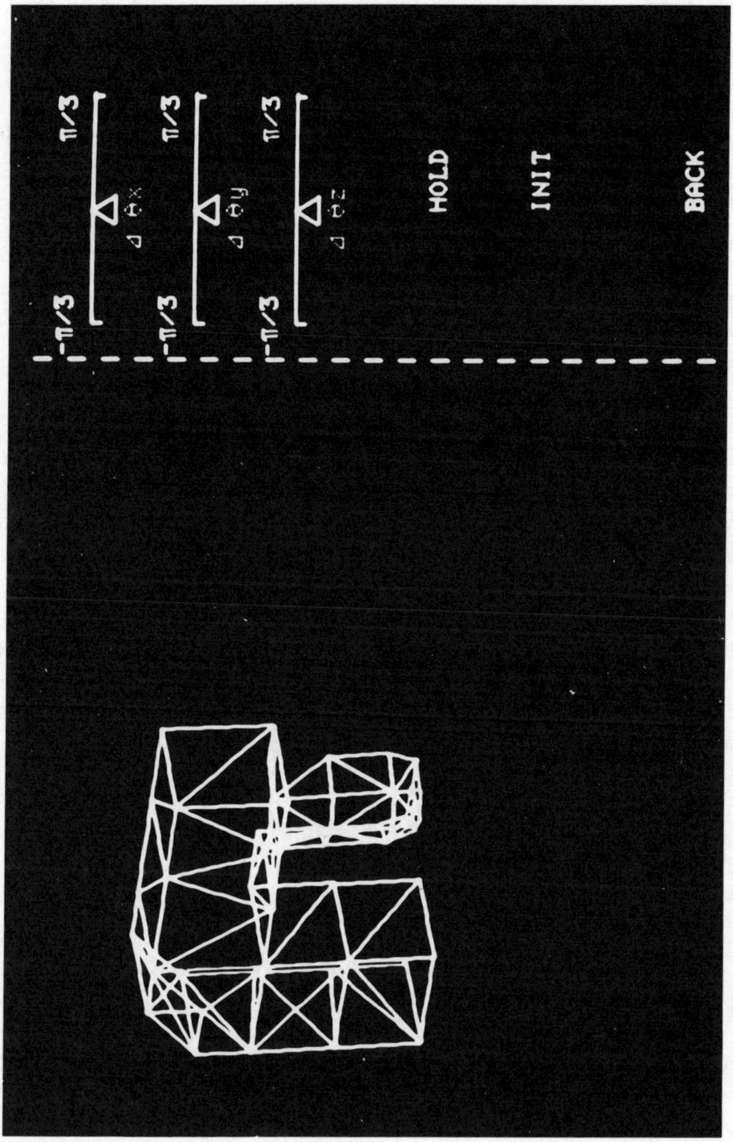

FIG. 4. CRT display, on a PDP 11 computer, of one-quarter of an electromagnet boundary element model. An error in data preparation, in this figure, is evident. (Computer graphics software developed by H. A. Brydon.)

quadrant, i.e., over one-eighth of the surface, only. Only 39 nodes were needed for this problem. It appears, judging by the parallel contours within the spheroid, that the solution is quite accurate. Error is evidenced by deviation from parallelism of the contours near the boundary.

A study of the magnetic field produced by an electromagnet is also to be reported (Jeng & Wexler, to be published). In this case the problem is treated in two parts in a manner described by Trowbridge (1979): the initial magnetic field, due to a pair of current carrying coils in free space, is computed by triple Gaussian quadrature; and, subsequently, the iron (having a relative magnetic permeability of 1000) is introduced and its effect upon the initial field is computed.

The boundary element layout is shown in Fig. 4. It is a photograph of the CRT screen of a PDP 11 computer. Software has been developed for data verification of existing elements and it is planned to extend the software to permit convenient construction of element assemblies at the console. The object can be rotated about one, two or three axes of an orthogonal coordinate system by light-pen adjustment of the cursors displayed at the right of the image. This allows examination of, say, the top of the magnet and then the underside of the pole. Motion of the object creates the illusion of depth without the need of developing complicated, hidden-line algorithms.

It turns out that the formulation equivalent to eqn. (118) is ill-conditioned, due to rounding error problems, insofar as the subsequent calculation of magnetic field in the air region is concerned. On the other hand, field calculation within the iron is well-conditioned. In cases of high relative permeability it appears advisable to use a two-field formulation, i.e., to solve for both $\sigma(\mathbf{s})$ and $\partial n(\mathbf{s})/\partial n$, rather than to eliminate one of them from the boundary integral equation.

9.7 ELECTROMAGNETICS

Most open region problems in electromagnetics pertain to scattering and radiation. From an algorithmic point of view, they share similar traits since it is possible to derive formulations sufficiently general to allow particularisation to one or the other. This section will deal mainly with electromagnetic scattering problems while recognising the importance of antenna theory. In retrospect, antenna engineers have been involved in integral equation solutions using quasi-analytic methods for quite a while. An abundance of literature—King (1956), Schelkunoff (1952), Mei (1965),

Tai (1955), to name but a few—bears this out. We shall first state a few preliminary remarks about antennas in general.

Antennas, used as transmitter–receiver systems, are commonly modelled as thin-wire structures and are frequently formulated as Pocklington's or Hallen's integral equations. For a clearer insight in these integral forms, the reader is referred to Thiele (1973). In essence, the Pocklington's integral equation represents the entire electric field on one side of the equation and subsequently does not restrict its application to antennas of wire-type configuration. On the other hand, the Hallen integral equation is derived explicitly for centre-fed, delta-gap voltage generator systems, which further add to computational difficulties in the evaluation of antenna input impedances. Some efficient numerical techniques for solving Pocklington's equation are discussed by Wilton & Butler (1976). Seven numerical schemes for solving both Pocklington's and Hallen's equations for thin-wire scatters, with emphasis on the convergence rate of the solutions, are also detailed by Butler and Wilton (1975). In the use of thin-wire antennas, the EFIE (electric field integral equation) is generally preferred over the MFIE (magnetic field integral equation). Through experience, most researchers have found that the MFIE caters particularly well for large solid-surface geometries.

For maximum efficiency, antennas are usually operated directionally with the aid of reflectors, thus altering the resultant **E** or **H** field through multiple scattering processes between the antenna and the reflector (for example, Hassan, 1976). Antenna gain is increased together with a decrease in backward radiation. In time-varying problems, frequency of operation is determined by the electromagnetic 'size' of physical dimensions in wavelengths λ. For example, practical antennas used in the high frequency range (2–30 MHz) are often fractions of a wavelength in size. Hence, for antennas mounted on, say, a ship or aircraft, the antenna may not be divorced from the main structure in terms of modelling. However at very high frequencies (above 30 MHz), the superstructure or fuselage will be much larger than λ, and the modelling philosophy is accordingly different. Generally, surface patches and filamentary currents are used to represent the above. Solutions have been often obtained by the method of moments with certain asymptotic techniques thrown in to cater for the respective frequency range.

9.7.1 Electromagnetic Scattering

Scattering problems may be posed by either the MFIE or EFIE formulation. A similar argument (previously used), is employed in choosing which to implement. Conventionally, scattering problems are initially solved for the induced currents or sources on the scatterer surface prior to the computation of the other physical observables. Classical solutions have depended on frequency as a criterion, being able to cater almost exclusively only to the low (Rayleigh–Born approximation) or high frequency (optical techniques) regions. The transition region generally referred to as the 'resonance' region is less well defined. Let $k = 2\pi/\lambda$ be the wave number and define 'a' to be, say, the radius of a sphere or the semi-minor axis of a prolate spheroid. Then $ka = 2\pi(a/\lambda)$ is used as a common factor to scatterer dimension and frequency. The Rayleigh region (low frequency) may then be characterised by $a/\lambda \ll 1$, thus covering the entire frequency spectrum.

The radar cross-section σ(RCS) of a scatterer may be defined as

$$\sigma = \lim_{r\to\infty} 4\pi r^2 \frac{|\mathbf{E}^s|^2}{|\mathbf{E}^i|^2} = \lim_{r\to\infty} 4\pi r^2 \frac{|\mathbf{H}^s|^2}{|\mathbf{H}^i|^2} \tag{119}$$

Where $\mathbf{E}^i$, $\mathbf{H}^i$ and $\mathbf{E}^s$, $\mathbf{H}^s$ denote incident and scattered fields respectively. For the perfectly conducting sphere subject to a linearly polarised incident plane wave, the normalised RCS $(=\sigma/\pi a^2)$ is seen to be proportional to λ^{-4} in the Rayleigh region and equal to πa^2 in the optical region (Harrington, 1961).

The scattering of electromagnetic waves by an obstacle depends on the configuration and material of the scatterer, the frequency, polarisation, angle of incidence of the incident field and the far-field observation angle. Depending on obstacle material, one may particularise the problem to that of scattering by a perfectly conducting boundary, dielectric interfaces, or an impedance boundary. Readers are referred to the IEEE Special Issue on Radar Reflectivity (Proc. IEEE, 1965). A tremendous amount of interest in RCS evaluation came directly from obvious military implications. With a sound knowledge of the RCS behaviour of pertinent obstacles, one may attempt to reduce it by camouflaging techniques such as the use of radar absorbing material. Further extensions of such scattering studies may be motivated by the acute interest in reflectivity signature or vehicle identification. Further relevant information may be found in King & Wu (1956), Oshiro & Su (1965), Waterman (1965), Andreasen (1965*a*, *b*) and Crispin & Maffett (1965).

The practical measurement of physical observables depends on parameters of the receiver–transmitter system. Principal among them are radar (or carrier) frequency, polarisation of the radar signal, transmitted waveform and wavefront characteristic of the incident field if target is close to the transmitting system. In short, the observed cross-section, for example, is a function of target body and is unique for a particular frequency and polarisation.

The monostatic (backscatter) cross-section is viewed as that from an isotropic radiator having a constant magnitude of scattered field on a large sphere with radius defined by the distance between the point of impact and the receiver location. For nose-on incidence with bodies of rotational symmetry, the backscattered field retains the same polarisation as that of the incident field. In general, a de-polarising effect is experienced at all other points of incidence. A lucid account of cross-section estimation is provided by Senior (1965).

The analysis of solid surface scatterers usually involved planar-patch representation and subsequent application of the method of moments. Oshiro & Su (1965), for example, used pulse expansion and delta-function testing. They obtained good results for the prolate spheroid for $ka = 1{\cdot}7$ even with a lack of accurate modelling. It seems that a better geometrical model, provided by the isoparametric algorithm, will lead to even more consistent results with a lesser number of elements. Analyses of wire RCS have been reported by Curtis (1967) using the point-matching method.

Some inroads into the inverse scattering problem have been made by Mittra (1973), where he examined remote probing and wavefront reconstruction methods. An enlightening discussion of transient scattering (that aids in the analysis of antennas or scatterers actively loaded with non-linear elements) is also given by Mittra (1976).

9.7.2 Integral Formulation

Consider a volume V in free space bounded by an inner surface S and a spherical surface at infinity S_∞. In the case of a classical radar problem, let the incident field be a plane wave travelling along the positive z-direction. For a linear, isotropic, source-free media, the fields are described by Maxwell's equations

$$\begin{aligned} \nabla \times \mathbf{E} &= -j\omega\mu_0\mathbf{H}, \qquad \nabla \,.\, \mathbf{E} = \frac{\rho}{\varepsilon_0} \\ \nabla \times \mathbf{H} &= j\omega\varepsilon_0\bar{\mathbf{E}}, \qquad \nabla \,.\, \mathbf{H} = \frac{m}{\mu_0} \end{aligned} \tag{120}$$

where $\mathbf{E}, \mathbf{H}$ are the electric and magnetic field vectors respectively, μ_0 and ε_0 are as defined previously, and ρ and m are the electric and magnetic charge densities. A time variation of $e^{j\omega t}$ is assumed. Taking vector products (120) may be shown to result in

$$\begin{aligned}\nabla \times \nabla \times \mathbf{E}(\mathbf{r}) - k^2\mathbf{E}(\mathbf{r}) &= 0\\ \nabla \times \nabla \times \mathbf{H}(\mathbf{r}) - k^2\mathbf{H}(\mathbf{r}) &= 0\end{aligned} \tag{121}$$

where $\mathbf{E}(\mathbf{r}), \mathbf{H}(\mathbf{r}) \to 0$ as $\mathbf{r} \to \infty$, and $k^2 = \omega^2\varepsilon_0\mu_0$. Define the vector free-space Green's function $\mathbf{G}$ given by

$$\mathbf{G}(\mathbf{r}, \mathbf{r}') = \psi\hat{\mathbf{a}} \tag{122}$$

where $\hat{\mathbf{a}}$ is an arbitrary unit vector, and

$$\psi = \frac{e^{-jk}|\mathbf{r} - \mathbf{r}'|}{4\pi|\mathbf{r} - \mathbf{r}'|} \tag{123}$$

is the scalar Green's function. Obviously, $\bar{\mathbf{G}}$ satisfies the inhomogeneous Helmholtz equation

$$\nabla \times \nabla \times \mathbf{G}(\mathbf{r}, \mathbf{r}') - k^2\mathbf{G}(\mathbf{r}, \mathbf{r}') = -\delta(\mathbf{r} - \mathbf{r}')\hat{\mathbf{a}} \tag{124}$$

Consider the homogeneous Helmholtz equation in $\mathbf{H}(\mathbf{r})$ as given by (121). Taking the inner product of $\mathbf{G}(\mathbf{r}, \mathbf{r}')$ with (121) and of $\mathbf{H}(\mathbf{r})$ with (124), then subtracting the first from the second and integrating over V, one gets

$$\int_V [\mathbf{G}(\mathbf{r}, \mathbf{r}') . \nabla \times \nabla \times \mathbf{H}(\mathbf{r}) - H(r) . \nabla \times \nabla \times \mathbf{G}(\mathbf{r}, \mathbf{r}')]\,\mathrm{d}v = \int_V \mathbf{H}(\mathbf{r}) . \delta(\mathbf{r} - \mathbf{r}')\hat{\mathbf{a}}\,\mathrm{d}v \tag{125}$$

The right-hand side of (125) is $\mathbf{H}(\mathbf{r}')$.

Applying the vector Green's theorem to the left-hand side of (125), we get

$$\gamma \int_{S+S_\infty} [\mathbf{H}(\mathbf{s}) \times \nabla \times \mathbf{G}(\mathbf{s}, \mathbf{r}') - \mathbf{G}(\mathbf{s}, \mathbf{r}') \times \nabla \times \mathbf{H}(\mathbf{s})] . \hat{\mathbf{n}}\,\mathrm{d}s = \mathbf{H}(\mathbf{r}')\hat{\mathbf{a}} \tag{126}$$

where the singularity at $\mathbf{r} = \mathbf{r}'$ is excluded and γ is as defined earlier. If, in particular, S is a perfectly conducting surface, then enforcing boundary conditions $\hat{\mathbf{n}} . \mathbf{H}(\mathbf{r}) = m = 0$ (normal magnetic field is zero) and $\hat{\mathbf{n}} \times \mathbf{E}(\mathbf{r}) = 0$, (tangential electric field is zero), (126) may be written as

$$\mathbf{H}(\mathbf{r}')\hat{\mathbf{a}} = \gamma \int_{S+S_\infty} [\nabla \times \mathbf{G}(\mathbf{s}, \mathbf{r}') . \mathbf{J}(\mathbf{s})]\,\mathrm{d}s \tag{127}$$

where $\mathbf{J}(\mathbf{r})$ is given by $\mathbf{J}(\mathbf{r}) = \hat{\mathbf{n}} \times \mathbf{H}(\mathbf{r})$. Interchanging source and observation location, one finally gets

$$\mathbf{H}(\mathbf{r})\hat{\mathbf{a}} = \gamma \int_{S+S_\infty} [\nabla' \times \mathbf{G}(\mathbf{r}, \mathbf{s}') \,.\, \mathbf{J}(\mathbf{s}')]\, \mathrm{d}s' \tag{128}$$

The radiation condition states that the integral over S_∞ can include only contributions from its outside.

In the case of a radar problem, the source is at infinity (so that we have plane wave incidence) and the integral over $\mathbf{S}_\infty$ therefore amounts to $\mathbf{H}^i(\mathbf{r})$. Hence (128) may be re-assembled as

$$\mathbf{H}(\mathbf{s}) = 2\mathbf{H}^i(\mathbf{s}) + 2 \int_S \mathbf{J}(\mathbf{s}') \times \nabla\psi\, \mathrm{d}s, \qquad \mathbf{s} \in S \tag{129}$$

To solve for surface current density $\mathbf{J}(\mathbf{s})$, we consider

$$\mathbf{J}(\mathbf{s}) = 2\hat{\mathbf{n}} \times \mathbf{H}^i(\mathbf{s}) + 2\hat{\mathbf{n}} \times \int_S \mathbf{J}(\mathbf{s}') \times \nabla'\psi\, \mathrm{d}s', \qquad \mathbf{s} \in S \tag{130}$$

Equation (130) is a Fredholm integral equation of the second kind and may be solved for $\mathbf{J}(\mathbf{s})$.

If the EFIE formulation is desired, (121) may be used to derive

$$\hat{\mathbf{n}} \times \mathbf{E}^i(\mathbf{s}) = \frac{-j}{\omega\varepsilon_0} \hat{\mathbf{n}} \times \int_S [-k^2 \mathbf{J}(\mathbf{s}')\psi + \nabla'_s \,.\, \mathbf{J}(\mathbf{s}')\nabla'\psi]\, \mathrm{d}s', \qquad \mathbf{s} \in S \tag{131}$$

in the same manner.

Here, ∇'_s denotes the surface divergence operator in source co-ordinates. Equation (131) is a Fredholm integral equation of the first kind in $\mathbf{J}(\mathbf{s}')$. A formal interpretation of the boundary conditions on the perfectly conducting surface S can be given as follows: The electric field $\mathbf{E}^s$ (scattered fields) on S due to the currents induced on it must cancel the tangential components of $\mathbf{E}^i$ at each point on S. Hence

$$\mathbf{E}(\mathbf{s}) = \mathbf{E}^i(\mathbf{s}) + \mathbf{E}^s(\mathbf{s}), \qquad \mathbf{s} \in S \tag{132}$$

and similarly,

$$\mathbf{H}(\mathbf{s}) = \mathbf{H}^i(\mathbf{s}) + \mathbf{H}^s(\mathbf{s}), \qquad \mathbf{s} \in S \tag{133}$$

The scattered fields in V may now be calculated since $\mathbf{J}(\mathbf{s})$ is known. In particular, from (129)

$$\mathbf{H}^s(\mathbf{r}) = 2 \int_S \mathbf{J}(\mathbf{s}') \times \nabla'\psi\, \mathrm{d}s', \qquad \mathbf{r} \in V \tag{134}$$

where

$$R = |\mathbf{r} - \mathbf{r}'|$$

and

$$\nabla'\psi = \frac{1}{4\pi}\frac{1 + jkR}{R}\frac{e^{-jkR}}{R}\frac{(\mathbf{r} - \mathbf{r}')}{|\mathbf{r} - \mathbf{r}'|} \tag{135}$$

RCS may then be evaluated by the expression given in (119).

It should be noted at this juncture that the scalar free-space Green's function used, does not cater for other than plane-wave excitation. Hence only far-field quantities are meaningful. For the source located at a finite distance, recourse must be made to the free-space dyadic Green's function $\bar{\bar{\mathbf{G}}}(\mathbf{r}, \mathbf{r}')$ (Tai, 1971) which may be expressed as

$$\bar{\bar{\mathbf{G}}}(\mathbf{r}, \mathbf{r}') = \mathbf{G}^{(x)}(\mathbf{r}, \mathbf{r}')\hat{\mathbf{i}} + \mathbf{G}^{(y)}(\mathbf{r}, \mathbf{r}')\hat{\mathbf{j}} + \mathbf{G}^{(z)}(\mathbf{r}, \mathbf{r}')\hat{\mathbf{k}} \tag{136}$$

The above equation has $\hat{\mathbf{i}}, \hat{\mathbf{j}}$, and $\hat{\mathbf{k}}$ representing unit normals in the cartesian co-ordinate system. In general, each component of $\bar{\bar{\mathbf{G}}}(\mathbf{r}, \mathbf{r}')$ indicates a vector Green's function with a source pointed in that direction. As a whole or in part, they satisfy the respective inhomogeneous Helmholtz wave equation and the radiation condition. Equation (130) may then be re-written in the form

$$\mathbf{J}(\mathbf{s}) = 2\hat{\mathbf{n}} \times \mathbf{H}^{i}(\mathbf{s}) + 2\hat{\mathbf{n}} \times \int_S \mathbf{J}(\mathbf{s}') . \nabla \times \bar{\bar{\mathbf{G}}}(\mathbf{r}, \mathbf{r}')\, \mathrm{d}s', \qquad \mathbf{s} \in S \tag{137}$$

9.7.3 Isoparametric BEM Implementation

Using expansion functions previously specified in (66), we may attempt to write (130) in Galerkin form as

$$\int_S \underline{\alpha}(\mathbf{s})\underline{\alpha}^{\mathrm{T}}(\mathbf{s})\, \mathrm{d}s\underline{\mathbf{J}} - 2\hat{\mathbf{n}} \times \int_S \underline{\alpha}(\mathbf{s}) \int_S \underline{\alpha}^{\mathrm{T}}(\mathbf{s}')\underline{\mathbf{J}} \times \nabla'\psi\, \mathrm{d}s'\, \mathrm{d}s = 2\int_S \underline{\alpha}(\mathbf{s})\hat{\mathbf{n}} \times \mathbf{H}^{i}(\mathbf{s})\, \mathrm{d}s \tag{138}$$

The column matrix $\underline{\mathbf{J}}$ will contain M rows where M is specified from (59). Each element of $\underline{\mathbf{J}}$ is itself a cartesian vector of three components. Matrix accumulation is carried out in sequential order, one element at a time. Realising that already a first degree ($N = 1$) approximation would result in a 9×9 matrix, it is easy to understand that a large portion of the computation time is spent on matrix generation. By trimming back on the number of elements required while going to a higher degree representation, it is possible to reduce the total number of variables. However, in and around the 'resonance' region, one may run into numerical instability or inprecision due to the excitation of interior resonance (which may be of the antisymmetric mode). This can lead to numerically spurious results which while non-radiating may, because of their numerical inaccuracy, lead to degeneration of the far-fields and consequently to pattern degradation. There seems to be an upper bound on the physical dimension allowable for an element (Poggio & Miller, 1973).

As to the numerical instability, Mitzner (1968) tried coupling both MFIE and EFIE together and obtained an integral representation valid for the entire frequency spectrum. Another method would be to physically impose the required condition that no field exists within the surface S. This can be achieved by setting a vanishing field at a few internal points.

Subsequent to the matrix accumulation, one has to be concerned with the matrix solution scheme. With double precision and complex arithmetic, this will still prove to be an arduous task even for a digital computer, as the matrix is almost always totally dense. A less accurate scheme would be to replace the small entries of the matrix by zeros. Such techniques are *ad hoc*, and may only be justified on an individual basis. A better scheme may be the implementation of diakoptics to create block-sparse matrices.

At the moment, meaningful quantitative results are not available. However, preliminary results have been very encouraging, and the authors intend to publish a complete report in the near future.

9.8 PICTURE-FRAME METHODS

The appellation 'picture-frame methods' refers to a number of procedures for the mutual constraint of boundary integral and partial differential equations in various combinations and ways (McDonald & Wexler, 1979).

The technique is of particular use in catering for the Finite Element Method (FEM) solution of subregions having finite extent within open or extensive regions. Regions containing inhomogeneous, non-linear or anisotropic media are conveniently solved by FEM, partial differential equation techniques. On the other hand, open-region problems are more conveniently handled by BEM, boundary integral equation techniques because the unknowns exist solely upon the surface. The number of unknowns is greatly reduced in the latter case. The basic principle—not always expressed in these terms—is to formulate the finite-region problem within a circumscribing boundary (or picture frame) by a partial differential equation and the open-region problem by a boundary integral equation over the picture frame, and then to mutually constrain these two solutions.

A number of different approaches, using finite difference and finite element methods have been employed to obtain the overall solution involving iteration (e.g., boundary relaxation), root-mean-square minimisation of picture-frame interface discontinuities, direct elimination of interface dependent variables, expansion of the exterior field in terms of spherical and cylindrical harmonics, etc. A number of pertinent publications are the following: Greenspan & Werner (1966), Cermak & Silvester (1968), Cermak & Silvester (1969), Sandy & Sage (1971), Silvester & Hsieh (1971), Williams & Cambrell (1972), McDonald & Wexler (1972), Mei (1974), Chang & Mei (1976), Csendes (1976), Zienkiewicz *et al.* (1977), and Zienkiewicz *et al.* (1978).

One approach for solving a coupled FEM–BEM problem is to consider the region within the picture frame to be treated as an inhomogeneous Neumann problem, i.e.

$$\left.\frac{\partial\phi}{\partial n}\right|_s = h(\mathbf{s}) = -\lambda(\mathbf{s}) \tag{139}$$

where the function $\lambda(\mathbf{s})$ is, in fact, unknown. Using a variational approach the appropriate functional for the interior region is

$$F = \iint_{V_i} \kappa\nabla\phi \,.\, \nabla\phi \,\mathrm{d}x\,\mathrm{d}y - 2\iint_{V_i} \phi f \,\mathrm{d}x\,\mathrm{d}y - 2\oint_S \phi h \,\mathrm{d}s \tag{140}$$

Stationarity of (140) occurs for the solution of the Poisson equation

$$-\nabla\,.\,(\kappa\nabla\phi) = f \tag{141}$$

for an inhomogeneous medium.

Consider the FEM interpolation method

$$\phi = \underline{\alpha}^{\mathrm{T}}\underline{\phi} \tag{142}$$

and

$$\lambda = \underline{\alpha}^{\mathrm{T}}\underline{\lambda} \tag{143}$$

where the column matrix $\underline{\phi}$ consists of node potentials throughout the region and the column matrix $\underline{\lambda}$ consists of values of the normal derivative at boundary nodes only. Substitution of (142) and (143) into (140) yields

$$F = \underline{\phi}^{\mathrm{T}}S\underline{\phi} - 2\underline{\phi}^{\mathrm{T}}\underline{b} + 2\underline{\phi}^{\mathrm{T}}G\underline{\lambda} \tag{144}$$

where

$$S = \iint_{V_i} \kappa\nabla\underline{\alpha}\,.\,\nabla\underline{\alpha}^{\mathrm{T}}\,\mathrm{d}x\,\mathrm{d}y \tag{145}$$

$$\underline{b} = \iint_{V_i} \underline{\alpha} f \,\mathrm{d}x\,\mathrm{d}y \tag{146}$$

and

$$\mathbf{G} = \oint_S \underline{\alpha}\underline{\alpha}^{\mathrm{T}}\,\mathrm{d}s \tag{147}$$

Differentiating (144) with respect to ϕ, one obtains

$$\mathbf{S}\underline{\phi} + \mathbf{G}\underline{\lambda} = \underline{b} \tag{148}$$

Substituting (142), (143) and

$$\left.\frac{\partial\phi}{\partial n}\right|_s = \lambda(\mathbf{s}) \tag{149}$$

where this unit normal is oppositely directed to that in (139), into

$$\phi(\mathbf{s}) = 2\oint_S \left(G(\mathbf{s},\mathbf{s}')\frac{\partial\phi}{\partial n'} - \phi(\mathbf{s}')\frac{\partial G(\mathbf{s},\mathbf{s}')}{\partial n'} \right)\mathrm{d}s' \tag{150}$$

we obtain

$$\left(\underline{\alpha}(\mathbf{s})^{\mathrm{T}} + 2\oint_S \underline{\alpha}(\mathbf{s}')^{\mathrm{T}}\frac{\partial G(\mathbf{s},\mathbf{s}')}{\partial n'}\,\mathrm{d}s' \right)\underline{\phi} - 2\oint_S G(\mathbf{s},\mathbf{s}')\underline{\alpha}(\mathbf{s}')^{\mathrm{T}}\,\mathrm{d}s'\underline{\lambda} = \underline{0} \tag{151}$$

Therefore

$$\left(\oint_S \underline{\alpha}(\mathbf{s})\alpha(\mathbf{s})^{\mathrm{T}}\,\mathrm{d}s + 2\oint_S\oint_S \underline{\alpha}(\mathbf{s})\underline{\alpha}(\mathbf{s}')^{\mathrm{T}}\frac{\partial G(\mathbf{s},\mathbf{s}')}{\partial n'}\,\mathrm{d}s'\,\mathrm{d}s \right)\underline{\phi}$$

$$-2\oint_S\oint_S G(\mathbf{s},\mathbf{s}')\underline{\alpha}(\mathbf{s})\underline{\alpha}(\mathbf{s}')^{\mathrm{T}}\,\mathrm{d}s'\,\mathrm{d}s\underline{\lambda} = \underline{0} \tag{152}$$

which may be expressed as

$$\mathbf{C}^{\mathrm{T}}\underline{\phi} - \mathbf{Z}\underline{\lambda} = \underline{0} \tag{153}$$

Equations (148) and (153) may be expressed in compound matrix form as

$$\begin{bmatrix} \mathbf{S} & \mathbf{G} \\ \mathbf{C}^{\mathrm{T}} & -\mathbf{Z} \end{bmatrix}\begin{bmatrix} \underline{\phi} \\ \underline{\lambda} \end{bmatrix} = \begin{bmatrix} \underline{b} \\ \underline{0} \end{bmatrix} \tag{154}$$

If sources exist in the exterior region, the bottom entry in the right-hand-side vector is non-zero. **G** and **C** are generally sparse rectangular matrices and **Z** is a square matrix and likely to be quite small.

Solution of matrix equation (154) could be attempted directly. However, since the augmented matrix is indefinite, it is probably safer to solve by matrix substitution in the following way: From (148),

$$\underline{\phi} = \mathbf{S}^{-1}\underline{b} - \mathbf{S}^{-1}\mathbf{G}\underline{\lambda} \tag{155}$$

Substituting $\underline{\phi}$ into (153),

$$\mathbf{C}^{\mathrm{T}}\mathbf{S}^{-1}\underline{b} - \mathbf{C}^{\mathrm{T}}\mathbf{S}^{-1}\mathbf{G}\underline{\lambda} - \mathbf{Z}\underline{\lambda} = \underline{0} \tag{156}$$

Solving for $\underline{\lambda}$,

$$\underline{\lambda} = (\mathbf{C}^{\mathrm{T}}\mathbf{S}^{-1}\mathbf{G} + \mathbf{Z})^{-1}\mathbf{C}^{\mathrm{T}}\mathbf{S}^{-1}\underline{b} \tag{157}$$

and, finally, substituting $\underline{\lambda}$ into (155), we obtain

$$\underline{\phi} = \mathbf{S}^{-1}\underline{\mathrm{b}} - \mathbf{S}^{-1}\mathbf{G}(\mathbf{C}^{\mathrm{T}}\mathbf{S}^{-1}\mathbf{G} + \mathbf{Z})^{-1}\mathbf{C}^{\mathrm{T}}\mathbf{S}^{-1}\underline{b} \tag{158}$$

A significant advantage of the algorithm described by (158) is that a number of sets of exterior constraint equations (i.e., the matrix coefficient of $\underline{b}$ in product form) can be retained within a FEM program for use if and when required.

The first term on the right-hand side of (158) represents the solution of the problem with homogeneous Neumann boundary conditions along the picture-frame boundary while the second term supplies the effect of the presence of the exterior region.

In Fig. 5, the electrostatic field about a ceramic high voltage insulator is illustrated. The demarcation between the FEM and BEM regions is

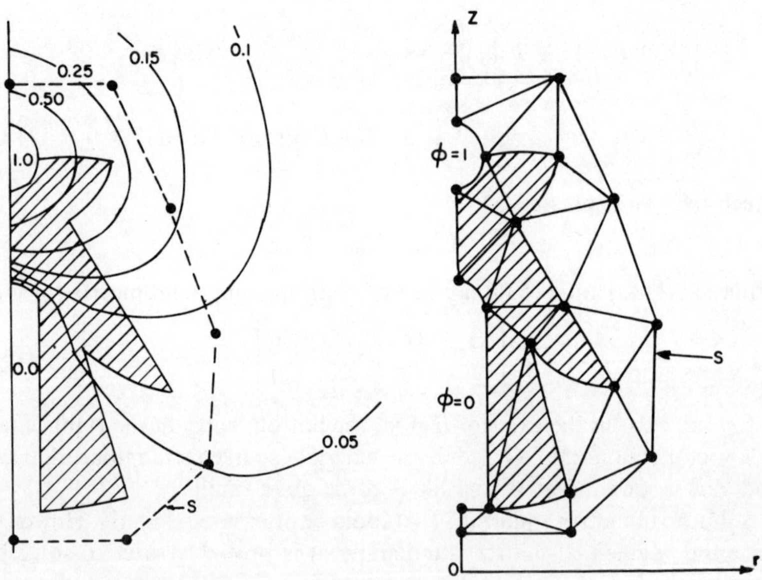

FIG. 5. A mixed PDE–BIE solution of the electrostatic field about a ceramic high voltage insulator. (From McDonald & Wexler, 1979.)

indicated. Contours of electrostatic potential appear unaffected by the presence of the picture-frame boundary.

A very similar technique can be employed to produce a discretisation of integral equations in a block-sparse rather than in a totally dense matrix form (McDonald & Wexler, 1979).

REFERENCES

ANDREASEN, M. G. (1965*a*). Scattering from cylinders with arbitrary surface impedance, *Proc. IEEE* (*Special Issue on Radar Reflectivity*), **53,** 812–17.

ANDREASEN, M. G. (1965*b*). Scattering from bodies of revolution, *IEEE Trans. Antennas and Prop.*, **AP-13,** 303–10.

BURTON, A. J. & MILLER, G. F. (1971). The application of integral equation methods to the numerical solution of some exterior boundary-value problems, *Proc. Roy. Soc. London, Series A.*, **323,** 201–10.

BUTLER, C. M. & WILTON, D. R. (1975). Analysis of various numerical techniques applied to thin-wire scatterers, *IEEE Trans. Antennas and Prop.*, **AP-23,** 534–40.

CERMAK, I. A. & SILVESTER, P. (1968). Solution of two-dimensional field problems by boundary relaxation, *Proc. IEEE*, **115,** 1341–8.

CERMAK, I. A. & SILVESTER, P. (1969). Analysis of coaxial line discontinuities by boundary relaxation, *IEEE Trans. Microwave Theory and Techniques* (*Special Issue on Computer-oriented Microwave Practices*), **MTT-17,** 489–95.

CHANG, S. K. & MEI, K. K. (1976). Application of the unimoment method to electromagnetic scattering of dielectric cylinders, *IEEE Trans. Antennas and Prop.*, **AP-24**(1), 35–42.

CLARK, J. H. (1976). Designing surfaces in 3-D, *Communications of the ACM*, **19**(8).

COONS, S. A. (1974). Surface patches and B-spline curves. In: *Computer Aided Geometric Design*, Eds R. E. Barnhill and R. F. Riesenfeld, Academic Press, New York.

CRISPIN, J. W. & MAFFETT, A. L. (1965). Radar cross-section estimation for simple shapes, *Proc. IEEE*, (*Special Issue on Radar Reflectivity*), **53,** 833–48.

CSENDES, Z. J. (1976). A note on the finite-element solution of exterior-field problems, *IEEE Trans. Microwave Theory and Techniques*, **MTT-24,** 468–73.

CURTIS, W. L. (1967). *Technical note: Radar cross-section of arbitrarily shaped wires*, The Boeing Company, Ant. and Prop. Unit, Z-5453-30-55.

GREENSPAN, D. & WERNER, P. (1966). A numerical method for the exterior Dirichlet problem for the reduced wave equation, *Arch. Rat. Mech. Analy.*, **23,** 288–316.

HARRINGTON, R. F. (1961). *Time-Harmonic Electromagnetic Fields*, McGraw-Hill, New York.

HARRINGTON, R. F. (1968). *Field Computation by Moment Methods*, Macmillan, New York.

HASSAN, M. A. Z. I. (1976). *Graphics aided projective method for plate-wire antennas*, Ph.D. Dissertation, McGill University.

JASWON, M. A. & SYMM, G. T. (1977). *Integral Equation Methods in Potential Theory and Elastostatics*, Academic Press, New York.

JENG, G. & WEXLER, A. (1977). Isoparametric, finite element variational solution of integral equations for three-dimensional fields, *Int. J. Num. Meth. Eng.*, **11,** 1455–71.

JENG, G. & WEXLER, A. (1978). Self-adjoint variational formulation of problems having non-self-adjoint operators, *IEEE Trans. Microwave Theory and Techniques*, **MTT-26,** 91–4.

JENG, G. & WEXLER, A. (to be published). Boundary element solution of magnetostatic problems.

KING, R. W. P. (1956). *The Theory of Linear Antennas*, Harvard University Press, Cambridge, Mass.

KING, R. W. P. & WU, T. T. (1956). *The Scattering and Diffraction of Waves*, Harvard University Press, Cambridge, Mass.

LACHAT, J. C. & WATSON, J. O. (1975). A second generation boundary integral equation program for 3-dimensional elastic analysis. In: *Boundary Integral Equation Method: Computational Applications in Applied Mechanics*, Eds T. Cruse and F. Rizzo, AMD 11, ASME, New York.

LACHAT, J. C. & WATSON, J. O. (1976). Effective numerical treatment of boundary integral equations: a formulation for 3-d elastostatics, *Int. J. Num. Meth. Eng.*, **10,** 991–1005.

LASKER, S. K. (1974). *A numerical approach to calculate capacity of conductors*, Unpublished report, Department of Mathematics, Assam Engineering College, Gauhati.

MCDONALD, B. H. & WEXLER, A. (1972). Finite element solution of unbounded field problems, *IEEE Trans. Microwave Theory and Techniques*, **MTT-20,** 841–47.

MCDONALD, B. H. & WEXLER, A. (1979). Mutually constrained partial differential and integral equation field formulations. In: *Finite Elements in Electrical and Magnetic Field Problems*, Eds M. V. K. Chari and P. Silvester, Wiley, New York, 1979.

MCDONALD, B. H., FRIEDMAN, M., DECRETON, M. & WEXLER, A. (1973). Integral finite-element approach for solving the Laplace equation, *Electronics Letters*, **9**(11).

MCDONALD, B. H., FRIEDMAN, M. & WEXLER, A. (1974). Variational solution of integral equations, *IEEE Trans. Microwave Theory and Techniques*, **MTT-22,** 237–48.

MEI, K. K. (1965). On the integral equations of thin wire antennas, *IEEE Trans. Antennas and Prop.*, **AP-13,** 374–8.

MEI, K. K. (1974). Unimoment method of solving antenna and scattering problem, *IEEE Trans. Antennas and Prop.*, **AP-22,** 760–6.

MIKHLIN, S. G. (1964). *Variational Methods in Mathematical Physics*, Macmillan, New York.

MITTRA, R. (1973). Inverse scattering and remote probing. In: *Computer Techniques for Electromagnetics*, Ed. R. Mittra, Pergamon Press, Oxford.

MITTRA, R. (1976). Integral equation methods for transient scattering. In: *Transient Electromagnetic Fields*, Ed. L. B. Felsen, *Topics in Applied Physics*, *Vol.* 10, Springer-Verlag, Berlin, Heidelberg.

MITZNER, K. M. (1968). Numerical solution of the exterior scattering problem at eigenfrequencies of the interior problem, *Proc. Fall URSI Meeting, Boston, Mass.*, 1968.

NEDELEC, J. C. (1976). Curved finite element methods for the solution of singular integral equations on surfaces in R^3, *Computing in Applied Mechanics and Engineering*, **8,** 61–80.

NOBLE, B. (1971). Some applications of the numerical solution of integral equations to boundary value problems. In: *Conference on Applications of Numerical Analysis*, Ed. J. L. Morris, Springer-Verlag, Berlin, Heidelberg.

OSHIRO, F. K. & SU, C. S., *A source distribution technique for the solution of general electromagnetic scattering problems*, Northrop Norair Rept., NOR, pp. 65–271.

PARIS, D. T. & HURD, F. K. (1969). *Basic Electromagnetic Theory*, McGraw-Hill, New York.

POGGIO, A. J. & MILLER, E. K. (1973). Integral equation solutions of three-dimensional scattering problems. In: *Computer Techniques for Electromagnetics*, Ed. R. Mittra, Pergamon Press, Oxford.

Proc. IEEE (1965). Special Issue on Radar Reflectivity, **53,** 805–48.

REITAN, D. K. & HIGGINS, T. J. (1951). Calculation of the electrical capacitance of a cube, *J. Appl. Phys.*, **22**(2).

SANDY, F. & SAGE, J. (1971). Use of finite difference approximations to partial differential equations for problems having boundaries at infinity, *IEEE Trans. Microwave Theory and Techniques (Corresp.)*, **MTT-19,** 484–6.

SCHELKUNOFF, S. A. (1952). *Advanced Antenna Theory*, Wiley, New York.

SENIOR, T. B. A. (1965). A survey of analytical techniques for cross-section estimation, *Proc. IEEE (Special Issue on Radar Reflectivity)*, **53,** 822–33.

SILVESTER, P. & HSICH, M. S. (1971). Finite-element solution of 2-dimensional exterior field problems, *Proc. Inst. Elec. Engrg.*, **118**(12), 1743–7.

STAKGOLD, I. (1968). *Boundary Value Problems of Mathematical Physics*, Vol. II, Macmillan, New York.

TAI, C. T. (1955). A new interpretation of the integral equation formulation of cylindrical antennas, *IEEE Trans. Antennas and Prop.*, **AP-3,** 125–7.

TAI, C. T. (1971). *Dyadic Green's Functions in in Electromagnetic Theory*, Intext, Scranton, Pennsylvania.

THIELE, G. A. (1973). Wire antennas. In: *Computer Techniques for Electromagnetics*, Ed. R. Mittra, Pergamon Press, Oxford.

TROWBRIDGE, C. W. (1979). Application of integral equation methods for the numerical solution of magnetostatic and eddycurrent problems. In: *Finite Elements in Electrical and Magnetic Field Problems*, Eds M. U. K. Chari and P. Silvester, Wiley, New York.

WATERMAN, P. C. (1965). Matrix formulation of electromagnetic scattering, *Proc. IEEE (Special Issue on Radar Reflectivity)*, **53,** 805–12.

WATSON, J. O. (unpublished report). The solution of boundary integral equations of three-dimensional elastostatics for infinite regions.

WEXLER, A. (1978). Some applications of the boundary element method to electrical engineering problems. In: *Recent Advances in Boundary Element Methods*, Ed. C. A. Brebbia, Pentech Press, London.

WILLIAMS, C. G. & CAMBRELL, G. K. (1972). Efficient numerical solution of unbalanced field problems, *Electronics Letters*, **8**(9), 247–8.

WILTON, D. R. & BUTLER, C. M. (1976). Efficient numerical techniques for solving Pocklington's equations and their relationship to other methods, *IEEE Trans. Antennas and Prop.*, **AP-24,** 83–6.

ZIENKIEWICZ, O. C., KELLY, D. W. & BETTESS, P. (1977). The coupling of the finite element method and boundary solution procedures, *Int. J. Num. Meth. Eng.*, **11**(12), 355–75.

Chapter 10

COUPLING BOUNDARY ELEMENT METHODS WITH OTHER NUMERICAL METHODS

D. W. KELLY, G. G. W. MUSTOE and O. C. ZIENKIEWICZ

Department of Civil Engineering, University of Wales, Swansea, UK

SUMMARY

In the numerical discretisation of a physical problem there are three primary solution techniques. These are finite differences, finite elements and boundary integral equations. A common basis for these methods has been identified for some time now, and today they can be presented to engineers and scientists as allied tools rather than distinctly separate methods. Therefore, it is logical to describe all the above procedures within a single methodology, and provide them in the same computer software as separate options that can be used simultaneously in the solution of a given problem. In this chapter we begin with a description of a weighted residual process which can include all three methods as special cases. Specific examples have appeared elsewhere in the text showing the advantage of a combined analysis approach involving the coupling of boundary solution procedures with finite differences (Chapter 4) and finite elements (Chapter 7). Here we try to summarise the basic theory behind these applications and to show that the coupling can be executed in a symmetric mode for self-adjoint problems.

10.1 INTRODUCTION

In general the physical characteristics of an engineering problem can be mathematically described in three different ways.

(i) The physical system's characteristics may be described by formulating the appropriate partial differential equation by considering some elemental portion of the domain and applying some physical law such as a conservation principle. The particular solution required is then found by applying the boundary conditions of the problem.

(ii) Another description of the physical problem can be achieved by the application of a variational principle or a weak integral form over the problem domain. In this type of formulation the boundary conditions can either be satisfied implicitly within the procedures or forced explicitly after the procedure has been applied.

(iii) Further formulations of the physical problem are also possible by utilising singular solutions of the governing differential equations (e.g., sources, doublets, etc.). This is achieved by either a superposition of the singular solutions whose distribution is determined by the boundary conditions of the problem, or by the application of a reciprocal theorem (e.g., the Somigliana identity in elasticity) used in conjunction with the singular solutions.

Historically the above analytical formulations of the physical problem have led to the development of three, apparently unrelated, numerical discretisation procedures. These processes are finite difference methods, finite element techniques and boundary solution procedures respectively. However it is shown later in this chapter that all these numerical solution techniques have a common basis and may be described by a single methodology.

The domain subdivision required in the finite element and finite difference procedures readily allows many types of complicated boundary and initial value engineering problems to be resolved as shown in Zienkiewicz (1977). However, difficulties can occur when attempting to solve an example which extends over an infinite domain. This is because the finite difference and finite element procedures were primarily developed to tackle problems within a finite domain.

Some of these limitations in the finite element procedure have been overcome by the introduction of 'infinite' elements as used by Bettess (1977) to solve many infinite wave problems.

On the other hand the boundary solution procedures require known singular solutions of the governing differential equations. This prerequisite means that problems which are linear and piece-wise homogeneous are the type most effectively dealt with by these methods. However, some solutions to problems with anisotropic and orthotropic materials have been obtained by Butterfield and Tomlin (Butterfield & Tomlin, 1972; Tomlin & Butterfield, 1974). Other areas where the boundary integral methods can be successful are problems extending over an infinite domain and examples which have a large volume/surface ratio. An example exhibiting some of these features is illustrated in Fig. 1. Many applications of the boundary solution procedures are given by Cruse (1973), Lachat & Watson (1976), Wexler (1978) and Banerjee (1976).

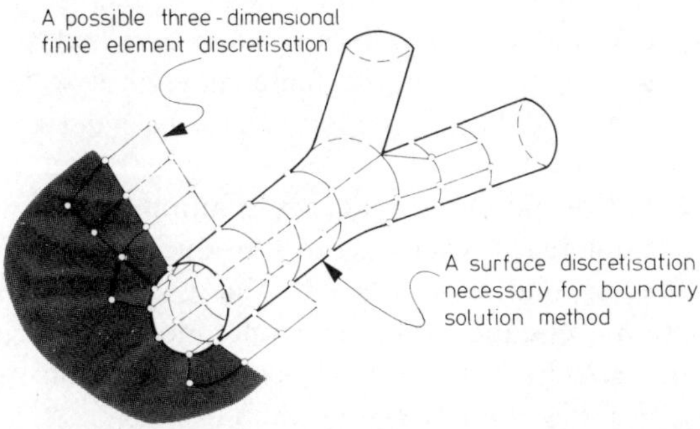

FIG. 1. Problem illustrating merits of boundary solution processes—and possible use of coupling with finite elements. Analysis of a tunnel in an infinite elastic and homogeneous medium with inelastic lining.

The most difficult class of problems for the boundary solution techniques to solve are examples containing material non-linearities and rapidly varying inhomogeneities, which cannot be approximated by a number of homogeneous subregions. This is because the appropriate singular solutions are not usually known. Solutions to these types of problems are possible by using similar iterative techniques to those employed by finite element and finite difference procedures. However, the viability and economy of these calculations requires further research before a critical assessment is possible.

In some engineering examples, the domain's physical properties are such that it could be naturally divided into a number of subdomains which are: (i) linear and homogeneous, (ii) non-linear or inhomogeneous, and (iii) semi-infinite. When this situation occurs, obviously the best discretisation would be of a mixed nature, so that the particular advantage of each numerical method could be used in an optimal manner. To summarise the various ways of coupling the numerical procedures; consider a problem where the domain is such that material non-linearities or inhomogeneities occur in a localised part of the region. In this case it is possible to divide the domain by means of an artificial boundary into an inner region in which there are non-linearities or inhomogeneities, and an 'outer' region where the problem is linear and homogeneous.

To discretise the 'inner' region, some weighted residual procedure is applied so that a finite difference or finite element system of equations is obtained which describes this zone. The 'outer' region on the other hand is easily approximated by deriving a boundary integral equation around the interface boundary of the 'inner' and 'outer' regions, in terms of the dependent variable and its derivatives. This integral relation is now a suitable boundary condition for the finite difference or finite element approximation to be linked to. This approach has been developed by Shaw (1978).

Another method of linking a boundary solution process with a finite element or finite difference scheme, is to introduce an energy functional which is modified so that its stationary value gives the solution of the governing differential equation in both regions, and enforces compatibility in some integral sense along the interface of the region as shown by Zienkiewicz *et al.* (1977*a*) and Chen & Mei (1974).

Finally, a coupling procedure may be performed by simply linking together the two sets of linear algebraic equations obtained from the boundary solution procedure and the other technique.

In Section 10.2 weighted residual methods are investigated to show that all three numerical formulations discussed above have a common basis. Coupling them in a combined solution process then becomes a natural rather than an incongruous marriage. Direct linking processes where either procedure provides boundary conditions for a region modelled by either the finite element, or the finite difference, or a boundary integral procedure are then discussed in Section 10.3.

The fundamental differences in the various coupling techniques occur in the treatment of the interface condition. In the symmetric formulations, compatibility can either be forced explicitly or satisfied in some integral sense by modifying the variational or weighted residual procedure being used. Equilibrium in this formulation is satisfied in a nodal force sense.

In the direct non-symmetric mechanical form of coupling, compatibility is satisfied explicitly at the nodes, and the equilibrium condition achieved by either a nodal force matching, or a linking of the nodal tractions or fluxes on the interface. The description of variational processes which lead to symmetric matrices follows in Section 10.4. A number of examples then follow in Section 10.5 which are intended to reinforce those of Chapter 4 and 7 in demonstrating that coupled solution schemes can be successfully implemented.

10.2 THE WEIGHTED RESIDUAL PROCESS

To set the stage we shall concentrate on the approximate solution of a continuum problem involving the determination of a function ϕ (Fig. 2) such that

$$A(\phi) = 0 \qquad \text{in the domain } \Omega \tag{1}$$

and

$$\bar{B}(\phi) = 0 \qquad \text{on the boundary part } \Gamma_1 \tag{2a}$$

$$\bar{\bar{B}}(\phi) = 0 \qquad \text{on the boundary part } \Gamma_2 \tag{2b}$$

In the above A and B can be linear or non-linear operators of a differential or indeed, integral type.

The algebraic equations for numerical determination of the solution parameters are formed as in Zienkiewicz (1978)

$$\int_\Omega W_j A(\hat{\phi})\,d\Omega + \int_{\Gamma_1} \bar{W}_j \bar{B}(\hat{\phi})\,d\Gamma + \int_{\Gamma_2} \bar{\bar{W}}_j \bar{\bar{B}}(\hat{\phi})\,d\Gamma = 0 \tag{3}$$

where W_j, $\bar{W}_j$ and $\bar{\bar{W}}_j$ are a set $j = 1, n$ of independent weighting (or test) functions. It follows immediately that the algebraic equations can be assembled by evaluating the various integrals for subdomains Ω^e, Γ_1^e and Γ_2^e such that

$$\Omega = \sum \Omega^e \qquad \Gamma_1 = \sum \Gamma_1^e \qquad \Gamma_2 = \sum \Gamma_2^e \tag{4}$$

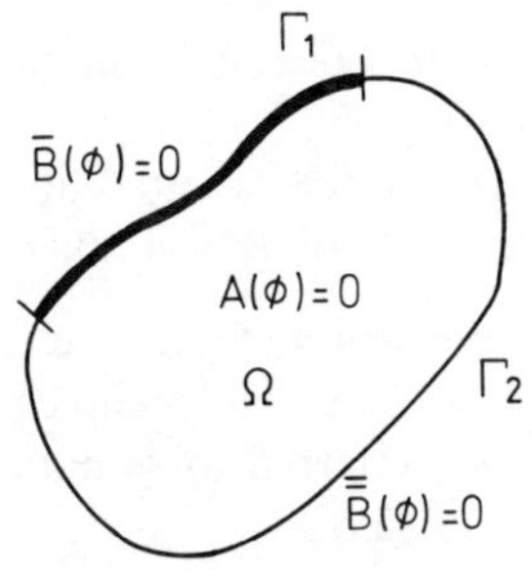

FIG. 2. Definition of continuum problem.

and adding the contributions. This is a corollary of the simple fact that

$$\int_{\Omega} (\quad) \, d\Omega = \sum_{e=1}^{m} \int_{\Omega^e} (\quad) \, d\Omega \tag{5}$$

and means that the equations of approximation behave as a 'standard discrete system' and obey its general rules. The linear system equations (3) then result in a matrix equation

$$\mathbf{Ha} = \mathbf{f} \tag{6}$$

which can be solved in a variety of ways.

Our generalised numerical process will be defined as an approximation in which the independent variable ϕ is expanded in a finite series.

$$\phi \approx \hat{\phi} = \sum_{i=1}^{n} N_i a_i \tag{7}$$

In this expression $N_i = N_i(\mathbf{x})$ where $\mathbf{x}$ are the independent co-ordinates and N_i are prescribed functions known alternatively as shape, basis or trial functions. The a_i are a set of parameters determining the approximate solution.

In conventional finite element analysis the parameters a_i are often identified with the values of ϕ at nodes i and the shape functions N_i have a narrow base being defined separately in each element. This leads to banded forms of eqn. (6) and this is particularly useful in solution but by no means essential.

Equation (3) can be interpreted as an exact statement equivalent to eqns. (1) and (2) as it is true for all functions W_j, $\bar{W}_j$ and $\bar{\bar{W}}_j$ and also for the exact form of ϕ. It follows therefore that the process is convergent providing

(a) The set of expansion functions (7) is complete, i.e., can represent the exact solution when $n \to \infty$
(b) Integrals of eqn. (3) can be evaluated at all points and
(c) Suitably independent sets of W are taken.

In certain cases it is possible to eliminate some of the boundary integrals involving the unknown ϕ by a suitable choice of $\bar{W}$. Corresponding boundary conditions are then referred to as natural.

It can be shown that the choice

$$W_j = \delta(\mathbf{x}_j) \tag{8}$$

where $\delta(\mathbf{x}_j)$ is the Dirac delta function, and where suitably localised shape functions are used, gives the standard finite difference approximation. Thus this approximation can readily be shown to be included in the general formulation.

On the other hand if we limit our considerations to linear operators

$$A(\phi) = L\phi = 0 \tag{9}$$

and choose shape functions such that

$$LN_i = 0 \qquad x \in \Omega \tag{10}$$

Then the domain part of the integral in expression (3) disappears and we have only to satisfy the boundary conditions. The general approximation form is given by

$$\int_{\Gamma_1} \bar{W}_j \bar{B}(\hat{\phi})\,\mathrm{d}\Gamma + \int_{\Gamma_2} \bar{\bar{W}}_j \bar{\bar{B}}(\hat{\phi})\,\mathrm{d}\Gamma = 0 \tag{11}$$

which reduces again on approximation to

$$\mathbf{Ha} = \mathbf{f} \tag{12}$$

Here the coefficients of **H** only involve boundary terms. Within this latest class we have the boundary solution procedures.

The differences between these procedures have therefore become obscure within the theory of weighted residuals. The differences in programming also all but disappear if iterative solution schemes are used with the explicit formulations of Pica (1978). Here, for example, in the simplest, the Jacobi, procedure only the diagonal of the coefficient matrix **H** in (11) is evaluated and

$$\mathbf{H}_{ii}\mathbf{a}_{n+1} = \mathbf{f} - \bar{\mathbf{H}}\mathbf{a}_n \tag{13}$$

where

$$\bar{\mathbf{H}} = \mathbf{H} - \mathbf{H}_{ii}$$

Note that $\bar{\mathbf{H}}\mathbf{a}_n$ can be evaluated as $\mathbf{f}^1 - \mathbf{H}_{ii}\mathbf{a}_n$ where $\mathbf{f}^1$ are the values of the parameters in **f** corresponding to the current $\mathbf{a}_n$. To evaluate $\mathbf{f}^1$ from $\mathbf{a}_n$ requires in finite elements a domain integration, and in boundary solution a boundary integration. Indeed these algorithms involve an extremely low core requirement and their future application to boundary solution processes could be promising if convergence of the iterative process can be guaranteed.

10.3 DIRECT COUPLING THEORY

A linking between the boundary integral procedures and finite element or finite difference methods is possible in principle simply by applying the appropriate interface conditions between the two regions.

For example in a potential problem the interface constraints are that the potential ϕ must be continuous and the flux $\partial\phi/\partial n$ equal and opposite. It should be noted that the flux condition can be interpreted in an integral sense, where the sum of the 'flux forces' at any node are balanced by the resultant exterior force applied at the node. The analogous interface conditions in elasticity are displacement continuity and traction cancellation. Here the traction condition may be replaced by the usual nodal force equilibrium condition.

The general interface conditions for a potential problem may be written as

$$\phi_{\mathrm{BI}} = \phi_{\mathrm{FE}} \quad \text{or} \quad \phi_{\mathrm{FD}} \tag{14}$$

$$\frac{\partial \phi}{\partial n_{\mathrm{BI}}} = -\frac{\partial \phi}{\partial \mathbf{n}_{\mathrm{FE}}} \quad \text{or} \quad -\frac{\partial \phi}{\partial \mathbf{n}_{\mathrm{FD}}} \tag{15}$$

Where the alternative statement of (15) is

$$\sum_{e=1}^{\text{No. elements}} \mathbf{F}_{\mathrm{e}}^{i} = \mathbf{F}_{\text{external}}^{i} \tag{16}$$

which is a force balance equation written at node i.

10.3.1 The Direct Boundary Integral Procedure

Let us now consider a potential problem where the exterior domain Ω_{e} is modelled by boundary integrals and the interior domain Ω_{i} discretised by finite elements or finite differences where Γ_{i} is the interface between the two regions which is smooth, and Γ_{e} is the remaining part of the boundary of the region Ω_{e}. (See Fig. 3).

In the usual notation the direct boundary integral equation for Ω_{e} is

$$\alpha\phi_i(p) = \int_{\Gamma_i+\Gamma_e} \frac{\partial G}{\partial n}(p,q)\phi(q)\,\mathrm{d}\Gamma_q - \int_{\Gamma_i+\Gamma_e} G(p,q)\frac{\partial \phi}{\partial n}(q)\,\mathrm{d}\Gamma_q \tag{17}$$

where α is a constant for each point p and $G(p,q)$ the Green's function for the problem.

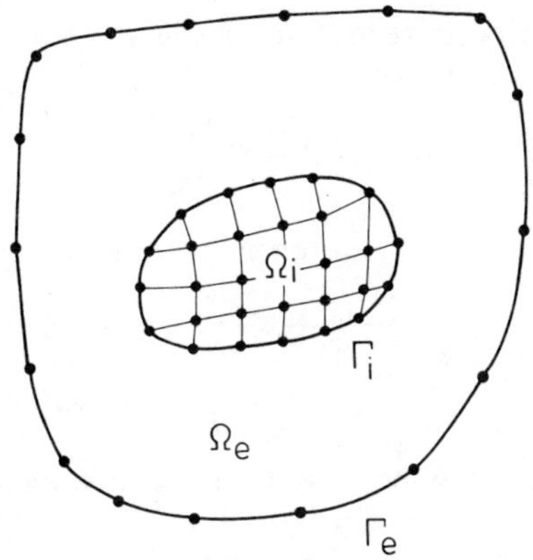

FIG. 3. Two region potential problem.

A boundary collocation and numerical discretisation of the above equation gives

$$\mathbf{A}\boldsymbol{\phi}_{\mathrm{BI}} = \mathbf{B}\frac{\partial \boldsymbol{\phi}}{\partial \mathbf{n}_{\mathrm{BI}}} \tag{18}$$

where $\boldsymbol{\phi}_{\mathrm{BI}}$ and $\partial\boldsymbol{\phi}/\partial\mathbf{n}_{\mathrm{BI}}$ are nodal values of ϕ and $\partial\phi/\partial n$ around Γ_{i} and Γ_{e}.

In the interior domain the finite element or finite difference equations are of the form as detailed by Zienkiewicz (1977)

$$\mathbf{K}\boldsymbol{\phi}_{\mathrm{FE}} = \mathbf{F} \tag{19}$$

where

$$\mathbf{F} = \int_{\Gamma_{\mathrm{i}}} \mathbf{N}^{\mathrm{T}} \overline{\frac{\partial \phi}{\partial n}} \mathrm{d}\Gamma = \mathbf{C}\frac{\partial \boldsymbol{\phi}}{\partial \mathbf{n}} \tag{20}$$

(Note: $\partial\bar{\phi}/\partial n$ are known values of $\partial\phi/\partial n$ on boundary of interior domain.)

If it is desired to couple the methods by using (14) and (16) then the boundary integral equation must be manipulated and integrated so that

$$\mathbf{F}_{\mathrm{BI}} = \mathbf{C}\mathbf{B}^{-1}\mathbf{A}\boldsymbol{\phi}_{\mathrm{BI}} = \mathbf{C}\frac{\partial \boldsymbol{\phi}}{\partial \mathbf{n}_{\mathrm{BI}}} \tag{21}$$

It is now easily seen that the equations can be linked in the usual finite element assemblage process, and then solved for the unknown $\boldsymbol{\phi}$s and **F**s.

Alternatively, it is possible to link the above methods together by using conditions (14) and (15).

To perform this procedure the finite element or finite difference equations are written as

$$\mathbf{K}\boldsymbol{\phi} = \mathbf{C}\frac{\partial \boldsymbol{\phi}}{\partial \mathbf{n}} \tag{22}$$

The coupling of (18) and (22) is then given by applying (14) and (15), which gives the result

$$\begin{bmatrix} \mathbf{K} & \mathbf{O} \\ \mathbf{O} & \mathbf{A} \end{bmatrix} \left[\boldsymbol{\phi} \right] = \begin{bmatrix} \mathbf{C} & \mathbf{O} \\ \mathbf{O} & \mathbf{B} \end{bmatrix} \left[\frac{\partial \boldsymbol{\phi}}{\partial \mathbf{n}} \right] \tag{23}$$

These equations can now be solved for the unknown $\boldsymbol{\phi}$ and $\partial\boldsymbol{\phi}/\partial\mathbf{n}$.

10.3.2 The Indirect Boundary Integral Procedure

The indirect boundary integral equations in the fictitious sources density $\mu(q)$ for the domain Ω_e are

$$\phi(p) = \int_{\Gamma_i + \Gamma_e} G(p, q)\mu(q)\,\mathrm{d}\Gamma_q \tag{24}$$

$$\frac{\partial \phi}{\partial n}(p) = \int_{\Gamma_i + \Gamma_e} \frac{\partial G}{\partial n}(p, q)\mu(q)\,\mathrm{d}\Gamma_q \tag{25}$$

Performing the usual numerical approximation of eqns. (24) and (25) results in the two matrix equations

$$\boldsymbol{\phi} = \bar{\mathbf{N}}\boldsymbol{\mu} \tag{26}$$

and

$$\frac{\partial \boldsymbol{\phi}}{\partial \mathbf{n}} = \bar{\mathbf{M}}\boldsymbol{\mu} \tag{27}$$

where $\boldsymbol{\phi}$, $\partial\boldsymbol{\phi}/\partial\mathbf{n}$ and $\boldsymbol{\mu}$ are the values at the boundary nodes of potential, flux and source density respectively.

To effect a coupling of the procedures using the interface conditions (14) and (16), the source density must be eliminated from (26) and (27) to give

$$\frac{\partial \boldsymbol{\phi}}{\partial \mathbf{n}} = \bar{\mathbf{M}}\bar{\mathbf{N}}^{-1}\boldsymbol{\phi} \tag{28}$$

Equation (28) is now integrated to provide the necessary forces required, so that (28) becomes

$$\mathbf{F}_{\mathrm{BI}} = \mathbf{C}\frac{\partial \boldsymbol{\phi}}{\partial \mathbf{n}} = \mathbf{C}\bar{\mathbf{M}}\bar{\mathbf{N}}^{-1}\boldsymbol{\phi} \tag{29}$$

Now eqns. (19) and (29) can be assembled by using the nodal force equilibrium technique as employed in the finite element method. Another form of linking can be achieved by using the interface conditions (14) and (15) with eqns. (22), (26) and (27) to give

$$\begin{bmatrix} \mathbf{K} & & \mathbf{O} \\ \mathbf{O} & \mathbf{I} & -\bar{\mathbf{N}} \\ \mathbf{O} & & \bar{\mathbf{M}} \end{bmatrix} \begin{Bmatrix} \boldsymbol{\phi} \\ \boldsymbol{\mu} \end{Bmatrix} = \begin{bmatrix} \mathbf{C} & \mathbf{O} \\ \mathbf{O} & \\ \mathbf{O} & \mathbf{I} \end{bmatrix} \begin{Bmatrix} \dfrac{\partial \boldsymbol{\phi}}{\partial \mathbf{n}} \end{Bmatrix} \tag{30}$$

This system of equations is now solved for the unknown values of source density $\boldsymbol{\mu}$. Back substitution will then recover all the boundary values of $\boldsymbol{\phi}$ and $\partial\boldsymbol{\phi}/\partial\mathbf{n}$.

In the above section it was assumed that ϕ (and $\partial\phi/\partial n$) were defined at the same nodal points on the boundary as the finite element or finite difference methods. If this is not the case then a slight change to the above procedure must be carried out before linking. This modification simply requires a relationship between the nodal values ϕ and $\partial\phi/\partial n$ for the boundary integral region and the corresponding nodal values for the finite element or finite difference procedure. This relationship is of the form,

$$\boldsymbol{\phi}_{\mathrm{FE}} = \mathbf{G}\boldsymbol{\phi}_{\mathrm{BI}} \qquad \text{and} \qquad \frac{\partial \boldsymbol{\phi}}{\partial \mathbf{n}}_{\mathrm{FE}} = \mathbf{H}\frac{\partial \boldsymbol{\phi}}{\partial \mathbf{n}}_{\mathrm{BI}} \tag{31}$$

The methods can now be coupled as described above.

10.4 SYMMETRIC COUPLING THEORY

In this section we consider the potential problem,

$$\nabla^2\phi = 0 \qquad \text{in } \Omega$$

subject to the boundary conditions

$$\phi = \bar{\phi} \quad \text{on } \Gamma_\phi \qquad \text{and} \qquad \frac{\partial\phi}{\partial n} = \overline{\frac{\partial\phi}{\partial n}} \quad \text{on } \Gamma_q$$

to illustrate the various symmetric formulations of the boundary integral procedures.

For the above problem in two dimensions the kinetic energy is given by

$$U = \frac{1}{2}\int_\Omega \left(\left(\frac{\partial\phi}{\partial x}\right)^2 + \left(\frac{\partial\phi}{\partial y}\right)^2\right) \mathrm{d}\Omega \tag{32}$$

To transform the domain integral (32) into an equivalent boundary integral we recall Gauss's divergence theorem which states that for any vector function **F**

$$\int_\Omega \nabla . \mathbf{F} \, d\Omega = \int_\Gamma \mathbf{n} . \mathbf{F} \, d\Gamma \tag{33}$$

where Ω is a region bounded by Γ, and **n** is a unit normal vector at $d\Gamma$.

Applying the above theorem for $\mathbf{F} = \phi\nabla\phi$ gives

$$\int_\Omega \nabla . (\phi\nabla\phi) \, d\Omega = \int_\Gamma \phi \frac{\partial \phi}{\partial n} d\Gamma \tag{34}$$

If we expand the left-hand-side integral we have

$$\nabla . (\phi\nabla\phi) = \nabla\phi . \nabla\phi + \phi\nabla^2\phi$$

where $\nabla\phi . \nabla\phi$ is immediately recognised as $(\partial\phi/\partial x)^2 + (\partial\phi/\partial y)^2$ in two dimensions. Therefore if ϕ is a solution of Laplace's equation then we obtain the integral identity

$$\int_\Omega \left[\left(\frac{\partial \phi}{\partial x}\right)^2 + \left(\frac{\partial \phi}{\partial y}\right)^2 \right] d\Omega = \int_\Gamma \phi \frac{\partial \phi}{\partial n} d\Gamma$$

which leads to

$$U = \frac{1}{2} \int_\Gamma \phi \frac{\partial \phi}{\partial n} d\Gamma \tag{35}$$

where Γ is the boundary of the region Ω.

10.4.1 A Symmetric Form of the Direct Boundary Integral Procedure

To generate a symmetric system of equations from the direct boundary integral procedure we require the minimisation of an energy functional of the form

$$\pi = \frac{1}{2} \int_\Gamma \phi \frac{\partial \phi}{\partial n} d\Gamma - \int_{\Gamma_q} \phi \frac{\overline{\partial \phi}}{\partial n} d\Gamma \tag{36}$$

where the first term demands ϕ to be harmonic, and the second term leads on minimisation of the functional to satisfaction of the boundary conditions (16). If we substitute the boundary interpolants for ϕ and $\partial\phi/\partial n$, which are of the form

$$\phi = \mathbf{N}\boldsymbol{\phi} \quad \text{and} \quad \frac{\partial \phi}{\partial n} = \mathbf{M} \frac{\partial \boldsymbol{\phi}}{\partial \mathbf{n}} \tag{37}$$

the functional π becomes

$$\pi = \frac{1}{2}\frac{\partial \boldsymbol{\phi}^{\mathrm{T}}}{\partial \mathbf{n}} \int_{\Gamma} \mathbf{M}^{\mathrm{T}}\mathbf{N}\,\mathrm{d}\Gamma \boldsymbol{\phi} - \boldsymbol{\phi}^{\mathrm{T}} \int_{\Gamma_q} \mathbf{N}^{\mathrm{T}} \frac{\overline{\partial \phi}}{\partial n}\,\mathrm{d}\Gamma \tag{38}$$

Boundary collocation around the boundary integral region gives the direct boundary integral equations (18), i.e.,

$$\mathbf{A}\boldsymbol{\phi} = \mathbf{B}\frac{\partial \boldsymbol{\phi}}{\partial \mathbf{n}} \tag{39}$$

In order to ensure that the rows and columns of the final system matrix will sum to zero we supplement the equations (39) with an equilibrium equation

$$\int_{\Gamma} \frac{\partial \phi}{\partial n}\,\mathrm{d}\Gamma \equiv \int_{\Gamma} \mathbf{M}(\Gamma)\,\mathrm{d}\Gamma \frac{\partial \hat{\boldsymbol{\phi}}}{\partial \mathbf{n}} = 0 \tag{40}$$

which ensures equilibrium is identically satisfied by the approximate solution based on the interpolating shape functions $\mathbf{M}(\Gamma)$ of (37).

Linking the direct boundary integral equations (39) with the equilibrium condition (40) we have

$$\begin{bmatrix} \mathbf{A} \\ \mathbf{O} \end{bmatrix} \begin{bmatrix} \boldsymbol{\phi} \end{bmatrix} = \begin{bmatrix} \mathbf{B} & \mathbf{Q} \\ \mathbf{Q}^{\mathrm{T}} & \mathbf{O} \end{bmatrix} \begin{bmatrix} \dfrac{\partial \boldsymbol{\phi}}{\partial \mathbf{n}} \\ \lambda \end{bmatrix} \tag{41}$$

where λ takes the role of a Lagrange multiplier.

Here

$$\mathbf{Q} = \int_{\Gamma} \mathbf{M}(\Gamma)\,\mathrm{d}\Gamma$$

where from (37)

$$\frac{\partial \phi}{\partial n} = \mathbf{M}(\Gamma)\frac{\partial \hat{\boldsymbol{\phi}}}{\partial \mathbf{n}}$$

The role of the Lagrange multiplier in (41) is therefore to introduce a controlled error into each of the equations (39) which can no longer be satisfied identically if the extra condition (40) is imposed. The ith equation becomes

$$\sum_j A_{ij}\hat{\phi}_j = \sum_j B_{ij}\frac{\partial \hat{\phi}}{\partial n_j} + \int_{\Gamma_i} M_i\,\mathrm{d}\Gamma \lambda$$

The error in satisfaction of the equation from collocation on the ith segment is therefore proportional to the length of the ith segment and is usually negligible because equilibrium without interpolation is well approximated by the solution of the standard integral equations.

If the solution for $\partial\boldsymbol{\phi}/\partial\mathbf{n}$ from eqn. (41) is

$$\frac{\partial\boldsymbol{\phi}}{\partial\mathbf{n}} = \mathbf{E}\boldsymbol{\phi} \tag{42}$$

where $\mathbf{E}$ is obtained by inversion and partitioning in eqn. (41) then

$$\pi = \tfrac{1}{2}\boldsymbol{\phi}^{\mathrm{T}}\mathbf{E}^{\mathrm{T}}\int_{\Gamma}\mathbf{M}^{\mathrm{T}}\mathbf{N}\,\mathrm{d}\Gamma\boldsymbol{\phi} - \boldsymbol{\phi}^{\mathrm{T}}\int_{\Gamma}\mathbf{N}^{\mathrm{T}}\frac{\overline{\partial\phi}}{\partial n}\,\mathrm{d}\Gamma \tag{43}$$

The minimisation of π which requires $(\partial\boldsymbol{\pi}/\partial\boldsymbol{\phi}) = 0$ leads to

$$\mathbf{K}'\boldsymbol{\phi} + \mathbf{F} = \mathbf{0} \tag{44}$$

where

$$\mathbf{K}' = \tfrac{1}{2}\mathbf{E}^{\mathrm{T}}\int_{\Gamma}\mathbf{M}^{\mathrm{T}}\mathbf{N}\,\mathrm{d}\Gamma + \tfrac{1}{2}\left[\mathbf{E}^{\mathrm{T}}\int_{\Gamma}\mathbf{M}^{\mathrm{T}}\mathbf{N}\,\mathrm{d}\Gamma\right]^{\mathrm{T}} \tag{45}$$

and

$$\mathbf{F}' = -\int_{\Gamma_q}\mathbf{N}^{\mathrm{T}}\frac{\overline{\partial\phi}}{\partial n}\,\mathrm{d}\Gamma \tag{46}$$

The matrix equation (44) is of the same form as the finite element stiffness equation (19), therefore the matrics $\mathbf{K}'$ and $\mathbf{F}'$ can be assembled into standard finite element systems as contributions from a new element. If the flux is required it can be determined directly from (42). Compatibility between elements is ensured if the boundary interpolants for the boundary integral region are identical to the interpolation functions of the adjacent finite elements.

Within the formulation it should be noted that the direct boundary integral procedure assumes independent interpolants for ϕ and $\partial\phi/\partial n$. This is not physically consistent, as definition of the boundary variation of one completely defines the other. Therefore the resultant energy distribution modelled by

$$\frac{1}{2}\int_{\Gamma}\phi\frac{\partial\phi}{\partial n}\,\mathrm{d}\Gamma$$

cannot be correct. It also follows that the derivatives $\partial\boldsymbol{\pi}/\partial\boldsymbol{\phi}$ are not calculated accurately and so eqn. (44) must contain an inherent source of error. It is interesting to note however, that similar matrices can be derived (see Brebbia, 1978), from a more direct procedure by rearranging (39) so that

$$\frac{\partial \phi}{\partial \mathbf{n}} = \mathbf{B}^{-1}\mathbf{A}\phi$$

which we can integrate to obtain

$$\mathbf{F} = \mathbf{C}\mathbf{B}^{-1}\mathbf{A}\phi \tag{47}$$

Now $\mathbf{CB}^{-1}\mathbf{A}$ is a stiffness matrix and should, from first principles, be symmetric. In fact it does not turn out to be so and it has been suggested that this equation should be modified to

$$\mathbf{F} = \tfrac{1}{2}[\mathbf{C}\mathbf{B}^{-1}\mathbf{A} + (\mathbf{C}\mathbf{B}^{-1}\mathbf{A})^{\mathrm{T}}] \tag{48}$$

to produce a symmetric form. This modification apparently has no valid foundation except that the energy approach led to a similar form in (45).

10.4.2 A Symmetric Form of the Indirect Boundary Integral Procedure

Once again the formulation of a symmetric form for a boundary solution procedure requires the minimisation of a functional based upon energy considerations which can be written as

$$\pi^* = \frac{1}{2}\int_{\Gamma} \phi \frac{\partial \phi}{\partial n}\,\mathrm{d}\Gamma - \int_{\Gamma_q} \phi \overline{\frac{\partial \phi}{\partial n}}\,\mathrm{d}\Gamma - \int_{\Gamma_\phi} (\phi - \bar{\phi})\lambda\,\mathrm{d}\Gamma \tag{49}$$

where the functions performed by the first two terms of the functional are as described in the preceding subsection, and the introduction of the third term ensures satisfaction of boundary conditions ϕ, or that compatibility of ϕ is maintained across the interface with a neighbouring region.

It should be noted that λ is a Lagrange multiplier to enforce the constraint satisfaction and is shown in Zienkiewicz *et al.* (1977*a*) to have the value $\partial\phi/\partial n$ so that an alternative, reduced, form can be written

$$\pi^* = \frac{1}{2}\int_{\Gamma} \phi \frac{\partial \phi}{\partial n}\,\mathrm{d}\Gamma - \int_{\Gamma_q} \phi \overline{\frac{\partial \phi}{\partial n}}\,\mathrm{d}\Gamma - \int_{\Gamma_\phi} (\phi - \bar{\phi})\frac{\partial \phi}{\partial n}\,\mathrm{d}\Gamma \tag{50}$$

where $\bar{\phi}$ can be the value of ϕ on a neighbouring region along Γ_ϕ. Expressions for ϕ and $\partial\phi/\partial n$ from the indirect scheme, via eqns. (24) and (25) can be written in the form

$$\phi(p) = \bar{\mathbf{N}}(p)\boldsymbol{\mu} \tag{51}$$

$$\frac{\partial \phi}{\partial n}(p) = \bar{\mathbf{M}}(p)\boldsymbol{\mu} \tag{52}$$

Furthermore if the variation of $\bar{\phi}$ is described by

$$\bar{\phi} = \mathbf{N}\boldsymbol{\phi} \tag{53}$$

then the substitution of eqns. (51), (52) and (53) into the functional π^* will yield

$$\pi^*(\boldsymbol{\mu}, \boldsymbol{\phi}) = \tfrac{1}{2}\boldsymbol{\mu}^{\mathrm{T}} \int_{\Gamma} \bar{\mathbf{N}}^{\mathrm{T}}\mathbf{M}\,\mathrm{d}\Gamma\,\boldsymbol{\mu} - \boldsymbol{\mu}^{\mathrm{T}} \int_{\Gamma_q} \bar{\mathbf{N}}^{\mathrm{T}} \frac{\partial \bar{\phi}}{\partial n}\,\mathrm{d}\Gamma$$

$$- \int_{\Gamma_\phi} (\boldsymbol{\mu}^{\mathrm{T}}\bar{\mathbf{N}}^{\mathrm{T}} - \boldsymbol{\phi}^{\mathrm{T}}\mathbf{N}^{\mathrm{T}})\bar{\mathbf{M}}\,\mathrm{d}\Gamma\,\boldsymbol{\mu} \tag{54}$$

Because π^* is a functional of $\boldsymbol{\mu}$ and $\boldsymbol{\phi}$, performing the variation we have

$$\delta\pi^* = \frac{\partial \boldsymbol{\pi}}{\partial \hat{\boldsymbol{\mu}}}\,.\,\delta\boldsymbol{\mu}^{\mathrm{T}} + \frac{\partial \boldsymbol{\pi}}{\partial \hat{\boldsymbol{\phi}}}\,.\,\delta\boldsymbol{\phi}^{\mathrm{T}}$$

which can be written in the form

$$\delta\pi^* = (\mathbf{K}^{\mathrm{e}}\boldsymbol{\mu} + \bar{\mathbf{K}}^{\mathrm{e}}\boldsymbol{\phi} - \mathbf{F}^{\mathrm{e}})\,.\,\delta\boldsymbol{\mu}^{\mathrm{T}} + \bar{\mathbf{K}}^{\mathrm{e}}\boldsymbol{\mu}\,.\,\delta\boldsymbol{\phi}^{\mathrm{T}}$$

where

$$\mathbf{K}^{\mathrm{e}} = \tfrac{1}{2} \int_{\Gamma_q} (\bar{\mathbf{M}}^{\mathrm{T}}\bar{\mathbf{N}} + \bar{\mathbf{N}}^{\mathrm{T}}\bar{\mathbf{M}})\,\mathrm{d}\Gamma - \tfrac{1}{2} \int_{\Gamma_\phi} (\bar{\mathbf{M}}^{\mathrm{T}}\bar{\mathbf{N}} + \bar{\mathbf{N}}^{\mathrm{T}}\bar{\mathbf{M}})\,\mathrm{d}\Gamma$$

$$\bar{\mathbf{K}}^{\mathrm{e}} = \int_{\Gamma_\phi} \bar{\mathbf{M}}^{\mathrm{T}}\mathbf{N}\,\mathrm{d}\Gamma \tag{55}$$

and

$$\mathbf{F}^{\mathrm{e}} = -\int_{\Gamma_q} \bar{\mathbf{N}}^{\mathrm{T}} \frac{\overline{\partial \phi}}{\partial n}\,\mathrm{d}\Gamma$$

Therefore minimisation of π^* for arbitrary variations of $\boldsymbol{\mu}$ and $\boldsymbol{\phi}$ requires

$$\frac{\partial \boldsymbol{\pi}^*}{\partial \boldsymbol{\mu}} = 0 \qquad \text{and} \qquad \frac{\partial \boldsymbol{\pi}^*}{\partial \boldsymbol{\phi}} = 0$$

which leads to the equations in matrix form

$$\begin{bmatrix} \mathbf{K}^{\mathrm{e}} & \bar{\mathbf{K}}^{\mathrm{e}} \\ \bar{\mathbf{K}}^{\mathrm{e}\,\mathrm{T}} & \mathbf{O} \end{bmatrix} \begin{bmatrix} \boldsymbol{\mu} \\ \boldsymbol{\phi} \end{bmatrix} = \begin{bmatrix} \mathbf{F}^{\mathrm{e}} \\ \mathbf{O} \end{bmatrix} \tag{56}$$

(Note: Where $\boldsymbol{\phi}$ may be specified boundary conditions.)

These equations can again be incorporated directly into finite element systems with coupling between regions coming through the parameters $\boldsymbol{\phi}$. As these parameters $\boldsymbol{\phi}$ are only evaluated as unknowns when they exist on the interface between regions we must, in general, resort to the standard indirect equations (51) and (52) to determine the unknown values of ϕ and $\partial\phi/\partial n$ on the boundary of the region or the interior, with the vector $\hat{\boldsymbol{\mu}}$ given by the solution of (56).

A direct analogy exists here with the hybrid finite element schemes (see Zienkiewicz, 1977, Chapter 12), where parameters internal to the element define the variation of stresses and displacements, and interface variables link the elements and try to enforce equilibrium and compatibility across the interface. The enforcement of these conditions across the interface of multiple regions discretised using this symmetric form of the indirect method is discussed elsewhere.

10.4.3 The Expense of Implementing the Symmetric Procedures

The ease of implementing the symmetric forms of boundary integrals in standard finite element procedures makes them attractive formulations if they can be shown to be cost-effective. An estimate of this cost-effectiveness can be determined here by comparison with a solution based on the standard direct boundary integral procedure.

Here we postulate the solution of a large problem using the *standard direct boundary integral* procedure with the domain divided into 3 subregions each with 800 degrees of freedom ($n = 800$). Using the non-symmetric 'block banded' solution scheme the solution of the final set of equations would require $3 \times (n^3/3)$ operations (since $n^3/3$ operations are required for non-symmetric Gauss elimination on n full equations). But the solution time represents about $\frac{1}{3}$ of the total cost with the remaining time taken up by the boundary integrations, the cost of input/output, etc. An approximate order of the cost is therefore $3n^3$.

Using the *symmetric form of the direct boundary integral procedure* proposed in Section 10.3.1 we must first invert n equations (inversion n^3 operations) on each region giving $3n^3$ operations. The solution of a symmetric set of n equations however requires only $n^3/6$ operations so that the solution of the assembled equations for the 3 regions requires something less than $3 \times (n^3/6) = (n^3/2)$ (less than this figure because the assembly process identifies each interface variable now only once so the order of the total set of equations should be considerably less than $3n$). However, the same $2n^3$ operations are required as in the direct method for the boundary integration and input/output. We therefore arrive at an approximate $11n^3/2$ for the total solution.

With this direct comparison the symmetric form would appear to be roughly twice as expensive to implement as the standard direct boundary integral procedure. However, it must be noted that the symmetric forms would become much more cost-effective if a larger number of subregions were considered. The penalty of introducing 2 additional parameters at each interface node is reduced to 1 so the subdivision into many more regions is not directly penalised.

The implementation of the *symmetric form of the indirect method* also involves an n^3 operation on each subregion. If the integrals in eqn. (55) are carried out by numerical integration using p integrating points on each segment, the formulation of $\mathbf{K}^e$ contains pn^3 multiplications. This arises because $\bar{\mathbf{N}}$ and $\bar{\mathbf{M}}$ are vectors and the operation $\bar{\mathbf{N}}^T\mathbf{M}$ involves n^2 multiplications. For n segments this is carried out pn times leading to the final figure of pn^3. If $p > 1$ the total expense should exceed that of the symmetric direct method because there must be three variables introduced on each interface segment between subregions (a source term μ for each region and the coupled ϕ_i value).

In the next section a number of examples will be solved to compare the accuracy of the symmetric and non-symmetric coupling formulations presented in this chapter.

10.5 EXAMPLES

In this section a number of problems are solved by the various proposed coupling techniques to illustrate the accuracy and feasibility of linking either type of boundary solution technique to a finite element or finite difference discretisation.

10.5.1 L-shaped Domain Problem

(i) Non-symmetric Coupling of Finite Element and Direct Boundary Integral Procedures

Using the discretisation shown in Fig. 4(a) which is comprised of 9 linear finite elements and a direct boundary integral region with 8 nodes that has linear continuous boundary interpolants, a numerical solution was obtained by the non-symmetric coupling described in Section 10.3 which is shown in Table 1. As the results show, this procedure gives a good solution with the very coarse discretisation used.

TABLE 1

POTENTIAL SOLUTIONS FOR THE L-SHAPED DOMAIN PROBLEM

Node	*Accurate solution*	*FE + DBI (non-symmetric)*	*FE + DBI symmetric (quadratic)*	*FE + IBI symmetric*	*FE + IBI*[a] *symmetric*	*IBI symmetric 3-region*	*DBI symmetric 3-region*
A	0·8640	0·8586	0·8621	0·8865	0·8591	0·8644	0·8587
B	0·6667	0·6667	0·6722	0·6266	0·6687	0·6733	0·6667
C	0·2972	0·3046	0·2994	0·3229	0·2955	0·2997	0·3023
D	0·2881	0·2913	0·2880	0·2751	0·2894	0·2883	0·2875
E	0·5680	0·5707	0·5686	0·5877	0·5694	0·5692	0·5706
F	0·8081	0·8005	0·8020	0·7900	0·8010	0·8039	0·8040
G	0·8514	0·8479	0·8514	0·8637	0·8512	0·8478	0·8488
H	0·9109	0·9082	0·9099	0·9041	0·9107	0·9063	0·9085
Max. % error		2·7%	0·8%	8·5%	0·7%	0·85%	1·6%

[a] A supplementary solution is added to the symmetric form of the IBI procedure.

Note: The maximum percentage error quoted above is calculated from all nodal points within the discretisations used.

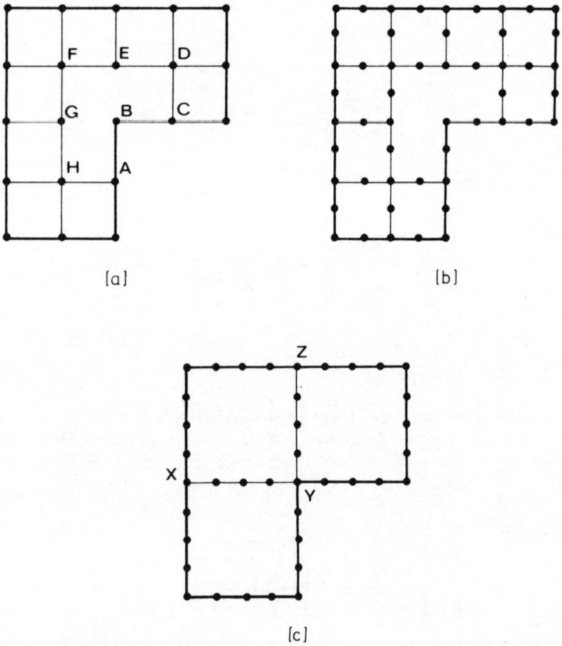

FIG. 4. L-shaped domain coupled discretisations. (a) Mixed discretisation for FE and DBI (linear), and also FE and IBI; (b) mixed discretisation for FE and DBI (quadratic); (c) discretisation for multi-region symmetric BI procedures.

(ii) Symmetric Coupling of Finite Element and Direct Boundary Integral Procedures (Quadratic Interpolants).

The symmetric formulation of the direct boundary integral procedure with quadratic boundary interpolants employed here does not require any finite difference approximation for $\partial\phi/\partial n$ in corners, because boundary collocation positions in corners are not used. Details of this procedure are given in Mustoe *et al.* (1978). A numerical solution obtained by using the discretisation illustrated in Fig. 4(b) is shown in Table 1. From this result it is seen that a very accurate solution to this problem has been found.

(iii) Symmetric Coupling of Finite Element and Indirect Boundary Integral Procedures

The symmetric formulation of the indirect boundary integral method employed here has a piece-wise constant source distribution with a linear boundary constraint on the potential wherever it is known or is an interface

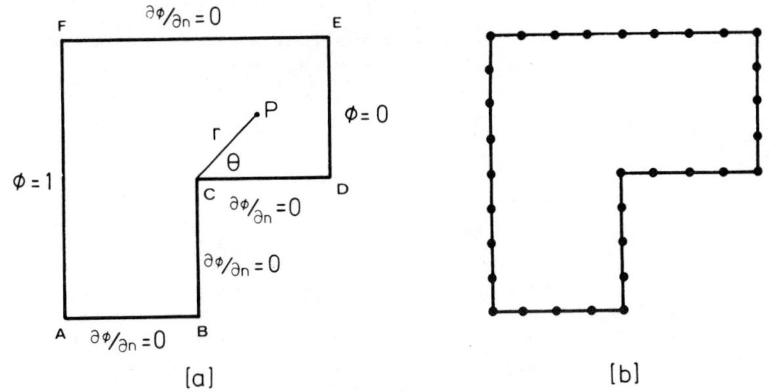

FIG. 5. L-shape domain boundary conditions, and single procedure discretisations. (a) Boundary conditions; (b) BI discretisations.

variable. Using the discretisation shown in Fig. 4(a) the resulting numerical solution is given in Table 1. This result can be improved by inserting a series approximation to the potential solution near the corner so that eqn. (51) becomes

$$\phi(p) = \sum_{i=1}^{n} \bar{N}_i(p)\mu_i + \mu_{n+1} + \mu_{n+2} r^{2/3} \cos\frac{2\theta}{3} + \mu_{n+3} r^{4/3} \cos\frac{4\theta}{3}$$

where r and θ are defined as in Fig. 5(a). However, because there is an unknown constant in the expansion for ϕ, the functional π^* must be altered as described in Kelly *et al.* (1978), so that

$$\pi^* = \pi^* + \lambda \int_{\Gamma_\phi} (\phi - \bar{\phi})\, \mathrm{d}\Gamma$$

where λ is a Lagrange multiplier required to determine μ_{n+1}.

The resulting numerical solution obtained with this modification, can be seen in Table 1 to be accurate.

(iv) Symmetric Direct Boundary Integral (Quadratic) Multi-region Solution

Using the discretisation shown in Fig. 4(c) a symmetric form of the direct boundary integral procedure with quadratic boundary interpolants for ϕ and $\partial\phi/\partial n$, where the $\partial\phi/\partial n$ approximants are discontinuous, is used to obtain the results shown in Table 1. These numerical solutions indicate that an accurate result has been found.

TABLE 2

INTERFACE $\partial\phi/\partial n$ VALUES FOR THE L-SHAPED DOMAIN PROBLEM OBTAINED BY SYMMETRIC MULTI-REGION DISCRETISATIONS

Node	*Symmetric DBI* $\left(quadratic\ \phi, \dfrac{\partial\phi}{\partial n}\ discontinuous\right)$		*Symmetric IBI*	
	$\dfrac{\partial\phi}{\partial n_1}$	$\dfrac{\partial\phi}{\partial n_2}$	$\dfrac{\partial\phi}{\partial n_1}$	$\dfrac{\partial\phi}{\partial n_2}$
X	−0·0030	0·0043	−0·0052	0·0052
	−0·0111	0·0119		
	−0·0237	0·0170	−0·0180	0·0180
	−0·0280	0·0362		
	−0·0452	0·0555	−0·0296	0·0296
Y	−0·0990	0·0692	−0·1111	0·1111
	−0·1095	0·1480		
	−0·1198	0·1075	−0·1616	0·1616
	−0·1115	0·1012		
	−0·0994	0·1077	−0·0984	0·0984
	−0·1045	0·1022	−0·0978	0·0978
Z	−0·1030	0·1015	−0·1088	0·1088

(*v*) *Symmetric Indirect Boundary Integral Multi-region Solution*

The symmetric indirect boundary integral technique using the multi-region discretisation in Fig. 4(c) gives the numerical results shown in Table 1. These results indicate that a good potential solution has been obtained.

(*vi*) *Interface Equilibrium Comparisons*

Table 2 gives nodal values of interface $\partial\phi/\partial n$ along the two sections XY and YZ (Fig. 4(c)). From these results it is easily seen that equilibrium is now being obeyed in the nodal force sense of the usual finite element procedure, and not in an interface matching of $\partial\phi/\partial n$. It is interesting to note that the symmetric form of the indirect boundary integral shows that the $\partial\phi/\partial n$ values are exactly equal and opposite along the interfaces. However, the potential values along the interfaces as given in Table 3, show that compatibility is not forced exactly.

10.5.2 Machine Component Problem

Figure 6(a) shows a geometrically complicated machine component with

TABLE 3
INTERFACE POTENTIAL VALUES FOR THE L-SHAPED DOMAIN PROBLEM GIVEN BY THE SYMMETRIC IBI MULTI-REGION DISCRETISATION

Node	*Symmetric IBI* ϕ_1	ϕ_2
X	0·9752	0·9700
	0·8888	0·8850
	0·8104	0·8073
Y	0·7439	0·7456
	0·6247	0·6234
	0·5792	0·5795
	0·5588	0·5590
Z	0·5617	0·5562

many re-entrant corners. The applied boundary conditions are also shown in Fig. 6(a).

In the following text the problem is solved by a selection of the proposed solution processes so that a detailed comparison of the numerical procedures can be made.

(i) Finite Element Solution

The finite element discretisation of this problem is illustrated in Fig. 6(a). The model is comprised of 95 eight-noded isoparametric elements. The total number of nodes within the model is 366, of which 140 are on the boundary The potential solution at various points around the boundary is given in Fig. 7. Additionally, in Fig. 8 a computer plot illustrates the equipotential contours throughout the machine component.

(Note: All solution procedures have 140 nodes placed on the boundary of the component.)

(ii) A Coupled Solution

The discretisation used by this mixed solution procedure is made up of 4 macro-regions modelled by the symmetric form of the direct boundary integral method with quadratic boundary interpolants and 19 eight-noded isoparametric finite elements. An illustration of this model is shown in Fig. 6(b). Here the coupling of the two methods is used so that finite elements model the slender part of the example where it is usually disadvantageous to use a boundary integral formulation because of the high boundary/surface ratio.

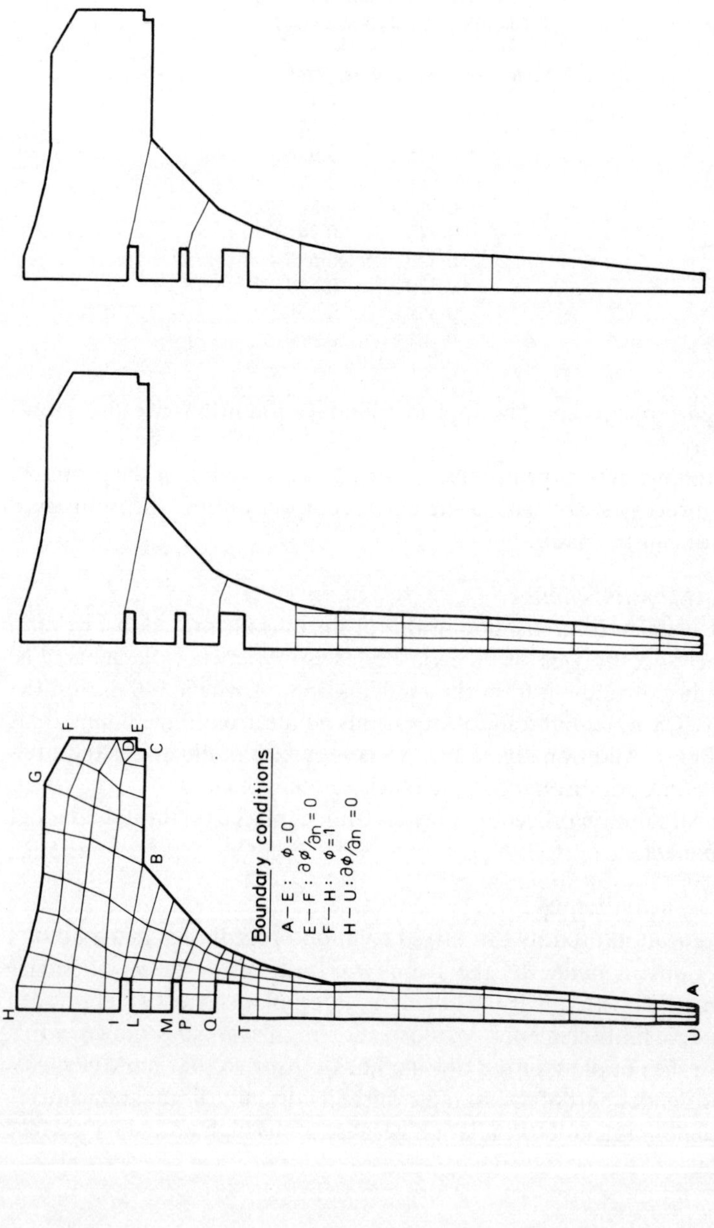

FIG. 6. Machine component problem, geometry, boundary conditions and various discretisations. (a) FE discretisation and boundary conditions; (b) FE and DBI discretisation; (c) multi-region DBI discretisation.

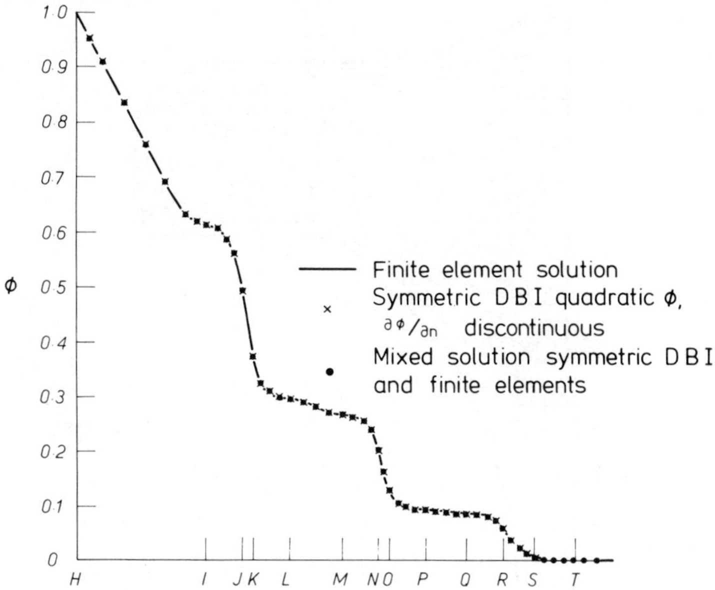

FIG. 7. Machine component potential solution given by the symmetric DBI (quadratic) procedure and the FE method, along the boundary H to T.

The results for the potential around a section of the boundary problem are illustrated in Fig. 7. It should be noted that this solution agrees very closely with the results obtained by the finite element model in part (i) of this section.

(iii) Symmetric Direct Boundary Integral Solution (Quadratic Interpolants)

The problem discretisation used by this procedure consists of 6 macro-regions modelled by the symmetric form of the direct boundary integral technique as described in Section 10.4. An illustration of this model is shown in Fig. 6(c). The boundary potential values are illustrated in Fig. 7 and as can be seen from this diagram, they are almost identical to those obtained by the mixed solution in part (ii) of this section.

(iv) Comparison of Results

The accurate solution obtained by the finite element method in part (i) is very similar to the results obtained by the mixed solution procedure and the symmetric direct boundary integral formulation with quadratic boundary interpolants.

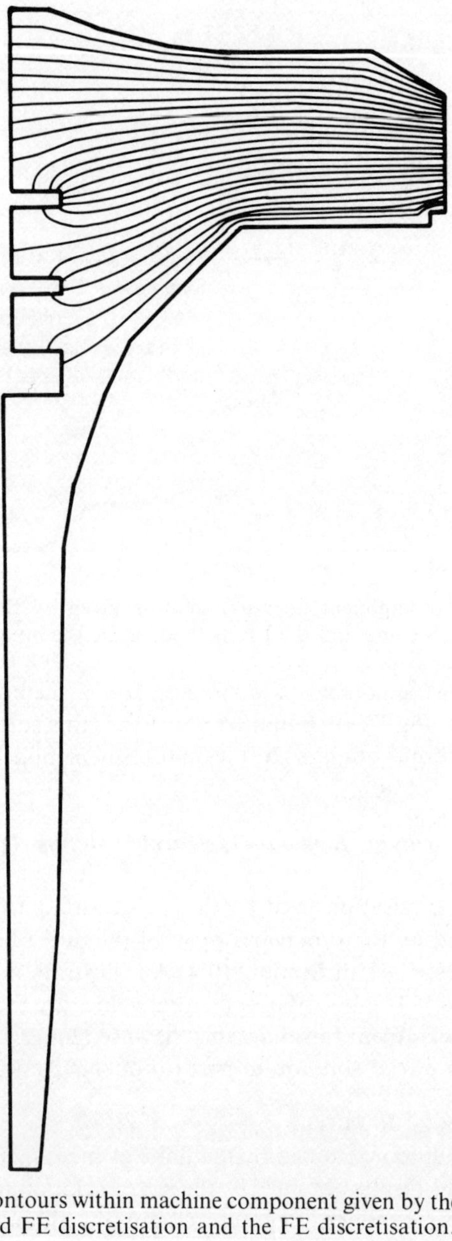

FIG. 8. Iso-phi contours within machine component given by the symmetric DBI and FE discretisation and the FE discretisation.

TABLE 4
$\partial\phi/\partial n$ VALUES AT THE UPPERMOST INTERFACE IN THE MACHINE COMPONENT PROBLEM

Symmetric DBI (quadratic) + FE	
$\frac{\partial \phi}{\partial n}_1$	$\frac{\partial \phi}{\partial n}_2$
−0·035 50	0·045 40
−0·032 20	0·029 82
−0·029 12	0·024 96
−0·023 83	0·025 99
−0·024 40	0·024 77
−0·026 32	0·022 86
−0·024 79	0·028 25
−0·027 52	0·031 80
−0·043 36	0·030 62

Note: $\partial\phi/\partial n$ results for symmetric DBI multi-region analysis are identical with those presented above.

An interesting feature in the symmetric coupled solution can be seen in Table 4. These results show that equilibrium across interfaces is no longer satisfied in the usual boundary integral sense where a continuous cancellation of $\partial\phi/\partial n$ occurs, but in a nodal manner requiring the sum of all forces at a node to be balanced by any resultant external force acting at that node.

The computing times for the various procedures are shown below. As expected the finite element procedure is appreciably faster than the other methods employed.

Computing times (CDC 7600)

Finite element solution	2·9 s C.P.U.
Finite element and symmetric direct boundary integral solution	8·6 s C.P.U.
Symmetric direct boundary integral solution	9·1 s C.P.U.

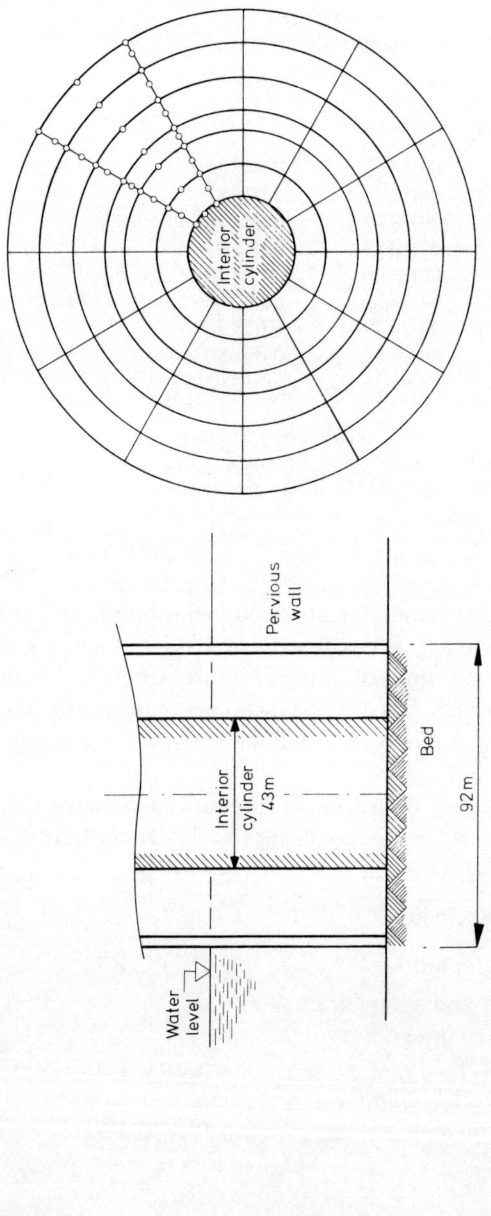

FIG. 9. Geometry and mesh for Ekofisk-type structure.

10.5.3 Waves Incident Upon a Cylinder with a Porous Protecting Wall

This problem, illustrated in Fig. 9(a), is based on the porous wall type offshore structures used in the North Sea. The problem defined is one where incident waves are impinging on an outer protective porous wall before hitting the solid cylinder core, with some reduced energy.

The coupled finite element and boundary integral discretisation of the problem is shown in Fig. 9(b). The model of this problem has been divided into 3 parts, which are modelled in the following ways.

(i) The region between the solid cylinder and porous wall is described with finite elements.
(ii) The region outside the porous wall to some circular boundary is modelled with finite elements.
(iii) The remaining semi-infinite region is discretised using the direct boundary integral scheme.

A novel feature in this problem is the modelling of the porous wall which is simulated by using a 6-noded finite element to produce the flow condition through the wall so that

$$\frac{\partial \phi}{\partial n} = k(\phi_2 - \phi_1)$$

where ϕ_1 and ϕ_2 are the velocity potentials interior and exterior to the porous interface respectively. This condition is enforced by adding the term

$$\int_{\text{wall}} k\frac{(\phi_2 - \phi_1)^2}{2}\,\mathrm{d}\Gamma$$

into the functional so that on extremisation the constraint is applied.

It should also be noted that the functional used in this problem must be modified so that the Helmholtz equation is satisfied. In this case the appropriate singular solutions used in the boundary integral formulation are Hankel functions.

A typical water surface for a permeability $\beta/\beta_{\text{crit}}$ for the porous wall of 0·5, is illustrated in Fig. 10. In Fig. 11 wave forces on the porous wall and cylinder are compared with the analytical solution for forces, acting on solid cylinders of the inner and outer size for various values of $\beta/\beta_{\text{crit}}$. Figure 12 illustrates the corresponding moments.

This example has been taken from Zienkiewicz *et al.* (1977*b*), where other incident wave problems are also studied.

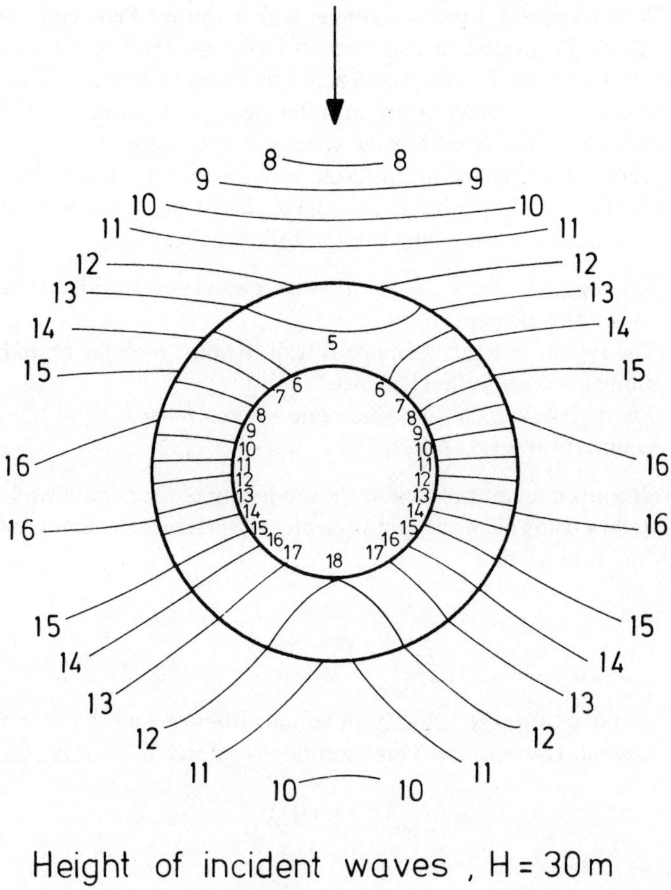

FIG. 10. Contours of a typical water surface.

10.5.4 Elastic Pile Group Problem

Figure 13 shows a capped group of 5 piles. The analysis here assumes that the pile group is embedded in a homogeneous elastic half space, and that the piles are also elastic.

Because the problem naturally can be divided into two parts, i.e., the soil mass and the pile group, the best numerical approach is to discretise each domain separately and then couple the discretisations together.

This strategy is performed by modelling the piles with a finite difference scheme, and the soil with an indirect boundary integral formulation. Here

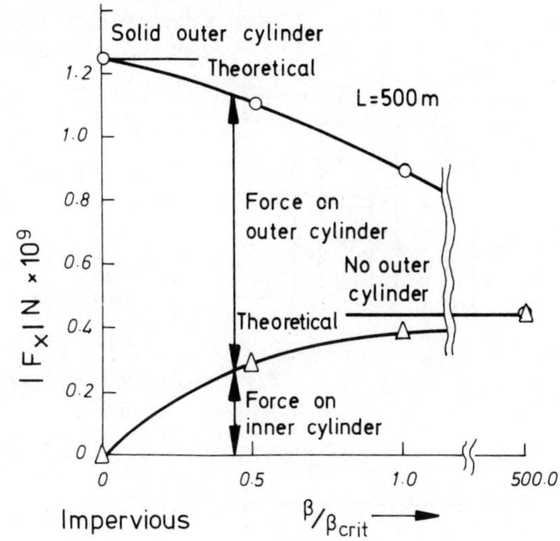

FIG. 11. Forces on cylinder with protecting wall.

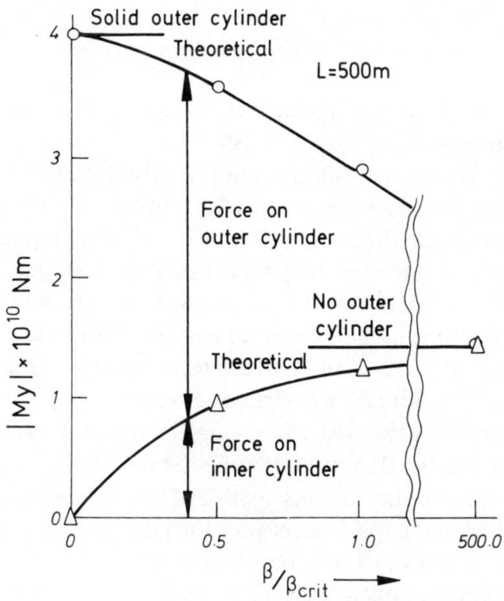

FIG. 12. Moments on cylinder with protecting wall.

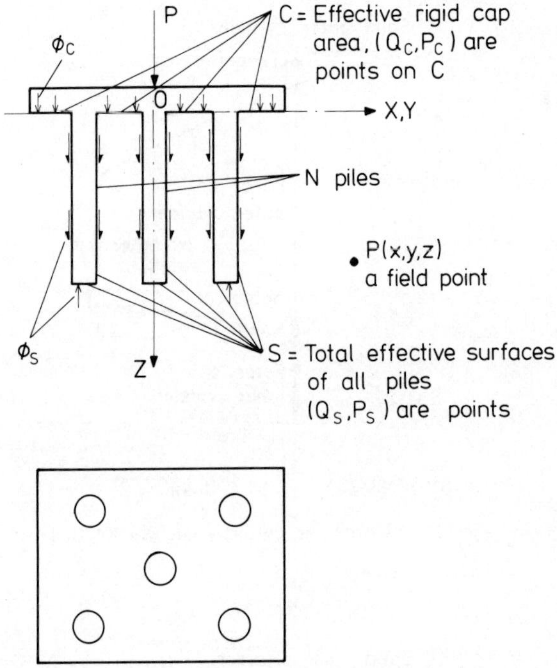

FIG. 13. General cap group system.

advantage is taken of the particular boundary conditions by using a singular function that naturally satisfies the stress-free condition at the ground surface, this is the Mindlin solution (Mindlin, 1936). This choice of discretisation means that the soil model 'boundary' only consists of the pile–soil interfaces and the pile cap. The method of linking employed here is similar to the non-symmetric technique detailed in Section 10.3. It should be noted however, that for this example the horizontal compatibility constraint across the pile–soil interface can be relaxed without introducing significant error into the result. This is simply because the external loading in the example is specified in a vertical direction.

The pile group is subjected to an axial load as shown in Fig. 13. The solutions obtained by Banerjee (1969) are presented in Figs. 14 and 15. Figure 14 illustrates the various load–deflection behaviours of the pile group when varying the spacing between the piles. In Fig. 15 the load shared between different piles in the group is shown.

Further studies of capped pile groups are given in Butterfield & Banerjee (1971).

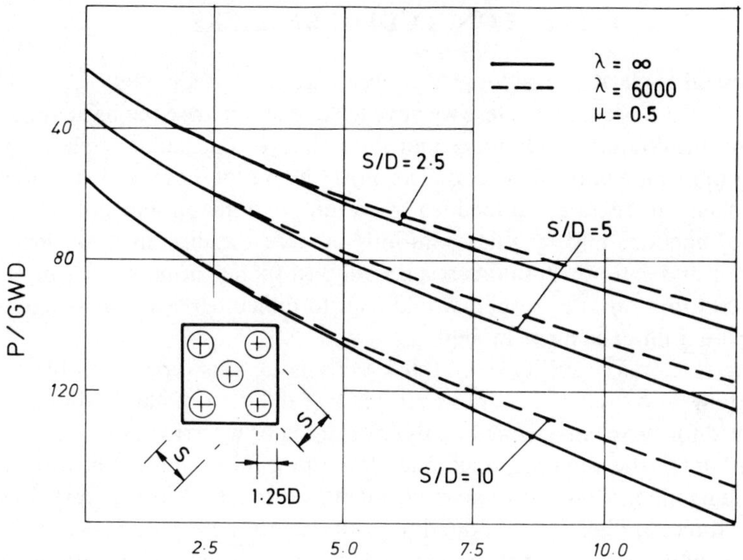

FIG. 14. Load displacement response of capped group of five piles.

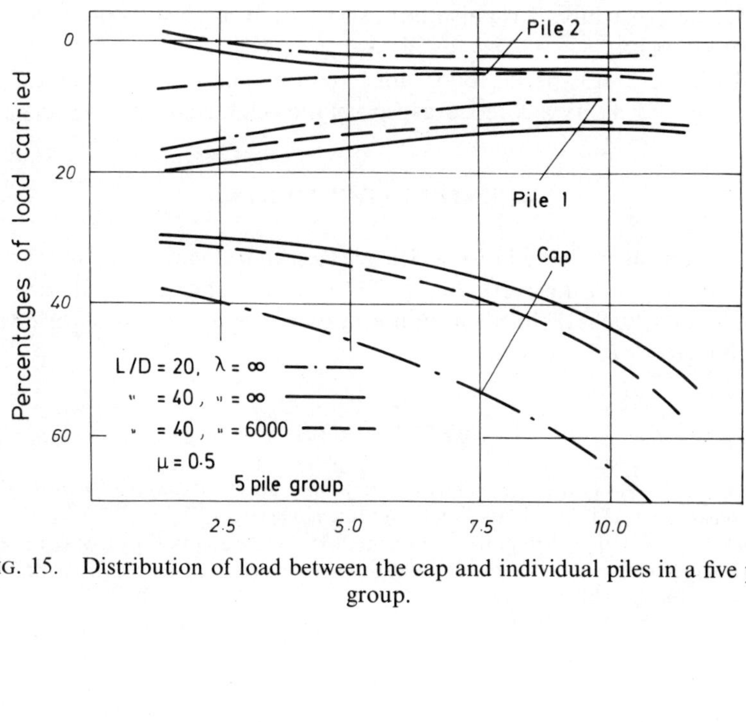

FIG. 15. Distribution of load between the cap and individual piles in a five pile group.

10.6 CONCLUDING REMARKS

Successful applications of coupled analyses are given in Chapters 4, 7 and 9 as well as in this chapter. Here we have tried to summarise the basic theory behind the coupling techniques and show that for particular applications the coupling is a natural process. The most obvious of these is where a non-homogeneous region is embedded in an infinite domain, and polynomial shape functions and singular solutions provide the discretisation in the interior and exterior domains respectively. In such a situation basing the discretisation on one method would lead to difficulties (such as where to truncate a finite element model).

The economic viability of coupled analysis for this type of problem is apparent. It is less obvious however whether different methods should be coupled for problems defined on interior domains with, for example, non-linear or plastic regions, since new developments in the application of boundary integral to non-linear problems, as reported in this text, may make direct application of boundary elements to these problems a viable proposition. However, the coupled approach with finite elements in the non-linear region remains a possible alternative and in this paper we have included an analysis of a machine component which shows that the coupled approach can be applied to complex problems defined on finite domains. The same procedures have also been successfully applied to problems of linear elasticity.

Much work still needs to be done to streamline the computational processes and clarify theoretical aspects of the weighted residual approach.

ACKNOWLEDGEMENTS

The authors are indebted to The Science Research Council for supporting the early stages of this work, and to The Centre Techniqué des Industries Mecaniques, Senlis, France for supporting Mr Mustoe during the period of much of this work.

REFERENCES

BANERJEE, P. K. (1969). *A contribution to the study of axially loaded pile foundations*, Ph.D. Thesis, Southampton University.

BANERJEE, P. K. (1976). Integral equation methods for the analysis of piecewise non-homogeneous three dimensional elastic solids of arbitrary shape, *Int. J. Mech. Sci.*, **18,** 293–303.

BETTESS, P. (1977). Infinite elements, *Int. J. Num. Meth. Eng.*, **11,** 53–64.

BREBBIA, C. A. (1978). *Introduction to boundary element methods*, Int. Symp. on Recent Advances in Boundary Element Methods, Southampton, July, 1978.

BUTTERFIELD, R. & BANERJEE, P. K. (1971). The elastic analysis of compressible piles and pile groups, *Geotechnique*, **21,** 43–60.

BUTTERFIELD, R. & TOMLIN, G. R. (1972). *Integral techniques for solving zoned anisotropic continuum problems*, Int. Conf. Var. Meth. in Engng. Southampton (9), pp. 31–51.

CHEN, H. S. & MEI, C. C. (1974). *Oscillations and wave forces in a man-made harbour in open sea*, 10th Naval Hydro, Symp., Department of Civil Engineering, M.I.T. Cambridge, Mass. 1974.

CRUSE, T. A. (1973). Application of the boundary integral equation method to 3-D stress analysis, *Comp. and Struct.*, **3,** 509–27.

KELLY, D. W., MUSTOE, G. G. W. & ZIENKIEWICZ, O. C. (1978). *On a hierarchical order for trial functions based on satisfaction of the governing equations*, Int. Symp. on Recent Advances in Boundary Element Methods, Southampton, July, 1978.

LACHAT, J. C. & WATSON, J. O. (1976). Effective numerical treatment of boundary integral equations; A formulation for three dimensional elastostatics, *Int. J. Num. Meth. Eng.*, **10,** 991–1006.

MINDLIN, R. D. (1936). Forces at a point in the interior of a semi-infinite solid, *Physics*, May, 195.

MUSTOE, G. G. W., KELLY, D. W. & ZIENKIEWICZ, O. C. (1978). The coupling of boundary integral and finite element procedures, Research Report prepared for CETIM, Senlis, France at University College, Swansea.

PICA, A. & HINTON, E. (1978). *Transient and pseudo-transient analysis of Mindlin plates: Theory and applications*, University College Swansea, Department of Civil Engineering, Internal Report C/R/329/78.

SHAW, R. P. (1978). *Coupling boundary integral equation method to other numerical techniques*, Int. Symp. on Recent Advances in Boundary Element Methods, Southampton, July, 1978.

TOMLIN, G. R. and BUTTERFIELD, R. (1974). Elastic analysis of zoned orthotropic contunua, *Proc. ASCE Mech. Div.*, **EM3,** 511–29.

WEXLER, A. (1978). *Some applications of the boundary element method to electrical engineering*, Int. Symp. on Recent Advances in Boundary Element Methods, Southampton, July, 1978.

ZIENKIEWICZ, O. C. (1977). *The Finite Element Method*, 3rd edition, McGraw-Hill, London.

ZIENKIEWICZ, O. C. (1978). The generalised finite element method and electromagnetic problems, *Compumag*, *Grenoble*, September.

ZIENKIEWICZ, O. C., KELLY, D. W. & BETTESS, P. (1977*a*). The coupling of finite element and boundary solution procedures, *Int. J. Num. Meth. Eng.*, **11,** 355–76.

ZIENKIEWICZ, O. C., BETTESS, P. & KELLY, D. W. (1977*b*). The finite element method for determining fluid loadings on rigid structures, two and three dimensional formulations. In: *Numerical Methods in Offshore Engineering*, Chapter 4 Wiley, Swansea.

INDEX